Digital Audio Processing

Doug Coulter

R&D Books
Lawrence, Kansas 66046

R&D Books
CMP Media, Inc.
1601 W. 23rd Street, Suite 200
Lawrence, KS 66046
USA

Cover art created by David Hamamoto, using images created by the code included.

Distributed in the U.S. and Canada by:
Publishers Group West
1700 Fourth Street
Berkeley, CA 94710
1-800-788-3123

ISBN 0-87930-566-5

Table of Contents

Preface

Acknowledgments

No book gets written without an author, and no author exists without parents that I'm aware of. So of course, I'd like to thank Mom and Dad. The former is a clinical psychologist, and the latter is a physicist, who became an engineer specializing in speech science and who was my first mentor. The head start they gave me has had inestimable value throughout my life. I'd like to thank all the other people who have mentored me over the years, either in person or through their own writings. Chief among the mentors have been associates such as Frank Harrison, longtime buddy and fellow audiophile, and Steve Kenyon and Tom Gamble, signal processing scientists I've worked with, who really accelerated my learning. Many other folks, whose books I'll be listing, have been most helpful to me in gaining understanding of various phenomena. My wife, Marlene, can't be left out here, and indeed, the odd meal and reminder to get back to work have been quite helpful in producing this work. Special thanks to the engineers at Valcom, who read several drafts of this book and paid me to teach a course based on it. The feedback from this was valuable indeed. Special kudos to the secretary at Valcom, Trish, who figured out how to print my drafts and made all the copies, besides catering in an excellent selection of food for us to eat during the courses. Of course, I'd like to thank my editor, Berney Williams, and the people at CMP Media who got me started on this adventure in the first place — it's been fun, guys, and I hope we'll be doing it again soon.

Thanks to Richard O. Duda, who allowed me to use his cool-looking and informative plots of the ear's frequency response versus direction. Guys like this are the type of academics I like.

Last but certainly not least, I need to thank "my kids." I don't have any biological kids, but I sometimes mentor the odd promising college student, and they repay me by writing some code, usually for beer, but sometimes for rent money. They keep me inspired, as well. Alan Robinson was a major help during my company's MusiCad project as we taught ourselves Windows UI programming. A little of his code also appears here. Troy Berg contributed more directly to this book by writing some of the code (see the header files) and by asking some not-so-dumb "dumb questions" about filters that in turn lead to new discoveries in the field when it turned out that the existing tools and explanations didn't cover the answers. Thanks for hammering on me until we discovered the truth, Troy.

Naks

During the final edit of this book, the following section was rewrote.

As I rewrite this, a judge has finally declared Microsoft™ a monopoly. Give the legal system just a few decades and it can figure out what just about everyone already knew! Now, I'm more or less grateful that we have a common platform; it means that I don't have to spend as much my life porting code and redesigning into the strengths of whatever system I'm going to port it to.

However, this excess market power leads to sloppiness. The toolset provided by Microsoft, which tends to be the only one that supports recent opsys developments, is full of bugs. OK, anything this complex is bound to have a few — even my own code probably has a bug here and there. But when I am informed of them, I fix them. I refer the reader to the last few years worth of *Windows Developers Journal*. Every month it seems, one or more bugs are reported and verified. When Microsoft is contacted about these, the response is uniform, "This is a known issue that will be fixed in a future release." In other words, they know their bug list and simply don't tell us about it so we can at least work around the errors — maybe this is because the bug list would take more paper to print than the manuals? The real problem here is that in fact, they don't fix them — some of these bugs have been around for quite a few releases. During the writing of this book,

I myself found quite a few bugs in the Microsoft C++ compiler and reported them to a contact I have over there, complete with code to reproduce them, and code to work around them. I pretty much got the same response that Ron Burk (editor of WDJ) gets, and one time was even referred to paid support, as if they could fix compiler product bugs! Sad to say, the time limitations involved in writing a book before the food runs out meant that there were a few places in the code where I ran into these bugs and simply didn't have time to find a workaround for each. This, in some cases, limited what I was able to do for you in the way of clean demo code for the concepts presented here. Now, I realize that Microsoft does very little floating point math in their products and so wouldn't run into many of these bugs themselves. But when I send demo code directly to their QA for tools representative, I guess I expect a better response than "we'll fix this someday, maybe" or "don't do that, then." It's not as though these tools are inexpensive. I find it reprehensible that they continue to release new versions without fixing old bugs; in effect, expecting us to buy a product that they know is defective in important ways. Other compiler products would be usable but for one little detail. Outside third parties have effectively no hope of keeping up with the flood of new technologies associated with the growth of Windows™ and Microsoft doesn't make the required information available to such without a stiff price and a long delay.

I find myself saddened by having to say this. At the beginning of the microprocessor revolution, I used and loved Microsoft tools and chose them over other possibilities because they were, in fact, the best out there. This may still be true in some sense, but now it seems to be because there is no effective competition for Windows programming, not because their tools have really improved.

Why did I rewrite this? Well, the first version I wrote might have been legally actionable, for starters, and I have better ways to be famous than being the target of a lawsuit. I was *mad* because some of the more subtle bugs caused me to have to write less than optimal code for you, the readers, and wasted many hours of my life. I also figure that a simple statement of fact may actually carry more weight than what appears to be a rant. I just have one request of Microsoft:

Why not just tell the truth, stop witholding information, and simply publish the known bug list if you can't or won't fix the bugs? The few hours it would take to read this would be amply repaid in hours saved by avoiding the coding constructs that create problems for the compiler.

Well, actually I have two requests. The other is, "When can we have color printout for listings that matches the onscreen display?"

Chapter 1

Overview

Virtually all sounds you hear, and especially any that come through speakers or headphones, have been either deliberately or accidentally processed and modified in some way. Even human speech falls into this category because the intelligibility is a result of the filtering a vocal tract performs on the original pitch impulses or air turbulence sounds. Special effects are a subset of this rather broad model, but may include deliberate processing that is intended to sound natural, as well as the more radical things that probably come to mind when the topic is considered, like fuzz tones or flangers. This book attempts to help the reader understand both, as well as give some information on how the human perceptual model gets involved in the overall process.

This is *important*, because if something can't be perceived, or will be perceived as other than what was intended, then why do it at all?

Where I'm Coming From

This work is intended to aid fairly experienced software developers and hardware designers to learn about audio special effects processing in computers for multimedia applications, with a big side order of digital signal processing (DSP) in general. If you're having trouble writing a hello-world application, figuring out how to unpack a wave file format, or wiring a flashlight, this book is not for you. Sorry. It may help you, but it will help you far more if you go and learn those other things first. Throughout, I stress practical understanding and techniques, giving the math where necessary or desirable. This is not intended to replace academic works on DSP, but rather to compliment them and fill in (in understandable English!) where they lack explanatory power or simply tend to obfuscate the obvious. It seems, at least in my experience, that most books about DSP with any useful content were written by people who were trying to impress dissertation advisors with their fancy math notation skills or their understanding of some unimportant esoterica, rather than trying to actually impart an understanding of the field to someone who doesn't already know it. In most cases, the "standard explanations" for things aren't truly explanations, but notational shorthand for people who get it already. I hope to fix that problem here, and where there's a standard explanation for some common concept, I'm going to deliberately avoid it, hoping that between my explanation and the standard, the "ah-ha!" will happen more easily for you. Thankfully, this bad situation is not true of all books in the DSP field, and where I can, I'll be listing some of the better works for further study. Out of the many thousands of dollars I've spent on books while learning this myself, these few stand out as definitely being worth the money and trouble to obtain.

Although I've done my best to arrange things so that each is the necessary prerequisite for the next, truly, this is impossible to do. Everything ties together in reality, and one understanding informs the other, so for example, I'll use filter terminology somewhat before I get into the in-depth explanations of it then refer you to the filter chapter for more detail. Effective special effects processing and invention just requires that you know a little something about a lot of different fields, and there's just no getting around the fact that some things are defined in terms of each other.

Academic versus Practical Understanding

I define (at least) two types of understanding here. One is what I'll call the academic flavor, where the author (e.g., the one who gets credit), history, and derivation, along with replicable proofs, are everything, and being able to generalize and actually do new things based on this understanding is less important. For instance, one ought to be able to derive all of chemistry and its implications, such as life and consciousness, from the basic quantum wave equations, but we're still waiting on that one many years later. Another type of understanding I'll call the practical flavor. This type is personified by the ability to get predictable results when faced with a task and by "feeling in your gut" or visualizing how a system works in general — something that's often completely missing from academic understanding. This second type is what I hope to impart here, but I'll present both as desirable and reasonably possible for each case. What I want to do is to hold these things up in front of your consciousness and show you how they jiggle when you shake them, so you can get that real practical insight into how they work and what they might be useful for.

What Else You'll Need to Know and How to Find It

This book is not going to cover all, or even most, of the large field of DSP. There will be little or nothing on statistical classifiers, neural nets, fuzzy logic, or other major DSP techniques. That will be another work. As I noted above, where a standard explanation exists for a widely used concept, I will come at it from a different direction, to avoid duplication. I'll limit myself to the major techniques used to process sound data in preparation for presentation to a listener. In fact, there's not even going to be enough space here to cover this subject fully, in the academic sense. Rather, I'll hit the high points then direct you to other sources for more complete coverage. I'll send you there and elsewhere as needed to keep this work at the level I want it to be; that is, more a compendium than total coverage of some narrow aspect. I find this is better for initially imparting some understanding of a field, and I hope you do too. The intent is to teach you enough to be dangerous right away and lead you toward a better ability to understand other works on the subject more easily.

A single book — at least not one of reasonable size — can't take someone from being a pretty good computer software programmer all the way to a DSP guru. If that were the case, the field would be trivial, and it's definitely not.

Here are some works I'll reference that I've found are "worth it," even on my limited budget. You don't have to go out and buy these to understand this book, but if I've stimulated your interest in this field, as I hope I will, here's where to go to get more, without wasting your money and time.

Acoustics   Beranek, American Institute of Physics, ISBN 0-88318-494-X

Beranek, for all practical purposes, invented and regularized the modern science of acoustics. Here, he generates the wave equations for sound in air, shows how to build circuit models of acoustic phenomena, and gives many practical measurements and valuable rules of thumb for acoustics.

Sound and Vibration   Morse, American Institute of Physics, ISBN 0-88318-287-4

As the title says, this is about sound and vibration. This work extends Beranek's to include wave equations in solids and covers harmonic and coupled harmonic oscillators well.

Signal Processing Routines in Fortran and C   Stearns and David, Prentice-Hall, ISBN 0-13-812694-1

In describing the *extremely* valuable code that comes with this book, the authors give a very good tutorial of all the DSP basics. This book of practical information is valuable for real-life practitioners of DSP. I'm almost embarrassed to admit how much money I've made consulting customers by repackaging the code that comes with this book for their needs. Very much a winner.

Musical Applications of Microprocessors   Hal Chamberlin, Hayden, ISBN 0-8104-5773-3

Hands-down the best DSP tutorial ever written. Many years later, I'm still using an FFT modeled on the one described in this book. His description is the one that "made the lights come on" for me. Although the book is somewhat outdated, it is extremely valuable for someone just starting to learn general DSP. It also covers analog design extensively for various functions. It seems to be out of print, but sometimes you can find a used copy, and it's worth whatever it takes to do so.

Signal Processing of Speech   Owens, McGraw-Hill, ISBN 0-07-047955-0

A compendium of current techniques used in speech processing. Although it doesn't go into great depth on most subjects, it fills in where most other books are lacking, and he has a unique take on many subjects that helps readers of the other works really understand what they say as well.

IEEE Reprints of Selected Papers on DSP, Volume I   Edited by Rabiner, Rader, IEEE Press ISBN 0-87942-017-0 (cloth) or 0-87942-018-9 (paper)

Professional papers, selected for clarity of exposition. You won't believe this after seeing the book, but it's what a lot of us had to learn from. The extensive coverage of most early DSP techniques, with brutal mathematical rigor, includes papers by the great innovators in DSP.

Selected Papers in Digital Signal Processing II   IEEE editorial board, IEEE Press ISBN 0-87942-059-6(cloth) or 0-87942-060-X (paper)

Edition 2 branches into 2d signal processing and covers "new" advances in 1-D stuff, again with brutal mathematical rigor. More papers by the innovators in DSP.

Practical Neural Network Recipes in C++   Masters, Academic Press, ISBN 0-12-479040-2

Signal and Image Processing with Neural Networks   Masters, John Wiley and Sons, ISBN 0-471-04963

Although the topics are not covered here, the Timothy Masters books on neural nets have been most helpful to me, and he is that most unusual of all beasts, a mathematician who actually writes well in English and C/C++. I read his books for the sheer intellectual enjoyment they provide, as well as the practical information and code. They cover all aspects of neural nets and statistical classifiers and quite a lot of signal processing in general. In the first book, regular and fuzzy logic neural nets are covered, as well as a lot of signal processing for data preparation for neural net use. In the second book, he advances the theory of neural nets to the complex number set and covers Morlet wavelet and Gabor transforms, general logic, and experimental design. These books cover how to think more so than what to think and do a very good job of it.

Psychology of Music   Carl E. Seashore, Dover Books, ISBN 0-486-21851-1

Written in 1938, this is a seminal work, which explores why people become musicians or simply love music. It seems to be the first attempt to *quantify* human perceptual abilities in musical areas and, despite the title, contains some good "hard" science and early signal processing techniques. Carl invented the Seashore test of musical ability, which accurately predicts success in various musical endeavors, and discovered that in most cases, it's natural ability rather than practice or hard work that really counts. In other words, if you can't hear your mistakes, there's no use trying to correct them. It turns out that many perceptual facilities are not very trainable — you either have them or you don't. Politically incorrect, I know, but true. I like these old books by people who basically began a type of science. They really impart useful information rather than simply try to impress. This book is still in print.

The Active Filter Cookbook   Lancaster, Sams Books, ISBN 0-672-21168-8

What can I say? No real engineer is without this one. Don Lancaster is very good at explaining things in general and filters in particular, which I suppose is why this book is in its seventeenth printing at last check. When writing some of the sample code for my book, I discovered that a few of his tabulated numbers are wrong, sad to say, but this is one of the very few works on the topic of filters where the author eschewed the hard-to-get standard explanations and derived better ways.

Professional MFC with Visual C++ 6   Blaszczak, Wrox Press, ISBN 1-861000-15-4

This is my all time favorite Windows programming book. The reader should consider it or a similar work as a prerequisite to the Windows programming parts of this book, where I avoid explaining Windows programming basics in order to concentrate on the DSP aspects of programming. More so than any other, *Professional MFC* gives the flavor of how MFC was designed, which is very useful when trying to make it do something the authors probably didn't anticipate.

Special versus Normal Effects

I should call this "processing" instead of "effects," at least for this paragraph. Effects is a vague word, but it has a common base of understanding in the context of audio processing. I define "special" as processing designed to sound unnatural — for example to make something sound different from a naturally produced set of vibrations in the air — and "normal" to mean processing that carefully duplicates various natural processes and, therefore, is undistinguishable by the human ear.

There is some overlap, since if done poorly or incompletely, one result may sound like the other. Also, some special effects have analogs in the natural domain. Flanging, for instance, is what you hear when a jet takes off, and that's natural in the sense that the effect is produced by varying reflection paths as the jet moves. You can duplicate this effect right now. Just close the book and say "sssssss" while moving it toward and away from your face with the cover parallel to the plane of your face.

You rarely hear this effect in music in most cultures, so your brain classifies it as something else. It's hard for a whole band to move around quickly enough for you to perceive flanging; however, it is part of what makes the Leslie speakers sound as they do (the rest is the Doppler effect).

Things Perceived as Normal

Certain sound modifications occur normally as sounds pass through the air, bounce off objects, and pass through acoustic resonators and rooms. Because these effects happen all the time, they are, in general, perceived as normal by the human ear. I'll go into much more depth in a later chapter on how the human ear works and the ramifications thereof, but for now, I'll just say that as long as you process sound in such a way as to mimic the effects of these modifiers, it will sound natural, which may or may not be what you want in a given case. In nature, reflectors aren't perfect, acoustic resonators have limited "q," and so on, so something natural can sound unnatural very quickly if you don't take certain rules of thumb into account. I'll give you the normal rules of thumb used in the business along the way.

Things Perceived as Special

Some sorts of sound modifications either never or rarely occur naturally. In this class are phaseless filters that affect frequency response without affecting the phase response, all pass filters that affect phase but not frequency, reflections from moving walls, some kinds of distortion, and such like. These usually sound "fake" and have some shock value. One could argue that the default processing used on most electric guitars is almost perceived as a natural sound nowadays, because we've all heard so much of it, or that the sound of the ubiquitous EV PL-20 microphone that produces those mellifluous voices on the radio is more natural than a more accurate microphone. For these cases, the listener's perception *is the reality*, so no other standard can be applied successfully. It's important to keep this in mind when processing sound.

When Are Effects Useful?

Effects have a wide variety of useful functions — from improving lousy recordings to improving real performances — and are used at several stages of the signal processing chain on the way to your ears.

Effects are generally used to enhance the listener's experience, although one could argue that they often fail in actual practice. Certainly you've heard the singer who doesn't think s/he sounds right without way too much reverb, or that guitar player who uses too much gain and distortion to attempt to cover up a lack of fingering technique. Although I can't seriously address taste issues here (after all, taste is, well, taste, and there's no accounting for it as they say), some things are pretty universal, and where it's appropriate (in MY version of taste), I'll mention these things as I go along.

Overcoming Recording Problems

Often in the recording business, there are problems. You may not be able to afford those nifty microphones you crave, you may not have the perfect room acoustics for a given performance, or you may not have a recording media that's perfect enough to satisfy you or your listeners. That's just for starters. So, how can you produce good results despite these things? With a good enough equalization system, you might be able to make your junky microphones sound better, eliminate some of the room effects, and pre-emphasize your recordings against tape hiss. With a more sophisticated set of effects in your toolbox, you might just be able to deconvolve (a fancy DSP word for "undo what's been done to the original signal" that I'll cover later) the microphone and room responses completely, pre-emphasize the highs so you can turn them and the tape hiss back down on playback, or compress and later re-expand the dynamic range so that the dynamic range of the performance fits better into your recording media's range, which is basically what a DBX unit does. Thankfully, most of the tape-related issues are history: I run two pro studios, and the only tape system is a cassette in my boom-box that's rarely, if ever, used for anything, except perhaps finding out how some particular mix will sound when degraded by a cheap tape system. Here, I go straight from disk to CD masters nowadays.

However, some folks think the tape sound is nice, and I'm one of them occasionally, so I'll tell you how to produce that sound, if you want it, on the more modern digital media.

Overcoming Performance Problems

On top of the basic problems of acquiring sounds, you often have less then perfect performances to begin with. This singer might hit a flat note; that guitar player might miss a chord and go "thuck" instead; a drummer might be late, have bad-sounding drums, or a squeaky bass drum pedal; the actor might cough between or during lines. All of this is expected to be fixed somehow before a product goes out. Anything less is perceived as amateurish by listeners who have heard nothing but "perfection" in the mass media for many years. The listener's assumption is that this is easy; in general it's not. Better tools are becoming widely available at last and are one of the things this book should promote the further development of, but you still have to use them, which can be tedious if there are lots of mistakes in the original. The pros simply make fewer mistakes in the first place, are usually more willing to do more takes, and of course, can afford the best tools and production help as well.

Enter efx. Now you can fix that one flat note, cut and paste another good chord over that thuck, and so on.

Even today and with professionals, these effects are useful and economical. Consider the "one flat note" case in an opera recording. Another take costs serious money for all those musicians, the hall, and so on. It's cheaper and not all that dishonest to just fix it, since it's such a small part of the overall effort and performance involved for this case. I'm sure you can imagine other cases where the principle applies.

Precorrecting Playback Problems

Often, you guess that a production will be played on reproduction hardware of poor quality, whether it be in a car or with cheap multimedia speakers on someone's computer. In a lot of cases you're pretty sure that it won't be played at the original level, which means the ear's loudness curve comes into play. I'll have a lot more to say on that subject later, since current methods for correcting this at either end of the chain aren't even close to perceptually or theoretically correct. When you hear one of those car sound systems that are apparently more powerful than the car's drivetrain, it's obvious that some listeners really like a lot of bass! Personally, I find that one-note-at-50Hz boomy stuff nauseating, but it must get people excited or they wouldn't pay for it. Although one can pre-emphasize, expand, or compress the dynamic range within frequency bands, I'll also cover some things you can do to fool the human ear into perceiving bass where little or none exists or that could be produced by the target system. The overall effect is far better than trying to get real bass out of a three-inch speaker and failing.

Noise Reduction

Although not a special effect, per se, I get more requests for noise reduction than for any other single processing effect. Noise reduction, done well, is popular and can save old or poor recordings from oblivion. There are many, many types of and techniques for noise reduction as a result of this popularity, and some work better in some cases than others. I'll cover most of the old hat as well as the newest, most advanced techniques to reduce actual and perceived noises.

All noise reduction techniques face a simple to define but hard to solve problem: Will the real noise please stand up? It can be very frustrating to realize that this is one of those times when human perception can readily separate the noise from the signal, but computers don't often do a very good job at it. In fact, the problem is similar in nature to continuous speech recognition, which is another thing computers don't do very well, so far, because the human consciousness uses *context* and *understanding* of the subject material to produce the results it does — something that is pretty hard to reduce to binary decisions at the current state of the computer art. Most people are surprised to discover that in individual word recognition tests, they only get 70 percent or so of the words right! In normal conversation, this is true as well, but your brain fills in the missing words and corrects errors based on your expectations and understanding of the speaker's topic and manner of speaking without you even realizing it. Perhaps, like that famous apple, this is just too obvious for most folks to recognize! In addition to these problems, information theory in effect "wraps around" at the extremes. What I mean by this is that a perfect information-containing signal is by definition completely unpredictable. If it were predicable, it'd be redundant, and therefore not perfectly information containing. Sounds like the definition of random noise, doesn't it?

Although music and other signals don't sound much like noise to the ear, they tend to be rather tough to predict perfectly. Again, if they were easy to predict, there wouldn't be much content involved. You get a little break because music, in general, is designed into human perceptual rhythms and, therefore, has some predictability. Usually, it's just those small differences between perfect predictability and what you actually get that make music pleasing; otherwise, it sounds machinelike, so you don't get that big of a break. However, given a point of attack, any problem can be solved eventually. Noise reduction, especially the single-ended type, where you don't have access to a signal before degradation, is an ongoing study. No one has solved it all.

The Power of Reality

I'll make various comments in different contexts about reality. In an awful lot of cases, the listener's perceived reality is what is most important — subjective reality, that is, and not yours either. However, I want to talk about what happens when you produce sounds that are so realistic that they are compelling for that reason alone, almost regardless of the actual content. Compared to most of what's out there, these created sounds have significant shock value. They get a listener's attention and keep it while providing a pleasant experience. Various techniques produce this effect even when the sound originates in a completely unnatural way, such as with a synthesizer. It's generally a bad move to compromise the reality effect with nonideal EQs (equalizers), for example, even if you seem to need them to get a nice frequency response. The result is often so much more compelling than just a good overall sound in many cases that it's better to have poor overall balance in a recording than to compromise the effect. Listeners quickly forget that the acoustic guitar sounds thin when it sounds like it's really in the room with them. Often you can balance a shortcoming by changing the arrangement or by mixing in other, realistic sounds. That said, I will show you techniques that let you keep the effect and achieve good balance at the same time.

Shock Value and Boredom — How Listeners Work

I shouldn't have to say this, but I will anyway, because it's obvious from what's out there that a lot of people still need to hear it: Many things have shock value, be they real or manufactured, but intelligent listeners get bored with it mighty quick, and it quits working. If you think your listeners aren't intelligent, even if it's borne out by marketing studies, you aren't going to sell very much by insulting them. Almost everything really popular strives to exemplify the basic. At the time, there were a lot of better guitar players than the Beatles or the Stones, for example, but who sold the most records? The guys who made it simple and strove for honest feeling in their expression, that's who. They used all manner of special effects, of course, but usually in subtle ways and not continuously "in your face." I'll have a lot more to say on this when I cover perceptual rhythms later. Then you can understand how to use effects in effective, rather than annoying, ways — or you can be maximally annoying if that's what you want. I don't care, if I don't have to listen to it! I also hope that this isn't read by the people who produce advertising, because the tricks *do* work, and I'd hate to be responsible for most of their stuff!

Chapter 2

PsychoAcoustics

What are PsychoAcoustics? The "acoustics" part ought to clue you in that it has something to do with sound, and indeed it does. The "psycho" part says that you will delve into that rather inexact, so-called science that involves human subjective perceptions. Or it could mean that this psycho (me!) is about to hand out some opinions about perception, I suppose. The overall idea here is that a listener's subjective reality may not precisely map onto objective reality. Surprised? You shouldn't be. Consider politics ... no, I won't go there, although it's perhaps the best phenomenon to use to demonstrate the differences between objective and subjective reality.

Anyone who's ever used a camera knows that humans aren't nearly as sensitive to color temperature as film is and that you have to train yourself to "see like film" to get a good result. What you're actually doing is untraining yourself to see all basic light sources as white, the way you normally do. Really, fluorescent lights are kind of green, incandescent lights are kind of orange, and only sunlight is pretty close to being truly white. In other words, your visual perception system has trained itself to compensate for variations in the color temperature of the source in order to save your brain some work. In fact, you have to untrain that reflex to see what's really there.

The same sort of untraining occurs with hearing and is perhaps even more complex. In this chapter, I explain how the hearing part of human perception works. Why do this in a book about practical audio processing? If you have no idea what the result of a particular process will be, you're reduced to casting about at random when trying to figure out how to get some "effect." In other words, the theory here has some very practical usage in guiding you toward whatever result you desire when processing audio. As it turns out, some things don't have to be done as precisely as you might think — a welcome situation when you're trying to save CPU cycles for best performance, or hardware for best profits. Just as your eyes quickly learn to ignore differences between white light and other light, your ears selectively process sound. You can take advantage of this fact to simplify processing.

Human Perception, Preprocessing, and the Lazy Brain

Human perceptual systems consist of several components. Obviously, there's a sensor of some kind that translates inputs of some kind and activates neurons. After this, there's some unconscious preprocessing of the signals by other neurons before they reach your brain, or at least your consciousness. This is a good thing, because if your conscious brain had to accept real-time input from every neuron in your body, there would be no time left for things like thinking or knowing your name. Imagine what it'd be like if you were constantly aware of every square millimeter of your skin, for instance. It might be fun for a few well-chosen minutes, but you'd not like it all the time! And that's just one sense.

As it turns out, some preprocessing has direct analogs in the signal processing and computer science worlds, and the mapping is pretty close to things you can understand quite well mathematically. This is a big plus when you're getting ready to deliberately fool the system, as one often does in audio signal processing.

The main point I want to make here is that your brain is and must be lazy when it comes to handling sense inputs, and that a system has evolved that allows this by preprocessing all input before the brain "sees" it, passing on only the interesting stuff, unless consciously trained or commanded otherwise. This point is absolutely crucial to understanding sense perceptions in general.

This process maps pretty well onto the design of conventional computer operating systems. In that model, you have low-level driver code for individual devices that handles time-critical or detailed operations and, in turn, presents a smoother and more uniform interface to the operating system and application programs. Thus, none of the high-level code has to worry about what brand of disk drive or video accelerator you have. Only when an operation is complete or there's a fatal error does the driver impinge on the upstairs stuff: that doesn't need to know the details of writing on a disk or putting out a pixel in most cases. The driver may not be very smart either. It doesn't know how to multitask programs or how the file or windowing system works. It just does a simple job and mostly stays out of sight, allowing the more complex, high-level code to ignore some details and time-critical issues and get on with its own part of the total job.

People work the same way, more or less, and it's a good thing because it allows sentience and independent thought to occur. Unless you're well out of the intended audience for this book, for instance, you're probably not aware of the reading process per se. You're not looking at each character, doing optical character recognition (OCR), sounding out the words, then figuring out what they mean. I hope not, anyway! You look at the page and just know the words and what they mean without thinking about that part of the process. The rest of your mental abilities can now be used for thinking about the content (or daydreaming) instead.

The Perceived Reality is What Matters

When you listen to a stereo, you know that the musicians aren't really there in the room with you. If the recording is good and your stereo is of high quality, though, you can probably fool yourself into imagining a soundscape, complete with the virtual locations of the musicians in your room. Although this may be slightly more likely to happen if the recording was done live with a pair of microphones in a misguided attempt to capture and then reproduce the original 4-D sound field accurately, the effect can also be produced artificially with *even more* apparent realism. It can be quite gratifying if you're a music lover, as I am, even though the objective reality is that those musicians aren't there and maybe aren't even still alive. *The perceived reality is what matters*! When processing audio for others' consumption, it pays to keep this in mind at all times. Your source material may consist of a number of mono tracks that weren't even recorded on the same day, or in the same room, but you can combine these into something that will produce the desired subjective reality, anyway — virtual soundscape and all! However, the target in this case is the listener's subjective reality; the objective reality is far less important. Because it's important to make the distinction between subjective and objective reality, I'll use subscripts when it matters — reality$_S$ or reality$_O$.

Human Hearing Process and Preprocessing

The normal human hearing equipment is quite complex and acoustically preprocesses input at several stages long before it gets down to the neurons that provide the brain with signals. First, there's the outer ear, or pinna — the things that hold your glasses on. Actually, this is a fancy acoustic structure that causes sounds coming from different directions to have different frequency/amplitude responses and that provides a certain amount of acoustic "gain" for some combinations of frequencies and directions. Figure 2.1 shows a frequency response of just this part of the outer ear, taken with a "standard" ear and head by my friend Richard O. Duda. Unfortunately for special efx people, everyone's ears are different, so it's hard to make a general "fake" transform that works for everyone. However, in the lab, measurements taken on a specific person's pinna serve to produce transforms that can fool that person quite convincingly. More work remains to be done in this area, and it seems that if you get "close enough" to a general transform, it's a lot better than doing nothing at all, especially when combined with other cues.

Figure 2.1   Frequency response of a standard pinna vs. horizontal angle.

(Used by permission, R. O. Duda)

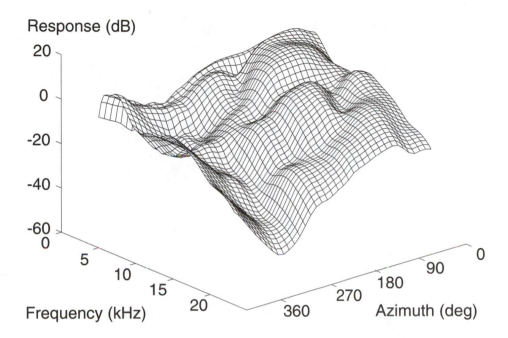

Notice that the human ear has a nonflat frequency response versus horizontal direction. This is partly how you can tell the direction a sound is coming from. Figure 2.2 plots data for different vertical angles.

Figure 2.2   Frequency response of a standard pinna, vertical angles.

(Used by permission, R. O. Duda)

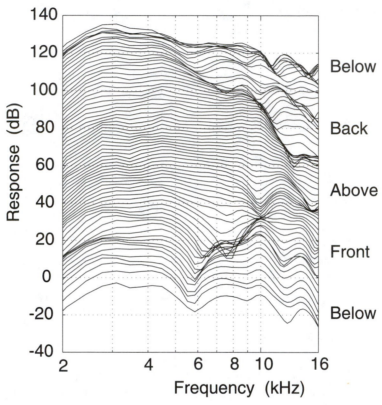

Interestingly, you don't usually notice these changes in frequency response; instead, they are usually perceived as directional cues. This is yet another example of preprocessing doing something that is not obvious and apparently hiding information to save the brain some work.

Following the outer ear is the eardrum and the small bones that transfer vibrations of the eardrum into the fluid in the cochlea. This system has a mechanical advantage built in to it that results in a gain of about 15×, which partly explains why the ear is so sensitive: An eardrum vibration $1/_{10}$ the size of a hydrogen molecule at 1KHz can be heard! The cochlea is an interesting acoustical/mechanical frequency analyzer. It's basically a tube wound into a spiral. It contains fluid, about 25,000 neurons line one side, and it reduces in cross section as it travels away from the excitation end. There's a longitudinal partition through the cochlea that is part bony and part basilar membrane, which is lined with the nerve endings. This mechanism is so arranged that by varying acoustic impedances, transit times, and reflections, a coarse frequency analyzer is formed. A particular frequency component produces a point of maximum excitation on the basilar membrane, with the bass frequencies producing it at the far end from the eardrum and the higher frequencies producing it nearer the front end. The relationship between the input frequency and the area of maximum excitation is fairly close to a straight line when graphed versus log frequency, which may explain why an octave, or a frequency doubling, is an octave in all human cultures and why, regardless of the tuning of a given musical scale, it always repeats ratiometrically in the next octave, instead of at a constant frequency spacing. There is a higher nerve density near the bass/far end of the cochlea where the basic frequency mechanical resolution of the structure is the worst, compensating for that effect somewhat. Humans must like bass! All this happens before a single neuron is fired. This is important to understand because by that time, the ear has already processed the data in ways that turn out to be useful to survival as well as pleasure. The extra resolution in the bass allows source size estimates (the tiger vs. the kitty cat). For each ear, a different resonance/reflection pattern is put on all incoming sound that is dependent on its direction, then the result is broken down into the frequency domain, mostly, with time domain information available up to the normal neuron firing rates — a few hundred Hertz or so.

At the next level, the neurons in the basilar membrane connect to each other on the way out in what looks like an FIR (Finite Impulse Response) edge enhancement filter. You can at least describe the neural behavior in this part of the system this way. The effect, if not the method, and the requisite numbers match the observations of actual sense abilities and mechanical neuron densities. This "structure" is common throughout the senses, and to a programmer, it looks like data preprocessing, such as a deblur and edge detection processor. In theory, FIR-mediated interpolation can improve the actual mechanical resolution by the factor required to explain things like perfect absolute pitch and other observations hard to describe by the apparent mechanical limits of the structures involved. The standard computer multilayer feed-forward neural net model, which is a weighted sum of inputs followed by a nonlinear function to "squash" the output range and in which the positive or negative weights are leaned, can do the same thing, as well as more complex functions, such as pattern recognition (see the Masters books), so I assume that the mechanism here is at least as good or powerful. Nature is more complex than this simple model, but the model does well to do all it does with a maximum sample rate in the few hundred Hertz range, at most. It remains to be seen whether this structure also can match the various reflection patterns caused by the ear's response to sounds from different directions. It shouldn't be too hard to do at this stage. Nature is generally economical, which would account for the frequency analysis approach as well: Simple physical shapes causing reflections in a sound field have simple shape effects on the cochlea excitation as well.

Another perceptual effect that seems to be based on the ear's hardware is that of "critical bands." Within a critical band, one loud tone can mask a softer one so that it is simply not perceived. Outside a critical band, a loud tone cannot mask a softer tone. For instance, a loud squeal will not prevent you from hearing low bass. You hear both. Although the rest of the ear's responses are almost purely logarithmic, this one does not match that curve well. Critical bands begin at about a 50 to 100Hz width at the low end and stay pretty narrow through a few kilohertz, after which they then begin to follow a log curve fairly well. This extra resolution at low and middle frequencies overcompensates the basic mechanical resolution of the ear's analyzing structure and must be partly why there are so many more neurons at the bass end of the cochlea. Knowledge of masking phenomena is often used to decide what's perceptually important, especially in cases where you've decided to accept some degradation of the original but want to minimize its perceptual effect. As such, MPEG-2 layer 3 audio coding and decoding, for instance, make extensive use of *a priori* knowledge of critical bandwidths to minimize the audible artifacts of throwing away some information in the process of data compression. If you're only throwing away information that would have been masked anyway, perhaps it won't sound too bad to the listener. Given the popularity of this method, I'd have to say the idea works pretty well.

So, before the information even gets to the brain, a lot has been done to it by the preprocessing "hard-coded" into the wiring as a fixed response to a particular ear shape and cochlear resolution. It has analyzed the frequency, saved the time information below the neuron's firing rates, and matched the shapes for directional and other information, but we still have a fairly high bandwidth signal going to the brain proper. In any case, I have found that you can use knowledge of the structure of the ear to fool the brain into believing something is happening that isn't.

Spatial Sense and Localization of Nearby Reflecting Surfaces

Try to find your way through a dark room in silence, then try it again while making clicking noises. You'll find that you can tell a lot about the space you're in. This sense isn't as good as a bat's — for one thing, the frequencies are lower and the wavelengths of the interference patterns are therefore longer and less precise — but it works, and without thinking about it much. You can do better with volition (e.g., by paying attention) because the times of flight of the useful signals are within the neural time response range, so you can compare times of flight for signals from each ear along with directional information from each ear and arrive at a direction and delay for each incoming sound. There appears to be a limit on this, however, at a certain level of complexity: It only works well for a certain number of directions and separate signals. This sounds like a hardware limitation and amounts to a resolution or memory limit, and it's one of the things that makes 3-D soundscape synthesis possible because, as it turns out, all you have to do is overload this sense with "reasonable" signals to produce the effect perceptually. I'll get into that in the practical section further on, but this is an example where theory illuminates and suggests what works in the real world.

At any rate, it turns out that this is a natural result of the architecture of the ear: simple pattern matching on cochlear input. Various reflecting surfaces create comb filters of varying shapes on the cochlea and, bingo, out comes the information. It's an easy shape to match in this domain and to describe with only a few bits. The notch spacing, not the location of each notch, is all you need in order to know the distances involved.

Speech Processing

Most neuroscientists believe there's a special place (or places) in the brain where speech is handled. There may be some sexual differences as well, with females using more brain volume for language processing than males, in our species anyway. These places also receive input from both ears and manage the time of flight and the fine time domain, low-frequency analyses. However, there's something special about how language and speech are processed. To prove this, try a simple test on a friend. Sit in a squeaky chair and squeak it randomly while recording the sound in the room. At some point, ask the friend where in a sentence a squeak occurred. They'll almost always get it wrong — even you might. Speech is processed differently, or another way of saying it is that the same input data is processed for speech information in parallel with other uses, and they don't always take the same amount of time to get the job done on a particular stimulus. Just as with the perception of color temperature mentioned earlier, this is something you've trained yourself to ignore.

It also makes sense that speech is handled in a part of the brain that has good time domain information available. One of the significant noticeable features of speech and singing — that "gravelly" sound, or diplophonia — is just the delaying of alternate pitch pulses. This adds a half-frequency component by repeating the same thing over and over, but a little differently every other (or nth) time, and by smearing out all the other components a little bit. This would be more difficult to detect in the frequency domain, mainly causing some slight spreading of the main harmonic components, and in fact, you don't hear it as well for very high-pitched speech, so speech or voice processing gets its own spot and hardware (wetware?). This makes sense as an extension of the preprocessing already discussed, because it's just more simple pattern matching (but a lot more, using more neurons) on the cochlear input. Again, various simple shapes, such as those in your mouth as you articulate speech, map onto various simple shapes of stimulation on the cochlea. Form follows function, something we usually expect from nature, especially as it applies to our well-developed (and egotistical) bad selves. Another layer also appears to act (not proven, my conjecture; new science) on the information — the cepstrum, which I'll describe in Chapter 3 as a spectrum of a log spectrum and which is a domain where the shape matching that the ear does so well is done simply, at least by computers. As always, Occam's razor should inform the analysis.

An important ramification of separate speech processing is that you can "trigger" it as part of a special effect, and by understanding some of the characteristics of the speech you're targeting, you can do this. Voiced speech is characterized as an impulse source feeding a time-varying formant filter (the vocal tract) with a few peak frequencies, or for nasal sounds, some peaks and a notch. The peaks generally lie in the range from 220 to 3, 500Hz or so. From two to four of these peaks are required to form a vowel, and most formant-based speech synthesizers use three variable peaks and one fixed peak. Vowels are perceived by matching the *ratios* of these formant frequencies to each other, more or less. There's some nonlinearity in this ratio comparison at very high or low frequencies that makes it not quite a true ratio and makes it hard to perceive vowels if the formant frequencies are all too high or too low. Smaller people generally have higher formant frequencies than larger people, or anything that uses a mouth to express sounds; that is, you can make a size judgment, and you often do, subliminally, based on the *absolute* values of formant frequencies in a sound. The normal human adult range of formant frequencies covers about 10 to 17 percent of the formant frequencies — which explains how you can change the apparent sex of a singer by simply shifting all the formants up or down.

If you've ever tried to talk in a little kid's voice, you'll notice that what you're doing is attempting to compress the size of your vocal tract. On the other hand, you can tell when something big is coming by the low frequencies big things generally create. Unvoiced consonants appear as noise mostly through less complex filters and are also very much characterized by their onset and tail-off (e.g., amplitude envelope) functions. I know all the speech experts out there are writhing over this oversimplification, but it's good enough for the purpose of making special effects. There are plenty of good books out there for those of you who want to go farther.

Pitch Processing

Pitch is like frequency, but different. Frequency is an objective quantity, whereas pitch is a subjective quality. In reality$_O$, you might define pitch as the repeat rate of a complex signal; that is, the rate at which peaks in the autocorrelation function occur. The ear uses a similar definition, with the caveat that things sound "flatter" or lower pitched as they become very loud. This is a problem for violin players and singers, particularly, because they can sound right to themselves and wrong to someone else who is hearing a different balance of the sound levels. However, the ear hears diplophonia as a gravely or rough sound in speech, not as a halving in pitch, which is what you'd have in the reality$_O$ situation, unless the diplophonia is quite extreme. Interestingly, this effect does not apply nearly as strongly in music processing, which is useful to know for audio processing. In fact, you can produce "apparent" frequencies that the channel in general can't pass and fool the ear into thinking they're there, which can be very handy for producing apparent bass out of speakers that can't radiate the real thing. It appears that the ear uses something very cepstrum-like to discover pitch, because it can find "false" pitches that aren't there. In exactly the same way, a cepstrum analysis might be in error, for instance, on bell-like sounds, which have a harmonic structure that implies a fundamental frequency that's not actually present but is perceived as the bell's pitch. It appears that for very low frequency signals, the ear can also use time domain processing, but this effect is relatively weak, and the ear can be fooled by producing a harmonic structure that implies a fundamental frequency that isn't actually present in the signal.

Conscious Postprocessing

You can add a bit to your perceptual abilities by applying consciousness to the task. For a very simple example, if you know that any nonlinear process produces sum and difference frequencies, you can match two frequencies well by listening for the "beat notes" produced. Many musicians have been tuning this way for quite a while, millennia perhaps. I have a very good 1930s book that explains how to tune pianos using several different methods, including the not simple 12th root of two tuning (which generates irrational numbers for ratios) used today, just by counting beats of various harmonics of strings. Long before you could do this via a frequency counter, all you needed was a single frequency reference and you could generate the rest of the notes using observation and intelligence alone. The accuracy is not too shabby for "low-tech." If you know that dispersive filters spread out a white impulse into a "boing" sound, with each frequency delayed differently, you can learn to hear even subtle ones by concentrating on any impulses in the source material. I've already mentioned the spatial sense, and you've probably strained to pick out a conversation in the middle of a party using context, automatic "speaker identification," the spatial sense, and whatever else you could bring to bear to understand that attractive person over there. So, once something interesting happens, the conscious system is awoken and can be brought to bear on the perceptual problem with better results. Often in signal processing, you'll want to either do this or avoid it, so again, this has many practical implications. You might want to deliberately invoke the "strange detector" to get someone's attention but then avoid doing it for a while so they won't get hip to the other tricks you're using to fool their perception. Get the picture?

Perceptual Rhythms and Their Importance

Timing is everything.

Again, your brain is, and must be, lazy. This seems especially true in smart people, and it makes some sense in this context: the more *effectively* lazy your brain can be, the more other thinking it can get done. One of the brain's apparent approaches to being successfully lazy is to do something that amounts to batch processing. You alternate paying attention and analyzing the results, in effect, and this creates a rhythm as you switch back and forth between phases. Perhaps this also makes it easier to process data that has different delays through the preprocessing channels. Could one suppose this is why all popular music throughout time is rhythmic in nature? It's pretty hard to play music arhythmically, even when you try; you tend to drop into a rhythm eventually. There must be something going on here that tends to force rhythm on us.

This has far-reaching effects on perception, reality$_S$, and how the user is affected by music, other events, and life in general. If something keeps making you pay attention when you're in an "analyzing the results" frame of mind, it becomes annoying, whereas if things fall naturally into your expectations or are predictable enough in some way, you can synchronize your perceptual rhythms to it and find it pleasing instead. This is one of those fundamental truths that isn't spoken enough, in my opinion, although I'm not the first to say it. When using any kind of special effect, you should especially be aware of this idea, otherwise you might be annoying when you want to be pleasing or vice versa. You can't flood the listener with constant indicators to pay attention and expect them to like it. They'll quickly learn to ignore you in self defense. Marketing professionals take note. You usually want to get a person's attention but not necessarily hold it too tightly. Let the listeners relax, rhythmically, and they can take in much more. They will like it if the timing's right.

Thus are all the expectations of popular music and entertainment built. You seem to know when the change to the chorus or when the punch line must come. A slight difference in rhythm may get extra attention, but total disregard of it gets the listener's disgust. I'd say this is a case of a hardware limitation influencing taste, but that's conjecture. It certainly maps well onto the data, though.

You may have noticed that jazz or modern classical music often is listened to by "intellectuals" who presumably have more mental horsepower to spare for things that can surprise their sophisticated expectations. However, far-out jazz or modern classical music resolves to a pattern eventually, or it's unpopular even in that world.

Human Hearing Capabilities

Human beings are so varied in their perceptual abilities that it's important not to overgeneralize. Averages can be very misleading in this area. One of the first people to accurately measure individual talents and track their effects on a person's life was Carl E. Seashore (see the bibliography in Chapter 1).

Seashore has found that, as far as musical training and ability go, nature is much more important than nurture. By following the musical training of people through their lives, he showed that ability is something one is simply born with. Practice and training have, at most, about a 10 percent effect on talent, far smaller than the overall range of acuities evidenced in the general population. Controversial, perhaps, but provable and proved: nature wins over nurture. If you can get your hands on some of these original studies, I'm sure you'll find them enlightening.

Time Discrimination

Time discrimination issues fall into several categories. At the simplest level is an ability to tell whether two sounds occur simultaneously or the question of how much later than an original sound an echo has to be, to be perceived as a separate echo. It turns out that there's quite a range for this "magic" number. A few percent can discriminate simultaneity at approximately 5 milliseconds (ms), and almost everyone can definitely tell that sounds are not simultaneous when they're about 35ms apart. This brings up an important point for anyone developing special effects or digitally processing audio for the consumption of others: Your hearing may be better or worse than various members of your intended audience, so you might not produce the subjective effect you're after if you don't take these variances into account *all the time*. In other words, just making it sound good to you may not be enough, as is commonly thought in the business.

At a higher level of time discrimination is rhythm, which is more difficult to describe. As a successful drummer, I think I have a good mental model of rhythm, which I define as anything that repeats, or nearly repeats. First, assume that some pattern, a musical bar perhaps, is repeated exactly over and over. Look at what goes on inside that bar before proceeding. Most music is written using simple integers and their ratios for timing purposes. Thus, you have whole notes, half notes, which are half as long, quarter notes, and so forth for the series that's based on the number two. Going up the prime integers, three gives triplets or 3/4 and 6/8 time, which are sometimes used together with 4/4 or the 2-based series of divisions in a bar. This has been done since well before Chopin "introduced" polyrhythms. It works because it's easy to make a bar come out even while using both at once. For instance, a bar divided into twelfths can handle both systems at the same time. This "pulse," or underlying division of the bar, is how we tell if someone is playing "on time" or not, although it's a bit more complex than just that.

Most people perceive this division into the common multiple without thinking about it, and indeed, playing exactly on the pulse sounds good to most people. The two division systems based on the integers two and three cover most classical and popular music, whereas "quints," based on the integer five, are also coming into use these days, particularly in hard rock and jazz. I'm sure you can understand the principle and extend this to any prime or nonprime combination of divisions that come out even.

Now I want to talk about that thing called a "groove." I need to define what it is because it's so important to most modern music. In my definition, a groove happens by playing slightly off the pulse as well as directly on it. Where the accents occur also helps define a groove.

No one, obviously, is perfect, so no one ever is quite exactly on the pulse. You could consider perfection, like that a computer-based rhythm machine plays, to be one kind of groove, but it's too obvious and quickly becomes boring to most listeners. A real drummer will almost *never* attempt to play perfectly on the pulse which is what makes the machines stand out. I'm going to argue that it's nonperfection that helps make music pleasurable. Before I go on, I have to say that a lot of things that sound like a groove are simply the failure of the listener to hear implied divisions, when in fact they're there and all actual notes are being played right on the pulse. This is done on purpose when a composer deliberately leaves out cues. I learned this the hard way when I programmed rhythm boxes to sound natural. A lot of things I thought were human imperfections or groove could be programmed by simply going to a finer level of division and playing precisely on the pulse, as many swing variations do.

That's not what I'm talking about here. What I *am* talking about is the deliberate (perhaps intuitive) use of timings that are *not* on the pulse — usually late but occasionally early — in that magic region of 5 to 35ms or so. It turns out that a slightly late timing in the middle of otherwise correct timings is generally perceived as an accent rather than lateness per se. I've measured timings very accurately in digital recordings, and it turns out that good percussionists playing fast-attack rhythm section instruments can repeat these slight delays to an accuracy far better than it's usually assumed the human ear can discriminate — within a couple of milliseconds at most — and can deliberately vary the exactness of repetition from bar to bar to achieve the desired effect of either grabbing your attention or lulling you, as desired. Current musical notation just can't handle this case, but that doesn't mean it's not an important effect when a musician interprets a score! It's why some musicians make Mozart sound joyous (as I think most of it should sound) and others make it sound mechanical and boring, leaving out for the moment those who just can't play well. This ability may be what determines whether you have rhythm or not and whether you appreciate certain types of music or not. Again, it points to the idea that different subjective impressions are produced in different people by the same stimulus.

Locale is another type of time discrimination. It is how you can tell what kind of acoustic space you're in, even in the dark. Without getting into some pretty hairy math and a brutal and rigorous exposition, I can't say a whole lot more here, other than that the spatial sense provided by hearing is obviously as affected by the basic perceptual acuity in time as the issues I've discussed above. It seems the human system is able to do a kind of inverse ray tracing, in which it builds a mental model of the surroundings and only builds logical models of physically realizable spaces. To map this onto 3-D graphics, it's as though you could infer all the objects in a room and all their light source properties from the illumination data of a single object. (I'd like to see someone write that software or come up with a good math derivation to make it possible!) This topic is pretty complex because it also is affected by frequency responses and directional perception abilities. In other words, this ability seems to use more than one of the channels of information the ear provides the brain: timing and frequency response information are used together. In fact, I'll argue that what comes in by these channels must be logically consistent or you can't produce a good effect when creating a virtual soundscape. At least this is what I find in practice with the techniques I'll describe later. Luckily, you don't have to understand exactly how the ear reverse engineers the space if you stick to creating logical models.

Frequency Discrimination

I'll divide this into two major areas for exposition. The first is the ability to discriminate pitch, one from another and flat from sharp, and the second is discrimination of a sound's texture or character via its spectrum. These are two very different things, but they come through the same perceptual channel to your brain. Again, there are significant differences among listeners in these areas.

Pitch Discrimination

Just about everyone can tell the difference between notes a single step apart in the universal 12th root of two scale. In fact, almost everyone can easily detect about half of this. It's probably one reason the scale is made up of 12 notes, instead of more or fewer notes, per octave. It is a lowest common denominator, if you will. When Carl Seashore measured the ability to differentiate pitch in both accomplished musicians and the general public, he found a wide variance in basic abilities. People at the low end of the scale tended not to be much interested in music, whereas people at the high end of the scale tended to be very attracted to music, often becoming musicians themselves. When he restricted his tests to highly talented and popular players of instruments where pitch is very important, such as the violin, he found that those people tended to land in the top first percentile of acuity. One interesting anecdote occurred when one member of a musical family was asked why he shunned music and hadn't become a musician like his siblings and parents, despite testing very highly on the standard acuity tests. His answer was that it all sounded so bad to him! He was *so* acutely sensitive that all music sounded like it was out of tune! Luckily, most people are not as sensitive as this poor fellow, falling somewhere in the middle of the range. It turns out that the most sensitive people can differentiate pitch so well that some pretty fancy math or hardware theory has to be considered to explain how they might do it. For example, consider two different ways of measuring frequency: You could count cycles for a second, which would take one second to get a 1Hz resolution (in fact, this is how most frequency counters work), or you could simply time the length of a single half-cycle (e.g., time the zero crossings) to an accuracy determined by the signal to noise ratio and know to a degree of high precision what the frequency is with a better time/frequency resolution product. Apparently, the ear can do this even without the constraint of a good signal to noise ratio, aided by the presence of other tones that are apparently used as reference pitches. This is another mystery I'd love to see worked out rigorously, because it would inform more than one science to know how this is done. Information theory and time/frequency resolution limits would indicate that the ear couldn't accurately tell what note a bass was playing if played fairly fast. The note wouldn't last long enough to be resolved in frequency that well. But some people can do it!

At one extreme, everyone (or nearly everyone) can hear half of a 12th root of two scale note, a few can hear as little as one percent of this, and most people fall in the range of 10 percent or so before they consider a tone either definitely sharp or flat. I would say, however, that even if a person can't say a tone is sharp or flat, they can often tell that something isn't quite right. This is borne out by the fact that virtually all popular music is tuned dead-on or it sounds amateurish somehow, even to those who don't know exactly why. Garage bands take note! Buy a tuner! Alternatively, if you're charged with helping someone produce music that is supposed to sound pleasing and professional, you obviously should have either very good ears or a mechanical aid to ensure true pitch. One notable exception, which is too complex to go into in great detail here, is when a solo instrument uses a different scale than the 12th root of two and sounds better to most people that way. These scale methods, known as "just," "natural," or "Pythagorean" tunings, generally are used only in isolation and with instruments such as the violin or voice. They tend to very slightly modify the 12th root of tuning notes so that they fall on simple integer ratios of the note that defines the key the instrument is playing in, rather than the irrational 12th root of two ratio, which represents a best average for all keys.

Sound Character

In this section, I discuss the range of abilities of the general population to hear sound textures or characters. For the purposes of this discussion, I'll define sound character as the ability to perceive spectral components, whether the harmonics of a sound are in tune with the fundamental, and whether the original sound has gone through some highly phase-dispersing media. Although the latter ability is more controversial, I can surely hear the effect myself. You could define other attributes that might be described as sonic character, but I'll avoid the temptation. Sonic character also begs the general question of whether everyone, for example, sees red the same way; that is, is my internal experience of red the same as yours? Who knows? Like many things, this is the result of hearing others describe things as red and mapping your internal experience to some average of all other descriptions for future use in analyzing it or describing it to others. This is just one of those instances in which you use some unknown internal reference to make a judgement. In some ways, this is advantageous, because those with high-frequency hearing loss can judge and enjoy sonic characters even without the same amount or kind of information that another person may get.

Most naturally produced sounds have a frequency at which most of the energy is radiated. At frequencies above and below this ideal level, the energy falls off. The reason is simple. Below some frequency, a source of a given size cannot radiate its energy efficiently. Because most sources consist of an imperfect impulse followed by some filtering, there's a high-frequency roll-off, resulting in a characteristic curve with a peak that usually falls off at around 6 decibels (dB) per octave on both sides of the maximum. Sounds that violate this behavior are usually perceived as unnatural, at least at first. One subliminally judges the source size (perhaps a survival instinct) by this peak frequency. Large, perhaps dangerous, things have a lot more very low frequency content than smaller sources do. Who has never cringed at thunder or at the approach of loud, thudding footsteps? You learn not to react that way later in life, but at first, nearly all people fear these things.

You subliminally judge the distance to a sound source using a similar algorithm. In air, the attenuation is a by-wavelength phenomenon, so high frequencies, which go through more wavelengths per unit distance, are attenuated more than low frequencies. Audio engineers long ago learned to add "presence" by boosting the higher frequencies a few decibels In that art, the boost usually began at 1 or 2KHz and became a 6 or 8dB peak at around 8KHz. Formerly, engineers didn't continue this boost all the way out of the hearing range in order to avoid boosting the various kinds of noise the imperfect audio equipment injected into the sound to the point of annoyance or to the point of destroying the perceived effect. With modern equipment, this is not as much of a problem. My experiments indicate that continuing to boost the high frequencies to the limit of audibility is a good thing to do to get a source to sound present or closer. Another way to judge the distance to a source is in the ratio of its direct sound to any reflected and delayed sounds that may be present. In this case, less reverb makes a source sound closer.

Source motion is sensed via several cues. Obviously, the source's relationship to the reverb will change, as well as its high-frequency response. But even more obvious is the Doppler effect. This well-known effect causes the source frequencies to be shifted upward if a source is moving toward you and downward if a source is moving away, keeping all the harmonic relationships intact. In a sonic environment with reverb, however, the effect is more complex. As the source moves toward you, it may be moving away from a surface off of which reflections are generated, so you hear a much more complex set of shifted frequencies.

Today, I think the reality of sounds may be judged based on something more subtle than the characters described above. A sound that is hard band-limited, such as that produced by a speaker with a sharp low-frequency or high-frequency cutoff, sounds reproduced because the usual 6dB per octave fall-off shape of most natural sounds isn't reproduced well — a telephone being an extreme case. Certain types of phase dispersion also make a source sound like it's been taped because all analog tape decks have a nonminimum phase response. They all have a high-frequency fall-off that isn't accompanied by the usual phase lag produced by a natural source. In addition, small amounts of white or nearly white noise added to a sound may make a stereo sound better to some ears (as revealed by double-blind testing), but the noise is also a dead giveaway to many that it is not the real thing.

Phase Effects

There has been some controversy in the audio business about whether people can hear absolute and relative phase in audio signals. I'm here to say they can, no doubt whatsoever. In an extreme case, consider a highly phase-dispersive all-pass filter that has more delay for high frequencies than for low frequencies. Such a filter would change the click sound of a perfect impulse into a "boing" sound, or a frequency sweep, so at the extremes, humans can indeed hear phase variations.

They can also hear smaller ones, such as those introduced by crossovers or equalizers, and after a little thought, it's obvious why this is true in some cases. If there is no phase dispersion, an impulse is a very narrow, very tall pulse. With phase dispersion, the energy is spread out into a longer time frame, with different frequency components occurring at different times. What's so different about this? It seems that the sounds with greater peak sound pressure level have different characteristic effects on the perceptual system. It is an unquestionable fact that they have different effects on nonlinear media, such as air. A bigger peak is, well, bigger after all, and shorter too, and all the frequency components are there at the same time. This implies that if there is any nonlinearity in the system (and all systems, including air, have some nonlinearity), intermodulation distortion among the components can be produced. This can't happen if the components are spread out over time, in which case the only distortion is harmonic in nature, only one frequency component is present at any one time, and no intermodulation distortion is possible among the components. I'm convinced that this is why certain people think analog sound is better. In some cases, it might be in some sense, although now that you understand the type of distortion, you can reproduce it digitally. Real audio is a lot more complex than an impulse, but phase shifts make a definite audible difference as well. I noticed the effect when working on an arbitrary EQ for a wave editor. At first I used the FFT/convolution/FIR method, which produces zero-phase filters, something quite different from the minimum-phase filters found in nature (a minimum-phase filter is one in which a certain rate of gain change versus frequency creates a certain phase shift, for instance 90 degrees shift per 6dB per octave). Although I could produce any filter shape desired, it just didn't sound right, especially when used for sharp filters. Later, I used IIR (Infinite Impulse Response) filters, and the resulting EQ effect sounded much better. In effect, having the natural phase shift implied by the EQ shape sounded more natural. Indeed, if the EQ has a minimum-phase property when used to remove naturally produced degradation, it will cancel the degradation's most probable phase error, as well as the amplitude error, which is a useful property. Another place this effect comes into play is with analog tape decks. Although most pre- and post-EQ used by tape decks to cancel one another are used to get a better perceived signal to noise ratio only, some pre- and post-EQ is used simply to attempt to correct for the high-frequency fall-off caused by a finite tape head gap. It turns out that this fall-off is phaseless, so the EQ used to compensate it has some phase that isn't cancelled by this effect, producing a part of the signature tape sound. The effect of just about any frequency-dependent phase shift is to lower the amplitude of the peaks, because now they are spread around temporally. This means a lot of other events in the chain will be operating out of clipping (i.e., without clipping the peaks), where they might have clipped the original peaks, and the time smear may make the sound appear smoother, especially to the analog sound lovers. Now that you understand why analog media sound more natural to some people, the "analog sound" is no longer strictly in the domain of analog-only implementations. You can do it even better in digital. You just have to stop arguing over which is best, look for the truth of what is actually going on, then replicate it on demand.

Amplitude Effects

Amplitude effects on perception can be divided into two major kinds: those that are time invariant (stationary) and those that change with time (dynamic).

The two most commonly discussed stationary amplitude effects are loudness curves and masking. Loudness curves, such as those published by Fletcher and Munson, illustrate the different sensitivities of the ear and different dynamic ranges at different frequencies. The masking phenomenon is one in which closely spaced tones can mask each other if one is significantly louder than the others.

I have a major rant to make about how most current schemes attempt to account for loudness curves with the "loudness" control found on almost all reproduction devices. Looking at the curves (shown in Figure 2.3), you can see that the ear is most sensitive around 1KHz or so, with decreasing sensitivity on either side of this frequency. It's weird that this happens to be approximately the nuclear magnetic resonance frequency of hydrogen in the earth's approximately one-third Gauss magnetic field, because the physical connection sure isn't obvious. At any rate, when you look at the curves of perceived loudness versus decibels of physical amplitude, notice that at all frequencies above 1KHz, the curves are more or less equally spaced;

therefore, no compensation is needed for varying sound levels here. After all, you're interested in the *difference* in perception at different levels, and *absolute* sensitivity has little, if anything, to do with this. Loudness controls that boost the extreme high frequencies may sound good subjectively; yet, they have nothing whatever to do with reality. However, in the bass region, the curve spacing is compressed until the amplitude gets quite loud.

Figure 2.3   Equal-loudness contours for pure tones by Fletcher and Munson. (*American Standard for noise measurement*, Z24.2–1942.) The dotted lines are equal-loudness contours for bands of noise, 250 mels wide as determined by Pollack. [*Pollack, J. Acoust. Soc. Amer.*, 24: 533–538 (1952).

Reprinted with permission from Beranek, L.L. 1990. *Acoustics.*, 3rd printing. Acoustical Society of America.

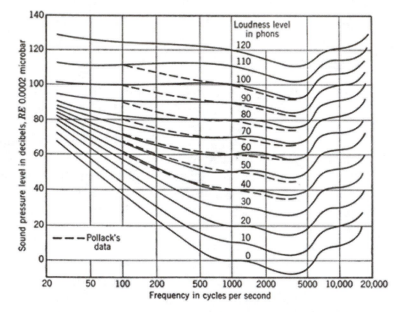

This means that every loudness control on earth, by simply boosting the bass at low levels, is doing the wrong thing, because in a performance recorded at high volume levels (virtually all rock and roll, for instance), a 10dB amplitude change in the bass sounds like a 10dB change; however, when played back at low volume levels, a 10dB change sounds like a larger change (see the curve's implicit compression). In the early days, controlling the dynamic range of low frequencies properly was difficult, but now it is not. With stereo equipment at stratospheric price levels, I wonder why no one has fixed this obvious defect. It should be simple to adjust the dynamic range in the bass appropriately to effect the correct perception. Boosting frequencies is better than nothing, but when someone charges $10,000 for a preamp and doesn't address this issue, I wonder whose pocket is being picked. It's not as though the parts would cost significantly more of the profit margin. Stereo manufacturers should want to compress the bass region at low levels, rather than simply boost it, to get the right effect, so that what sounded louder at a high volume level would sound equivalently louder when played back at a lower level. In real life, you can't control both the recording level and the playback level, so an adept system should at least account for the possibility of playback at high and low volume levels, especially when you're using equipment that costs as much as an automobile. The stereophiles all seem to argue about things that are essentially baloney to make sure there's enough controversy to sell the high-priced systems; yet, they miss the obvious problems. It sure isn't based on science.

What IS Normal, Anyway?

What do I mean by normal? I'll define this as sounds that have received no deliberate processing before you hear them, or naturally produced sounds. These tend to have certain identifiable characteristics that the perceptual system keys on to know whether a sound is normal, or has been modified since its production. Common characteristics of natural sounds include

- impulse excitation/filter model,
- minimum phase,
- radiation tilt,
- logical reflection structure, and
- exponential decay with time.

I will take each of these in turn. The impulse excitation/filter model is more thoroughly described in , where I cover analysis of sounds, but for now, what I mean here is that most natural sounds are produced by some impulse that is then processed by various filtering mechanisms. The sharp impulse might be a hammer strike, vocal chords slapping together, a reed in a woodwind slapping against the mouthpiece, a violin string releasing from the bow periodically, or any of a number of other things I'm sure you can think of. The important characteristic of an impulse for this discussion is that it is a harmonic-rich waveform — a "perfect" impulse has equal energy at *all* frequencies. Most natural impulses aren't perfect but are approximations that may have a significant nonflat frequency content that helps determine the overall character of the resulting sound. Perfect impulses are actually very uncommon in real life, although "good enough" approximations are fairly common. Once an impulse has been produced, nature generally contrives to filter it in various ways. The filter can arise from resonant mechanical structures, such as a piano string or an anvil, or it can arise from resonances produced by a confined blob of air, such as inside your mouth. Filters also arise due to whatever radiation mechanism moves the energy from the mechanical structure to the air. In general, big things radiate better than small things, big and small being defined in this context in terms of the wavelength of the frequency involved. The result is that most natural sources pick up a radiation-resistance defined "tilt" or a filter that passes higher frequencies better than low. A very important characteristic of virtually all natural filters is that they have the minimum phase shift possible for their sharpness, which appears to be something that the ear can often detect. Once a sound has been radiated into the air, and often even before this, it inevitably reflects from things, producing multiple copies at various delays from the original. As I state elsewhere, this also produces a filtering effect, though whether it's perceived as such or as reverb or ambiance is mostly determined by how long the delay in question is. When these delays reach the order of 10ms or so (corresponding very roughly to an extra 10 feet of flight in air), the reflections tend to be perceived as reflections or reverb rather than as filters. Your perceptual system then attempts to internally "draw a picture" of the shape and location of reflecting surfaces that would produce such reflections. If this does not result in a logical picture, the perceptual system immediately identifies the sound as an unnatural one.

Although this is not a property of all sources, after the excitation stops, the echoes or reverb of any natural sound tends to decay exponentially with time. Most sources follow this rule, and virtually all acoustic enclosures definitely do. This is yet another cue the ear uses to notice whether a sound or sound/environment combination is natural or not.

Given the above, you now have some clues on how to generate or process sounds such that they sound either natural, or unnatural, depending on what you want. If you want the sound to be perceived as natural, you need to ensure that you start with an impulse of some kind and filter it in a minimum-phase manner. If any reflections are present, they should correspond to those creatable by some possible group of reflecting surfaces in 3 space. If you want the sound to be perceived as unnatural, you would use a nonminimum (or zero) phase filter or filters and add delayed versions of the signal in such a way as to generate a nonlogical apparent group of surfaces. By knowing this, you can trigger the perceptual mechanisms the way you want, to produce a desired effect in the listener.

Fooling the Ear

Given the above information, you can plan intelligently how you might fool someone's perception. Because of preprocessing, the lazy brain, and perceptual rhythms, it's actually pretty easy once you figure out what you're after in doing so. The basic things you can fool are the spatial sense, the ability to pick one sound out of a group to tell it was "fake," and the sense of pitch. The first two can be overloaded by simply providing enough distinct stimuli to overload the short-term memory chunk limit for the "sense" in question. The latter is accomplished by providing fake harmonics of the pitch you want perceived.

Overloading the Spatial Sense

You can overload the spatial sense in either a realistic or unrealistic manner to achieve different goals. If you want to simulate something reverb-like, all you may have to do is ensure that the number of echoes builds up very rapidly after their onset so that the ear cannot really build a good internal model of the space. This can sound good, if not realistic, and is the basis of many widely used inexpensive reverbs. Another often-used technique is to have a number of delay lines with feedback, in a chain. One of the ways this falls down in terms of realism is that delays with feedback only produce precisely periodic echoes, which are not the only or even main kind that occur in real life in real rooms, where the sound does not simply bounce between pairs of walls, but takes all sorts of oblique paths through the space. Another way the delay-chain falls down in terms of realism is that it produces a relatively simple, effective comb filter which the ear picks up on easily. Real reverb produces such complex effects on total frequency response, often with several peaks and dips per Hertz, that the ear doesn't perceive it as a filter.

If you want to simulate either a feeling of being in a large reverberant "real" space or a 3-D soundscape, you must use accurate delay taps that simulate what would actually happen in such a space, at least for as long after initial onset as is required to allow the ear to perceive the virtual walls, after which you can use the easier multitapped simple delay and feedback strategy, feeding it the "real" front end echoes so it'll be repeating something more complex than just the original sound and not give itself away so easily. Since this is only used for the lower level water echoes, it's also easier to fool the ear with them, since much of the content is masked by the louder direct paths and shorter delays.

You might also want to put a head-related transfer function approximation on each echo separately depending on what direction it's supposedly coming from, to enhance the perceived reality, as well as low- and high-pass filter successive echoes more severely to simulate what happens when sound bounces from surfaces, where bass frequencies tend to pass through the surface rather than reflect and higher frequencies are absorbed both by surfaces and the air preferentially. It turns out that nearly all acoustic tile designed to quell excessive reverb and noise has a peak of absorption at around 1KHz, so an accurate scheme might want to simulate this as well. Although all this more or less eliminates the simple tapped delay or convolution as a contender for producing an accurate 3-D soundscape, I've discovered a method that is relatively computationally efficient, which will be given in Section 2 of this book, as well as in a code example.

Chorusing and Multiple Sources

If you provide the listener with enough slightly different versions of the same original sound, you can fool their perception into believing that there really are that many unique sources, by overloading their ability to pick them out individually and determine that they are just copies of a single source. The popular chorus effect is based on this idea. The more different copies you have, the less convincing each has to be, especially if there is also a dominant, obviously real sound in the mix. Elsewhere I provide some examples of this and several different techniques for modifying each copy slightly differently so as to produce either a convincing or merely pleasant effect.

False Pitches and Fundamentals

Although amplitude and, to some extent, harmonic content influence the perception of pitch, these are not often used as part of special efx to produce apparent pitch shifts, since the amplitude of a sound is usually determined by other needs and the harmonic content determines its basic character, which you usually don't want to change much. However, another effect is commonly used to produce a false apparent fundamental pitch when no energy at that frequency is actually present. A large part of the pitch detection mechanism of the ear works by looking at harmonic *spacing* and assuming the pitch equals this frequency, so if you have a 30Hz signal you know can't be reproduced, you may add some harmonic distortion to it to add all its harmonics (or accomplish this some other way) and fool the ear into hearing 30Hz *even though you've filtered all of that away* to leave more headroom in the signal for frequencies that *can* be produced by the channel. One simple trick for producing the "octaver" effect that seems to add a note an octave below what you're playing is to generate the pitch divided by two, but instead of *adding* this to the output, use it to *modulate*, or multiply the original signal, in effect inverting every other pitch period. This has the same effect as harmonic distortion in that it adds sidebands to the existing harmonics in such a way as to produce a series spaced at half the original pitch, but without altering the overall energy balance versus frequency as much, and with less intermodulation distortion among the original components. Although nothing sounds (or feels) quite like the real thing here, it can be very convincing if you haven't just heard the real thing for comparison.

Summary

Let me summarize what I've tried to impart in this rather long chapter. First, one must understand the characteristics of a given receiver if one is going to be a successful transmitter, and while most of the receivers are about the same, there *are* significant variances to take into account. Second, the way a lot of these receivers work is to compare what is being heard now against a template of the characteristics of most common natural sounds, and slight differences from these templates are attention-getting. Third, receiver attention is a rhythmic phenomenon. Fourth, the receiver has some limitations, which allow it to be fooled into perceiving something that isn't really there. All these are very important concepts to anyone who is going to be programming (in either sense) for the general population. Perhaps most important of all is the concept that there is definitely a large range of hearing abilities, which becomes very important when you try to make something sound natural that didn't originally. You cannot assume that because it sounds good enough to you, it will fool everyone out there. The only cure for this is to engage other ears and to have some analytic knowledge of natural sounds so you can replicate their characteristics more exactly.

Chapter 3

Practical DSP Primer

What is Signal Processing?

I can't really cover all of DSP in a chapter or two — even in its current, fairly primitive, yet very complex state — so I'm going for just the hot spots and the implications for special effects work in a practical, rather than an academic, sense. Digital signal processing (DSP) is a subset of General Signal Processing, which is a very broad topic.

For the purposes of this discussion, I'll define a Signal Processor as something that takes one or more inputs; does something to them, perhaps using and modifying some internal state information; and produces one or more outputs that are more useful than the original inputs for a particular purpose. At least, you hope so! I suppose that for the most general definition, the last part is not necessary. You can process a signal and turn it into garbage and still be a processor of it, after all. It all depends on what you wanted. One could make a pretty good argument that consciousness itself is a signal processing phenomena, since the input comes through the senses, gets processed using and modifying the internal state (otherwise known as memory and emotions), and produces behavior (output!). Of course in this case, too, the output may not be exactly what you wanted; but indeed, it fits the general definition of signal processing given above quite well. All this electronics and computer stuff is just scratching the surface of the real topic of general signal processing. Digital signal processing is just another way of doing the same old stuff, with some different caveats and capabilities than for the analog signal processing case. In fact, there's really not too much theoretical difference between analog and digital signal processing, except that certain things are much easier to do in a particular domain. Therefore, much work has gone into the development of the techniques that are easier in a particular domain, separated by domain, which tends to confuse the issues. For instance, there are tons of books on analog filter design and few on analog techniques for statistical operations, or things that need random signal access or long-term memory to work, whereas the reverse is true for the digital domain. Once you have an analog filter design, a simple substitution equation will make it into a digital design, so there's little need for more books on this topic — the old ones suffice, mostly. Because statistical or random access sampling is easy to do digitally and quite difficult to do in analog, you mostly see books and papers on doing those things digitally. But this confuses the issue for many readers who are led to think that analog can only do some things and digital can only do other things. In reality$_0$, you can do almost anything in either domain. Digital is popular because it's now cost competitive, and once you're in the digital domain, certain classes of things become far easier to implement.

So, a lot of things do signal processing. Here, I'm mainly concerned with a small subset, that is, signal processing of audio. A large number of signal processing techniques and combinations thereof find application here, so in covering these, I'll actually be covering quite a lot. The most basic technique, at least in terms of its frequency of use and the length of time it's been around, is to filter a signal. Normally, a filter means a processor that alters the amplitude and probably phase (time) response of a signal versus its frequency. A "null" filter would be a piece of wire in analog or the identity transformation in digital, out[n] = in[n], and would have no effect whatever. It's the non-null filters that are interesting, of course. These alter the amplitude and phase versus frequency curves of a signal in the most common usages. Thus, the tone controls on your stereo are filters, the speakers are unintentional but nonetheless *actual* filters, and so on.

The second common technique involves using some delayed signal, and adding it back into the signal as an echo, which sounds like, well, an echo if the delay is perceptually long, like flanging if the delay is short, or perhaps like reverb if you use many delay taps and filter each tap separately to mimic the frequency response changes of sounds as they bounce from objects in the real world. This technique also produces filter-like effects, since adding a delayed signal causes cancellations at frequencies where any odd multiple of half-periods are equal to the delay time and causes additions at the even multiples. In other words, at any frequency where the delay produces a 180-degree phase shift (or any odd multiple of 180 degrees), you get a notch, and where the phase shift is 0 degrees or any multiple of 360 degrees (i.e., even multiples of 180 degrees), you get a peak. If the delayed and original signals are the same amplitude, the notch has infinite depth, called a zero, and the peaks are just 6db higher than the original signal; that is, twice as big as the original signal, which seems reasonable since $1 + 1 = 2$ and $1 - 1 = 0$. This is called a comb filter because a plot of its response looks like a comb with its teeth pointing down.

The third major technique I'll be covering involves various nonlinear things one can do to a signal. These could be lumped under the word distortion, or perhaps modulation, since they always produce new frequency components that weren't in the original signal. A guitar amplifier turned up so far that the peaks of the signal are "clipped" due to insufficient amplifier power for the gain used is the classic example of this, but there are other interesting things that one might do, such as multiplying the whole signal by another signal, or by itself, or using just about any nonlinear curve as the input to output transformation, rather than simply limiting the signal to some level. This class of techniques also includes compression and expansion, which cause distortion of the original amplitude envelope characteristics of the signal, and can produce spurious frequencies in noticeable amounts if the gain control signal you're using to multiply by has any significant higher frequency components.

Taken together, these three classes of techniques comprise virtually all signal processing of audio. The more complex and interesting efx use more than one filter, delay, or nonlinear element in various combinations, but you have to understand some of the basics and their implications first — after that, the rest becomes obvious pretty quickly.

What is DSP?

DSP is a subset of General Signal Processing where the "D" seems to mean that digital hardware is used. It further implies that the input signal(s) are discrete, meaning that they have been reduced to samples of the original input at some sampling interval that is not necessarily regular, but almost always constant for audio processing. Thus DSP, in general, *usually* works with quantized samples of what was originally a continuous signal, in both the amplitude and time domains. Some signals are inherently discrete already, such as switch closures, which is why I say *usually* here, to be correct. This implies that from the beginning, you're losing some information about the signal. The trick is to make the losses minimal or effectively nonexistent.

Actually, this is true with analog processing too, where some noise and other inaccuracies always creep in, and the frequency response isn't infinite, either. In the digital domain at least, you can reduce these errors as much as you want, or are willing to pay for, with increased converter resolution and processing horsepower. As long as you don't get too pedantic about having "zero" errors, they can indeed be made vanishingly small by using faster sample rates, more bits of resolution, higher precision math, and so forth. A fundamental theory by Nyquist states that as long as you sample at a rate more than twice the highest frequency component in the signal, all the original information is still there. This seems counterintuitive at first, so I'll go over it in some detail a little later.

Comparing ASP and DSP

To explain what's special about DSP, one has to understand what analog signal processing (ASP) does well and what it doesn't do well, or easily, and the same for DSP. In analog, things are hard to set very precisely, tend to drift, and add noise and distortion. In analog, it's pretty hard to have a long delay and good fidelity at the same time. For analog delays, you're usually limited to using tape or some other mechanical medium, which have a lot of problems of their own. The so-called analog delays in stomp boxes are actually bucket brigade devices (a shift register made up of sample and hold elements) that sample the input signal, and although they store the amplitude as an analog value, since they sample the signal at discrete times, they are really digital already. In analog, it's tough to design hardware that can be dynamically reconfigured with any real level of generality. On the other hand, in analog one can filter a many-megahertz signal with a tuned circuit — a 20-cent part — whereas this would take a pretty expensive digital processor. In general, this sort of thing costs more to do in digital than in analog because of the speed required by the Nyquist criterion.

What's so special about DSP that everyone is adopting it nowadays? The answer's pretty simple. When doing DSP you have random access memory, and you usually have a relatively general-purpose processor. The benefits of these two things can be enormous in certain situations. For instance, if you need a single piece of hardware that can be reconfigured on the fly, digital is the way to go, period. Although analog can do this too, it's much harder, and never as general as what you can do simply by changing the program in a DSP situation. Rather than confuse you with a lot of spurious verbiage here, I'll just list common types of designs and their major points that will help you decide which processing technique to use for a given set of specifications.

Property	Digital	Analog	Hybrid
Dynamic range	more than needed for most uses	above ~80dB is hard work and expensive	same as pure analog
Distortion	usually .0026 percent or less	below ~0.01 percent is difficult	same as pure analog
Filter tuning precision	as good as the number system in use allows	better than ~0.1 percent is difficult and costly; high q's difficult	can be good if d/a's used to set filter parameters
Flat-phase filters	easy, using symmetrical FIRs	hard extra phase compensation sections	hard
Random signal access/delays/out-of-time-order processing	trivial; this is digital's strongest single point	very difficult	very difficult
Settable presets	trivial; a user interface issue for digital machines	very hard/costly, as in, for example, switching between potentiometers	possible using d/a's to set analog parameters
Reconfigurability	trivial; load different code into the same hardware	possible using analog switches, but never very general or without some problems	same as analog, but uses uP to handle stored configuration/ switch info
Frequency range	usually less than analog for a given cost	DC to daylight	DC to daylight
Power consumption	higher than analog	less than digital	in between
Hardware design cost	low; designs often reusable for other tasks	high; several revs usually needed if a tough problem	easier than pure analog because you can tune after design
Software design cost	the major design cost of DSP	none	low; usually simple stuff
Chainability	easy, with no losses	losses at each stage	usually the same as pure analog

If you don't need precision much better than about one percent, dynamic range greater than about 80dB, significant memory/internal state, random access to past signal values, reconfigurability, or a lot of presets and adjustments, analog will cost less and use less power for a given job. Low distortion is fairly easy to achieve in analog these days.

If you do need high precision, huge dynamic range, significant memory/state, random access to a signal, or all the benefits of random access memory — reconfigurability, presets, adjustments and so on — digital wins, unless the signal frequency range of interest is very high. Even then, you might use a combination of techniques: analog processing to translate the signal down to a low frequency, digital processing, and a little more analog to translate the signal back up to a high center frequency, if needed. A pretty serious digital signal processor for audio signals can be built on a cost-competitive basis with analog gear for most nontrivial types of functionality as this is written in early 1998, and it's only getting better.

A hybrid of the two techniques is often better than either alone. An inexpensive microcontroller can provide analog circuits with presets and adjustments, for instance, by using d/a converters to set analog tuning parameters or to drive analog switches or relays for reconfiguration. A hybrid radio might be a good example here. You'd design this so that the hardware that handles very high frequencies are analog, and you'd only digitize the signal after you'd converted it to an easier-to-handle frequency or "baseband" for further processing. In general, this is practical because a radio signal has a relatively low band*width* compared to its center frequency. You might also use the on-board processor to "tune" the analog parts — in this case, literally. Those analog parts could be made even less expensive and using the digital portion to retune or recalibrate them as necessary. As a simple example, suppose you know the analog section's drift versus temperature. You could then cancel this out by measuring the temperature and sending the analog parts different tuning values to compensate temperature drift to arbitrary accuracy. When done well, hybrids combine the best of each technology to do a job both better and less expensively than either can do alone.

A Short DSP Glossary

It's convenient to define terms here so I can use them later without breaking up the flow of things to stop and define them in place. Some definitions may be slightly different or less robust than those used by the math crowd, but they won't lead you too far astray and are the most common usages in the

DSP field. These are alphabetical; there is no best order because many terms reference one another.

Bin When you take the spectrum of a digital signal, you get a result that has finite frequency resolution; that is, you have a finite number of frequencies you can resolve from a given sample of the signal. With most common techniques, such as the FFT or DFT, described below, you get output bins that have equal spacing and width and contain the energy at each frequency you can resolve.

Canonical When applied to an IIR filter, this means that it uses all of the possible coefficients for its order and can produce all the possible shapes for the order specified. The short meaning is "by the rules," just as Bach's canons were written in a rule-based way. Some filters are implemented in non-canonical ways to reduce computation time when all the possible shapes aren't needed.

Complex Number (domain) In DSP it's often convenient or necessary to "phrase" certain classes of problems in the complex number domain. If you remember your school algebra, a complex number is really a pair of numbers, one describing the real part and the other describing the imaginary part, usually written as (A, Bi) where i signifies the $\sqrt{-1}$. To add complex numbers, add the real and imaginary parts separately; thus, $(A, Bi) + (C, Di) = (A + C, (B + D)i)$. The rules are different for multiplication: $(A, Bi)(C, Di) = (AC - BD, (AD + BC)i)$ You get the magnitude of a complex number as the square root of the sum of the squares of the real and imaginary parts and the angle as $\tan^{-1}(\text{imag/real})$. The multiplicative identity value is $(1, 0i)$. Dividing two complex numbers is a real bear. You do this by taking the inverse of one of them then doing the complex multiply. The inverse of (A, Bi) is $(A - Bi)/(A^2 + B^2)$. The most common complex number pair used in DSP is just the analytic sinusoid, or $(\cos(\omega nt), i^*\sin(\omega nt))$. The Fourier transform is phrased in the complex number domain, although a modification that can efficiently handle real-only data is also widely used. Some engineering texts use j instead of i to represent imaginary numbers, and I have used both in this book as well. Usually, the context should make it clear. In general, complex numbers are useful almost any time samples come in pairs, such as when simultaneously measuring the voltage across and the current through a circuit.

Convolution describes a multiplicative transform of some type. "Straight" convolution in the math sense means that you multiply some impulse response vector by past signal values pairwise and accumulate the results to produce an output sample of a filter. This is why many DSP processors boast of a fast multiply–accumulate (MAC) speed, and is exactly how FIR filters are computed. Because this straight convolution needs operations that go up as $M \times N$, where M is the filter length and N is the signal length, you usually use the convolution property of a fast Fourier transform to compute a large convolution that only goes up as $\sim 5N \log_2(2N)$ for an efficient implementation. A filter can be said to convolve its impulse or frequency response with the signal, for instance (depending on whether we're thinking in the time or frequency domain), or you might say that a microphone produces a signal that is the original sound pressure convolved with the microphone's impulse response. Deconvolution is often possible as well, so that in the above example, you might recover the original sound pressure values without the microphone's contamination by deconvolving the microphone's impulse response from its output signal. This is done by doing another convolve with the *inverse* of the microphone impulse or frequency response. One class of cases where deconvolution is either not possible or not a good idea is when the thing you're trying to deconvolve has zeros in its response (e.g., zero output at some frequency). This causes the deconvolution to try to multiply 0 by infinity, which doesn't reproduce the original signal very well at all but sure boosts any noise that was added after the original convolution.

Damping is the amount of implied friction in a filter; that is, how fast it rings down after an impulse, which is the inverse of the filter's Q factor. A *critically damped* second-order filter has a damping factor of $\sqrt{2}$ or a Q of $1/\sqrt{2}$.

Domain Often you'll find it convenient to specify a domain that you're working in or talking about. This generally means the variable you're using for the x-axis or global properties and rules that are assumed when talking about operations in a domain. This saves lots of words. For example, in the time domain you use time as the x-axis variable, whereas in the frequency domain you use frequency. Other types of domains might include the continuous domain, where signals have values at every instant of time (not just during a sample) and can take on a continuous range of values that would take a number of infinite precision to specify, or the discrete domain, where you only have samples of a signal at discrete times, which are quantized to less than infinite precision. I'll also be mentioning the *S*-domain and the Z-domain, which are used in describing filters mostly in the continuous and discrete domains, respectively, as well as the ever-necessary complex number domain, with its own set of rules for multiplying, dividing, and so on. In general, specifying a domain is just a handy way to say "the rules here are thus'" to avoid having to repeat them over and over. This is sort of similar to a name space (or a scope) in C++, for instance; the underlying idea is the same.

Euler's Relation   states that *e* (the base of natural logarithms) raised to an imaginary power produces a complex number that is the cosine and sine of that power. Thus, e^{jnt} is the complex number $(\cos(nt), j\sin(nt))$. This wave has the special property of having only a positive frequency; that is, it tells which way time runs. If you negate the exponent, you get a backward-running wave pair. A single sine or cosine wave cannot indicate the direction of time — it's the same either way. You see this a lot in DSP texts as a notational shorthand, but in reality no one actually generates sines and cosines this way. It keeps things confusing for the neophyte, to be sure.

Filter   is defined as a device (or a software simulation of one) that can reduce or increase energy at specific frequencies or ranges thereof from a signal presented to its input. It may also have a time delay or phase response that varies with frequency. This is a more specific definition than is currently used in general software development, but this definition came first, and you can see how a "filename filter" might have been derived from this concept. Here, I'll use filter in its older meaning: a linear transform that removes or boosts some components from its input, usually a range of frequencies, but doesn't add any new frequencies. In this sense, a filter is a linear transform. It can also be reversible if the filer doesn't reduce any input component to a zero output.

Fourier/Fourier Transform   Fourier was a mathematician who invented what's known as the Fourier transform. This transform allows one to move data from the time domain to the frequency domain and back. The essence of the concept is that any signal can be reproduced by specifying enough frequencies, amplitudes, and phases of component sine waves. It can also be used for convolution and several other purposes. It is probably the most widely used transform in the DSP business. The fast Fourier transform (FFT) is a discrete version of the Fourier transform, which was originally defined in the continuous domain. Cooley and Tukey usually get the credit for this modification of the original transform, which avoids many of the multiplications and additions that would be needed in a straight conversion, by eliminating duplications. You will also hear of a discrete Fourier transform (DFT), which is equivalent in results, but needs N^2 operations instead of the approximately $N \log_2(N)$ that the FFT uses. The nice thing about the DFT is that it's very easy to explain, and if you don't need to compute a full spectrum, rather only a few values, it can actually be faster than the FFT, which always computes all possible bins.

IIR and FIR stand for infinite impulse response and finite impulse response, respectively. An IIR filter never reaches zero energy, no matter how long you wait after providing the network with a unit impulse, whereas an FIR implies that the impulse response is truncated or reaches zero energy after some point in time. These terms are used to describe the most common filter types used in DSP. An IIR filter has feedback from its output to its internal state, so that its impulse response may never completely die out. In practice, it usually does eventually, if for no other reason than eventual truncation loss by the number system used. An FIR filter uses convolution of a finite-length impulse response with past signal values, so its impulse response must eventually die out. IIR filters usually are more efficient computationally, whereas there's more freedom of some design parameters for FIR filters. IIR filters are usually minimum phase; that is, they have an amount of phase shift versus frequency associated with an amount of slope in the frequency domain plot they produce, whereas FIR filters can easily be constant delay, and thus have *no* phase response if the fixed delay can be canceled by a matching delay elsewhere. Loosely stated, an FIR filter corresponds to the numerator of the general S-domain or Z-domain filter equations, which only works with previous inputs, whereas an IIR filter may have terms in both the numerator and denominator and so works with past outputs as well as past inputs.

Impulse/Unit Impulse is defined as an infinitely narrow, infinitely tall spike surrounded by silence in the continuous domain. It therefore has unit energy and no duration, at least in the continuous domain. For the discrete domain, a a single non-zero sample of maximum height is usually substituted, to be practical. Impulses have equal energy at all frequencies — in other words, an impulse is "white." The phase of every component is zero, unlike white noise, which has a flat frequency spectrum but random phases. Impulses can be used to measure all the characteristics of any linear transform or filter. For instance, if you acquire the impulse response of a filter by inputting an impulse and recording the resulting output, you can determine its frequency and phase response by using a Fourier transform to translate this data into the frequency domain.

Impulse Response   is defined as what comes out of a device or transform when you apply an impulse to its input. You can use a Fourier transform to convert an impulse response to the frequency domain and therefore learn all there is to know about the frequency and phase/time responses of a linear network quite conveniently. You can also generate an impulse response going the other way by specifying a frequency/phase response and inverse Fourier transforming this to get the impulse response of a linear network that would have those frequency/time characteristics. This is one of the many things that makes Fourier transforms so useful in DSP. You can, for instance, get the desired impulse response for an FIR filter by putting the desired frequency and phase response into an FFT and inverse transforming back to the time domain.

Linear   is a qualifier used to describe some process or transform. In DSP, this means that a linear transform has some linear input–output function and thus adds no new frequencies to the input. Most filters are linear transforms since they only remove or boost existing frequencies from the input, for instance. Linear transforms are usually invertible unless they produce a zero output for some non-zero input.

lsb, msb   are shorthand for least significant bit and most significant bit, respectively, in a digital binary number.

Nonlinear   is a qualifier used to describe some process or transform. The outstanding feature of a nonlinear transform is that it will add new frequency components to an input signal. Nonlinear transforms tend not to be invertible, but some are. If the transform has a unique output for every input, it can usually be inverted. If it has the same output for more than one input, you cannot unambiguously invert it. Any transform that generates this many to one mapping is said to be noninvertible.

Nyquist   is the name of the fellow who popularized the sampling theorem and proved that analog and digital systems can be equivalent, but it has become a more general-purpose noun in DSP usage. His theory states that as long as you sample at over twice the highest frequency component present in a signal, you can get all the available information. Thus, the

Nyquist limit, or just Nyquist, is the highest frequency that can be reproduced at the current sample rate.

Order when applied to a filter describes how complex it is. In the general case, it's the highest power of some variable used. Most IIR filters can be specified in direct form, or can be factored into first- and second-order sections to make implementation and description easier and to reduce math precision errors and coefficient sensitivities that tend to occur in direct-form filters.

Phase defines where you are in a sine wave cycle. It is often a more convenient way to specify a frequency-dependent delay than simply specifying a time delay, because often what matters is how much a particular frequency component is phase shifted, rather than how many whole cycles it might be delayed. After all, whole cycles all look alike, and in cases where things are added, whether the addition adds or cancels depends only on the phase of the components. Phase, like sine waves, conveniently "wraps around" or repeats for each cycle, so it is always between 0 and 360 degrees, or sometimes it is stated as between ± 180 degrees — it depends on what you want 0 degrees to mean. In radians, this is the range from 0 to 2π or $\pm\pi$ radians. Phase is therefore effectively time-delay autonormalized to the frequency in question, with excess whole cycles tossed out, like a modulo operation or the % operator in C. Sometimes the ambiguity implied here is good, since it saves bits and maps directly to what you need. Sometimes it is not good, since it makes taking invertible complex logarithms more difficult, if not impossible, because of the ambiguity the modulo operation introduces. Some heuristics that do "phase unwrapping" thus exist to allow this sort of thing. Although it cannot be done unambiguously, often just adding 360 degrees every time a phase takes a big jump, if that would reduce the jump size, is good enough.

Q/q is the quality of a filter. A high q represents a filter with low damping or losses compared to the energy it stores (Q = 1/D), a narrow passband, and a long ring time. It used to be hard to make a high-Q filter at audio frequencies. The necessary high Quality inductors and/or capacitors were expensive. This is why the term Q arose. In practice, a high Q does not necessarily mean a filter is better for some particular application. In some cases, it's more useful to use the inverse of Q, or D (damping factor) in analysis, so both are defined here.

Quantize   is the act of moving from the continuous domain to the discrete domain. In doing so, you let samples at discrete times and levels represent a continuous signal. As long as the sampling is above Nyquist, you capture all the signal information about time variations. Quantization of amplitudes into discrete levels adds noise to a signal, because the signal is continuous and perfect and you are in effect limiting its range to be right-on certain levels chosen to divide that range into a finite number of different levels, as in approximating a sloped line with a stair-step function. The more levels you use, the less quantization noise you get. For audio, you usually use a sample rate of 44.1KHz for a Nyquist limit of 22,050Hz and a 16-bit quantizer for 65,536 unique levels or a quantization noise of approximately −93dB.

Transform   is generally something that takes input and produces output. This term is usually used with qualifiers such as "linear," "null," and so on. A transform might also change the input signal from one domain to another, as well, as a Fourier transform does, for instance.

Common variable names and super- and subscripts are defined as follows.

F, f   are frequency variables: F_c for center frequency and F_s for sample rate. The Nyquist cutoff frequency is $F_s/2$.

i   is $\sqrt{-1}$. Some people use j instead. I use both.

N, M   define the size of the block of numbers being used, and

n, m   are indexes into these blocks, as in $y(m) = x(n) - x(n-1)$.

t   is a sample time; for instance, $\sin(\omega nt)$ would generate a frequency of ω. "Tick couchnt in samples."

ω is a frequency in radians (e.g., $f_c = 2\pi\omega_c$).

The Time Domain

Virtually all analog, and quite a bit of digital, signal processing occurs in the time domain. This domain is concerned with some value of the signal versus time. It's the way the signals usually exist in the first place, value by value, one after another, as they are generated by the source. The most common y-axis in the time domain is amplitude or instantaneous voltage, although other things can be defined and used. In analog, most signals are continuous in both time and amplitude; the analog time domain is a continuous domain as well. In the digital world, it's not practical to handle continuous signals either in time, which would imply an infinite number of samples per non-zero unit time, or in amplitude, which would imply an infinite number of bits per sample. Thus, digital processing occurs in the discrete domain. As you'll see below, this does not imply that you're losing much information about the signal at all — the loss can be made arbitrarily small.

Figure 3.1   Time-domain plot of speech

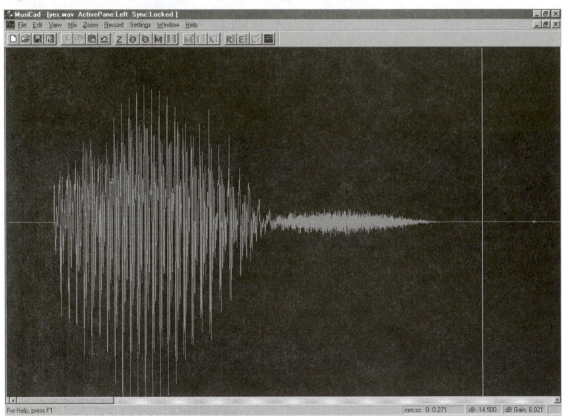

Figure 3.1 is an example of a time domain signal of, in this case, the word "yes." The plot shows the instantaneous air pressure versus time as

corrupted by the microphone and room noises. You can see the lower energy "sssss" at the end, but in general it takes a lot of practice to interpret much about a complex signal when looking at it this way.

The Frequency Domain

Especially in digital, but also in analog, it often is convenient to look at or process signals in the frequency domain. In this domain, frequency is the horizontal axis and amplitude or phase is usually the vertical axis. An important implication of this is that the plot "covers" some finite time span, since for zero time, no frequency can be defined, and in general, the lowest frequency you can deal with in the frequency domain accurately is one for which at least one full cycle is included in the time span under consideration; all frequencies below this are lumped into the DC category by all common analysis techniques, whether they are in fact DC or not. (Think about this — there's really no such thing as DC in a universe with a limited life span, just some very, very low frequencies!) The famous French mathematician Jean Baptiste Fourier defined a very useful theory and transformation to describe how to get from the time domain to the frequency domain and back. The reversibility and information conservation properties of this transform are crucial to many DSP techniques, so it's going to get its own section shortly. Basically, Fourier proved that any continuous time domain signal could be broken down into sine wave components and that the information about these components could then be used to regenerate the signal, exactly, in all cases. You might want to make some changes to the signal while it's in the frequency domain before changing it back, because quite a lot of things, such as filters, become very easy to accomplish once your signal is in this domain. As it turns out, this is also true in the discrete domain as well. This is very powerful stuff, which is why you'll hear his name a lot in virtually all books on signal processing of whatever type. His original formulation was in continuous domain calculus, allowing closed-form solutions to many trigonometry problems of the day, but it can also be formulated in the discrete domain in a way that's computationally efficient; hence, the term "fast Fourier transform," or FFT. This formulation was made popular by Cooley and Tukey, although it appears they might not have been the first to (re)invent it. Their ground-breaking paper, as well as many other papers on this important technique and its various uses, is reprinted in *IEEE Reprints of Selected Papers on DSP*.

One technique is just to look at signals to understand them. Figure 3.2 is a frequency domain of a little piece of the middle of the word "yes."

Figure 3.2   Frequency domain plot of voiced speech

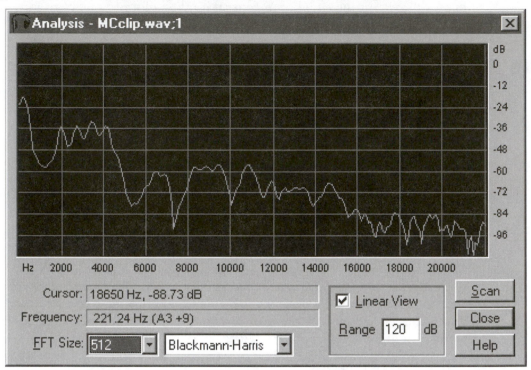

Figure 3.3 is a sample sonograph plot of the word "yes." Spectra are taken at each horizontal pixel of the original time-domain plot and displayed vertically instead, so the evolution of the signal over time can be seen.

Note the time–frequency resolution is good enough to resolve individual pitch pulses as well as the resonances in my vocal tract as I spoke the word. You can see the effect of my lips and tongue moving in this plot, which is pretty tough to see in the time domain plot. In this plot, time is from left to right, frequency is the vertical axis (0 to 5KHz), and the intensity at each frequency is mapped to darkness. The same plot is shown on the cover in color. Although I'm looking at frequency domain data, I'm plotting it in the time domain again. Many practitioners can actually read speech content from a plot like this with a little practice, and this technique is the basis of many attempts at computer speech recognition.

Figure 3.3   Sample sonograph plot of the word "yes."

You can see how an arbitrary filter might be created easily in this domain, such as setting all unwanted frequency bins to 0, or "white," and then transforming back to the time domain. This rather crude technique is called FFT-based convolution, and there are many refinements that make it work better, to arbitrary accuracy if you have the computer time.

To make this approach practical, you have to work on the entire data set at once. If not, you have to do the convolution on a lot of overlapping subsets to avoid the periodicity assumption errors the FFT will make then take the correct part of each little subset taken to get the answer.

Good Old Nyquist

An enormous number of words exist in print that cover this topic, and I'm going to try not to duplicate most of them here. Originally, they were needed to convince the skeptics that you could indeed to anything digitally that you could do in analog. The main player in this game was Nyquist, who propagated a theory of discrete systems, and under certain simple assumptions, he proved that a discrete system could model any continuous system to arbitrary accuracy. This is pretty well accepted these days, so I'll skip the derivations and cruft here, and get on with the interesting aspects and implications. Nyquist's theory, simply stated, is that if you sample a signal at discrete times fast enough, you can capture all the information in a continuous signal and convert it to the equivalent discrete one. He defines "fast enough" as at least twice the highest frequency component in the original signal. In other words, the analog signal must be band-limited to less than half the sampling rate, or information is lost. Some people find this counterintuitive and wonder about signal wiggles between samples that they'll never see. The simple answer to this objection is that if the signal is indeed band-limited to less than half the sample rate, the only kind of wiggle there *can* be between samples is a continuation of the basic sine wave shape, period. Because we know this, no information is lost. There's a caveat here though, that is often not mentioned. It turns out that although no information is actually lost when sampling above twice the highest frequency present, you may have to do some interpolation to really get it all back for some uses, and I'll point these out later. For the normal cases involved in audio signal processing, this is not a problem, and the interpolation required occurs in the final analog low-pass filter that is usually part of the output d/a converter. When you need to interpolate, for instance for resampling, you will usually use an FIR filter, since you can get any sample rate ratio this way. I'll talk about this usage in a later chapter. Figure 3.4 is a sample plot that shows the effects involved.

Here, the circles represent the samples at 40KHz, and the two sine waves are alias pairs for this sample rate. One is a 7KHz signal, which is accurately represented, and the other is a 33KHz signal above the Nyquist limit. Notice how you can't tell one from the other by the samples because they always have the same value at a given sample. Also note that although the samples seem pretty sparse on the 7KHz wave, all the information is still retained and can be recovered via interpolation. The important point here is to demonstrate that you can't tell 7KHz from 33KHz at a 40KHz sample rate!

Figure 3.4   Aliasing, 7 and 33KHz waves sampled at 40KHz. (Circles are sample times)

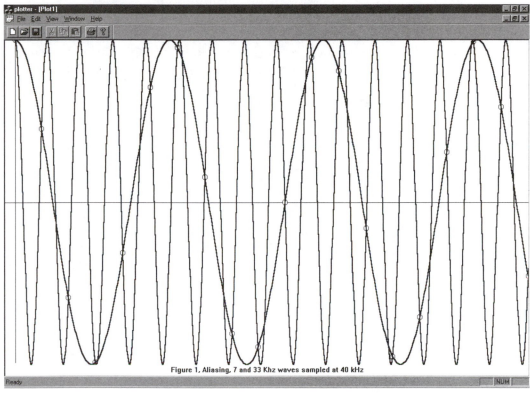

Figure 1, Aliasing, 7 and 33 Khz waves sampled at 40 kHz

Quantization of Time and Amplitude

The word "quantize" means that some continuous quantity is being converted into a discrete quantity. A continuous quantity is defined as one that can take on *any* value within some possibly infinite range, whereas a discrete quantity can only take on *certain* values within a *given* range. For time, this means that instead of having a representation for each possible instant, you have a finite number of samples for any given time interval. This is obviously a *very* important implication of Nyquist's theorem, because if you couldn't quantize in time and still have all the necessary information, you couldn't process the resulting infinite number of samples for any non-zero-length signal; you'd never get your job done! For amplitude, it means that instead of having an infinite-precision and -range number for a voltage, say, you have some finite range of numbers, and each sample is converted to the nearest number in your available set. Obviously, the latter case means that there is some accuracy limit implied by the quantization of continuous levels into a finite number of different values. As it turns out, the effect of this quantization accuracy limit is that some quantization noise is produced, limiting the dynamic range you can handle. As it turns out though, this limit is not serious these days since there are inexpensive analog to digital converters that have better dynamic range than most ears or stereos. For instance, a lowly 16-bit converter (65,536 possible values) gives a dynamic range of about 93dB, or about 20dB better than most stereos or listening situations. (Some marketing types claim that you get 6dB per bit, and therefore claim that a 16-bit converter gives a 96dB range — this is not quite true! Toggling a single bit gives you a square wave, which *could* represent a square wave perfectly if it were an exact submultiple of the sampling rate, but an input 1-bit sine wave would show 50 percent distortion, so I use the more conservative numbers here.) Some people think 16 bits isn't really enough for high quality audio, but I disagree. For example here in the boonies, it rarely if ever gets below 35dBa (absolute level, on a scale where the threshold of acute hearing is 0dBa); something is always making some noise, be it wind, wildlife, or the fridge.

Typical values for city dwellers are more like 45 to 55dBa if extreme care is taken to isolate an environment from outside noises; otherwise, the numbers are about 10dB more. So if you set your stereo so that any sound it produced was lower in level than this, you'd never hear it anyway: it would be masked by background noise in your listening room. If you add 93dB to 35dB, you get a maximum sound pressure level of 128dB — 8dB above the threshold of pain for most people. In the city, there's more reason to increase the maximum volume but less chance that you can produce a higher level without hearing a knock on the door from the local gendarmes. Higher resolution converters also exist for cases where this is really needed, although I might argue that in the audio case, more than 16 real bits of resolution is gilding the lily. The higher resolution stuff is mainly a marketing/specsmanship ploy anyway. One reason they get away with this is that there are many 16-bit implementations that don't really have 16 true bits of accuracy because of noise or nonlinearities in the converter or input amplifier, and since it's obvious to most people that more bits must be better somehow, they buy into the ploy. In reality, the problems are not due to a lack of bits, but rather poor design of the overall hardware. However, there's no excuse for this nowadays, except perhaps in very cheap sound cards.

Note that once you've acquired a signal to process, you can and should use higher precision math to do so: using 16-bit math to process 16-bit samples implies some other losses. For instance, you can't add two 16-bit samples and put the result in a 16-bit number in all cases. Multiplication is worse yet; every time you multiply, twice as many bits are needed in the result to contain all the result bits without any loss. Fortunately, some small loss is usually permissible, so you don't have to grow all your numbers by double the bits on each multiply or divide. I should mention that a dirty little secret of DSP is that much of the math formulation work is done simply to avoid the need for any divides, which not only need more bits to represent the results, like multiplies, but also have other troubles, such as low speed and the potential for dividing by zero. For most high quality audio processing, single-precision IEEE `floats` will suffice, having more than enough dynamic range and a 24-bit mantissa that limits the math quantization noise to very low levels — around −141dB. In cases where it's practical, one keeps more bits than this in partial results, using the 53-bit default internal precision of the Intel FPU. For lower quality applications, careful processing in 16- and 32-bit integers is often good enough, using the 32-bit integers only when needed. This is important since many available and inexpensive DSP processors are integer-only and handle 16-, 24-, or 32-bit numbers natively. `Floats` are almost always used in multimedia work on Wintel platforms, which actually compute faster than with integers and eliminates the need to worry about various overflow or number size issues until the final output is generated in some limited-range format.

Common Pitfalls

Something to remember as you process or generate a signal is that you must not violate the simple assumptions made by Nyquist's theory. The naive practitioner often forgets that some types of transformations, such as clipping, gating, or multiplying one signal by another, create new frequencies. If any resulting frequency is above the Nyquist limit of $F_s/2$, you'll get what's known as aliasing, which is usually an unpleasant side effect, since the new frequency created is not harmonically related to what was intended. An example of this might be squaring a sine wave input signal to double its frequency. If the new frequency is above the Nyquist limit, it will "bounce off Nyquist" or alias back down into the passband. For instance, if you try to create a 22KHz sine wave at a 40Hz sample rate, you will get is an aliased 18KHz sine wave as the result — probably not what you wanted at all. Because you went 2KHz above the Nyquist limit, your result folded around or bounced off the Nyquist frequency. This happens for all frequencies higher than the Nyquist limit, however generated. As you keep going higher and higher, the alias will eventually bounce off DC and start back up again, and so on, over and over. This effect can be useful in some cases other than audio processing: you could deliberately alias a high-frequency, but limited band*width* signal down to something you could handle easily via this effect. Simply gating a signal off or on instantly between samples will cause the same problem as generating a too-high frequency — the Nyquist limit implies that no rise times can be infinitely fast. Because these would generate extra high-frequency components that must land somewhere and because they'll always land between DC and Nyquist in a sampled representation, they usually manifest themselves as a "click."

Another common pitfall is blindly assuming that because all the information is there when sampling above the Nyquist limit, that every type of thing you might do can use just those samples alone. I give as an example the case of trying to find the exact area under a particular zero-crossing hump in a sampled signal. If you simply sum the samples involved, you'll get a wrong answer. This is because the zero crossing may actually have occurred at some unknown time *between* the samples and because you don't have an infinite number of samples to sum across. In calculus, you can set the delta to lim → 0, whereas in sampled data, you have samples that are some finite distance apart. You *can* recover all the information to whatever desired accuracy by interpolation to a higher effective sample rate internally; the thing is, you just have to remember to do so in these cases.

A third common pitfall, mostly among those who have done analog work and are now doing DSP with integer numbers, is that you must remember what happens when a value exceeds the available range of the number system in use. In analog, the signal simply "clips," or limits, which although probably not intended, at least leaves things stable most of the time. In computers, the numbers usually "wrap around," which is a very different behavior than clipping and can have undesired effects, such as making stable filters unstable, adding energy from "nowhere," and so forth. It's so much trouble to handle every case of this in integer math that many special-purpose DSP CPUs have a clipping mode built into their number handling to give a more analog-like behavior for these cases. This is, in general, a good thing; however, one thing about it that has bitten me is that often you *want* a number simply to wrap around on overflow, such as when indexing into a waveform table, so you have to remember to turn that mode on and off as required.

Dithering to Increase Effective Dynamic Range

Often the case comes up where you want to convert, say, 16-bit samples to lesser bits to save space or transmission bandwidth. The simple-minded way is to use just the top 8 bits of the 16-bit samples for your 8-bit output (converting the format as necessary if you're working with .wav files — these use two's complement signed numbers in 16 bits, but 128-offset numbers in 8 bits, just to make things interesting). This works "at all," but is nonideal in various ways. If the 16-bit samples contained some signal that didn't get up into those top 8 bits, it'd be completely lost in the result or appear to shut on and off as other signal components bring the small one (the component) up into the top of the waveform, causing unnecessary distortion. One approach to this problem is to compress the 16-bit signal first, which helps if you don't find the other artifacts of compression even more objectionable. However, you're still faced with this limit on low-level signals.

As it turns out, the ear can hear some sounds even if they're below the noise, because a particular sound may have more energy in a narrow frequency range than white noise of the same total power does, and this is where dithering comes in. Basically, dithering adds noise to the larger bitness signal before the truncation. The desired root mean square (rms noise amplitude) is usually on the order of one-third of the lowest bit that will make it into the result. This has the effect of causing the lowest bit of the result to always toggle, even in the complete absence of a signal. Now, a very low-level signal can affect when, on average, this toggling occurs, so some of its information can still make it into the output at the expense of adding a little noise. It turns out that the least significant bit (lsb) of any digital sample stream is almost always toggling anyway, unless it's been digitally noise-gated, because of infinitesimal noises that cause the sign of the input to appear to change. This is taken into account when one says that a 16-bit sample has a 93dB dynamic range and an 8-bit sample has a 45dB range. In effect, there's nothing to lose by doing this.

White noise is generally used because its energy is spread evenly across the spectrum and therefore less likely to swamp the desired low-level signal at any particular frequency, whereas a more correlated dithering signal would pile up in a narrower frequency band and be more objectionable or distracting. In addition, people are more used to hearing signals "under" white noise, more so than other types of noise, so that's what is most commonly done now. This technique of "fuzzy quantization" is also useful when going back from high-precision floating-point numbers to 16-bit samples, for instance. Although it extends the effective audible dynamic range somewhat, it has no effect on distortion whatsoever because noise is just added to, not multiplied by, the signal.

QuickMath — And Why Are These Sine Waves Really Cosines?

At this point, I will explain a couple of details about some of the math and terminology used throughout DSP. A lot of times I'll say, "multiply this by a frequency" or "complex multiply by a probe signal," and it would help if you knew what I'm talking about, where the frequency or probe comes from, and so on. First, I'll deal with the real domain (e.g., noncomplex numbers). In general, you're dealing with a discrete signal, call it $X(N)$, where the individual samples are $x(n)$, $n = 0, 1, \ldots, N - 1$.

The index n corresponds to the number of the sample after starting or its index in a memory buffer, for example. To normalize this to absolute time, use $x(nt)$, where t corresponds to the sample interval (in time units). Thus, nt takes on the dimensions of pure time at discrete values according to the sample count and has a slope determined by the sample rate. To generate a frequency, ω, use $(A \cos(\omega nt + \phi))$, where A is the desired amplitude of the wave, ω is a scaling factor to give a particular frequency, and ϕ adjusts the phase of the wave as required. The frequency generated is simply a function of the slope of ωnt. Every time ωnt increases by 2π, you've gone through another complete cycle. Any other factors you see are there just to allow the frequency to be specified in units that are more natural to use.

Why in the world, when speaking of sine or sinusoidal waves, is the cosine function used for this? There's a very simple, intuitive reason that no one ever seems to mention. Suppose you want to express DC or zero frequency in this form. This implies that $\omega = 0$. If you use the sine function (A $\sin(0) = 0$), you couldn't generate any DC! However, A $\cos(0) = A$, so for zero frequency you get a DC output equal to A in amplitude, which is just what you want. Cosines are used to extend the range of frequency generation all the way down to DC. For complex numbers, you can finally use the sine function for the imaginary part. In this case you might specify a complex sinusoid (r denotes real) by

$$(A \cos(\omega nt + \phi_r), i \times B \sin(\omega nt + \phi_i)),$$

which is the form used for a probe wave in DFTs and many other things, usually with $\phi_{r,i} = 0$ and $A = B = 1$. With complex sinusoids, the idea of negative frequency becomes a possibility. For a real-only wave, the symmetry of sinusoids creates a situation where you can't tell whether the frequency is positive or negative; time could have either a plus or a minus sign, and the sinusoid would appear precisely identical! However, when you have a complex wave, you *can* tell which sign time has by which direction the vector it specifies goes around the origin. For a positive frequency, this gives a unit vector that moves around the x and y axes counterclockwise, starting on the positive x-axis at $(1, 0)$, as time proceeds at a rate of ω.

Math guys, you just gotta love 'em. Who else could comfortably assert that the best way to look at time is by starting at 3 o'clock and then going counterclockwise? You can generate what's called a negative frequency by flipping the sign of the sine component, which is otherwise known as complex conjugation; ω or t or n could have been made negative, instead, to get the same result, but it's usually not done that way. Real life complex signals will not generally have this exact ±90-degree phase between the components or maximum values that are exactly ±1.0. Rather, A and B specify amplitudes that can vary separately for the real and imaginary components. By convention, the complex sinusoids used as probes nearly always have an amplitude of 1.0, except in wavelet transforms where the probe is windowed instead of the data.

When the amplitude or phase differences aren't exactly 1 or 90 degrees respectively, you still have a vector tip that rotates around the x and y axes, but it isn't a unit circle anymore — it may be an ellipse or some other more complex shape. It can still represent either positive or negative frequency, though, depending on which way it goes around. Believe it or not, this sometimes does matter, and knowing is often helpful! The DFT uses either negative or positive frequency probe waves depending on whether you're going "forward" to the frequency domain or "backward" to the time domain, for instance. In fact, the sign of the probe frequency is the *only* difference between the forward and reverse transforms, other than scaling!

Another way of specifying a complex sinusoid goes back to Euler's relation, which is used as a notational shorthand. It turns out that e (2.718...) raised to an imaginary (i) number generates a complex sinusoid. Therefore, you'll often see the intimidating $e^{i\omega nt}$ in various descriptions of Fourier and other transforms, but this only means $(\cos(\omega nt), \sin(\omega nt))$; it just takes fewer characters to say it that way, and it keeps the riffraff out. If the exponent of e is negative, simply flip the sign of the sine part to generate a negative frequency instead.

Perhaps other DSP writers favor this form to keep the unwashed from ever understanding it. I don't know, but it really confused me at first, so I'm explaining it up front for your benefit because it sure helped me once I got it.

Linear Transforms and Their Properties

A linear transform is defined here as one that adds no new frequencies to the signal transformed, and whose inputs and outputs are superposable, which means that the output generated for two summed inputs is the same as the sum of the outputs for each input separately. Linear transforms may or may not be reversible because there may not be a unique output for each input if the transform blocks some input component perfectly; that is, it has one or more zeros in the transfer function.

Multiplying a signal by a constant non-zero number, for instance, would qualify as invertible since this only changes its amplitude. Filtering a signal only subtracts or boosts existing frequencies so it is a linear transform, as are various applications of delays, except the time-variable delay. Transforms that change from one domain to another and are invertible can also be called linear transforms if the inverse is always applied later or if proper consideration is taken that you're in a new domain where things mean something different. The most notable of these is the Fourier transform.

The big difference between linear and nonlinear transforms in DSP is the issue of whether new frequency components are generated that are not present in the original signal. A good example of a nonlinear transform is multiplying a signal by some time-varying other signal, such as a sine wave. This produces new frequencies at the sum and difference of the input frequencies and is thus a nonlinear transform. This effect is how frequency translation is performed in all superheterodyne radios, for instance. With some notable exceptions, most linear transforms are easily invertible, and most nonlinear transforms are not. The invertibility trait comes in handy in audio processing, where often some uncontrolled linear transform, such as a room acoustic or a microphone response, has been applied to the signal before you got it and you'd like to remove it and have a "pure" signal again.

The Fourier transform described below can help do either convolution or deconvolution, as well as help define the unwanted convolution function to be deconvolved. This is done by acquiring the result of the unwanted convolution with an impulse, transforming to the frequency domain, doing the same for the signal to be "fixed," and multiplying the signal's transform by the *inverse* of the unwanted convolution's transform before inverse transforming the combination to get the deconvolved signal back again. It's a very powerful, sledgehammer technique, and an FFT can dramatically reduce the computation required to do large (de)convolutions.

The Fourier Transform Pair

Perhaps nothing would serve to explicate all this business about domains better than describing the most-used transform for moving between the time and frequency domains and the modifications from its original formulation for the continuous domain to handle the discrete domain. The Fourier transform pair is a pair because it's really two things: one converts data from the time domain to the frequency domain and one goes the other way.

The Fourier transform (FT) pair have perfect invertibility within math round off errors, and they have very few of those when formulated as an FFT, which turns out to be an extremely important and handy feature.

The FT is linear in the sense that a superposition of inputs creates a superposition of outputs in response, but it is nonlinear in the sense that if you use it to transform a time series into a frequency series then treat that frequency series as though it were a time series, you get "new" frequencies. This is never done. Usually you go from the time domain to the frequency domain, do something, then convert back. It is a linear transform or, more correctly, used as a part of one.

The basic original formulations of the Fourier transform pair for the continuous domain are

$$\mathbf{X}(\omega) = \int_{-\infty}^{+\infty} x(t)e^{-i\omega t} dt \, ,$$

to go from the time domain to the frequency domain, and

$$x(t) = \frac{1}{2}\pi \int_{-\infty}^{+\infty} \mathbf{X}(\omega)e^{i\omega t} dt \, ,$$

to go from the frequency domain to the time domain.

$\mathbf{X}(\omega)$ is the complex energy at the frequency ω in the frequency domain series; $x(t)$ is the value of the time domain signal at time t; and the complex exponential term $e^{\pm i\omega t}$ generates a complex sinusoidal probe wave, by Euler's relation, which is $e^{i\omega t} = (\cos(\omega t), i\sin(\omega t))$ — a notational shorthand, if you will. $\mathbf{X}(t)$ and $\mathbf{X}(\omega)$ are generally complex numbers, which has some interesting implications. One is that a frequency can, in fact, be considered negative when the frequency in question is composed of two orthogonal functions, for example (cosine, sine), and the sign of the sine part has been flipped. You can use this symmetry for frequency shifting, which will be worthwhile later. There is a very important problem with Fourier's original formulation, where it was assumed that the limit $dt \to 0$, but you have a noninfinite sample rate to deal with in any discrete system. Also, those integrations from plus to minus infinity are kind of hard to do in real life because it assumes your signal has infinite duration and gives infinite frequency resolution; too many numbers to handle as output. For cases where you can't use these equations to solve some problem in closed form by algebraic substitutions, they have to be reformulated as discrete sums, rather than continuous integrals. This requires an important new assumption about the signal that affects *all* uses of these sums, which is *the signal (or any component of the signal) is periodic at the block length or a submultiple thereof*. This is the result to moving to a non-infinite-length discrete input signal from an infinite-length continuous signal. They are called the discrete Fourier transform (DFT) pair and are

$$x[nT] = \frac{1}{N} \sum_{k=0}^{N-1} X_{(k\delta\omega)} e^{\frac{i2nk}{N}} \quad n = 0, 1, ..., N-1$$

for time to frequency domain and

$$\mathbf{X}(k\delta\omega) = \sum_{n=0}^{N-1} x[nT] e^{\frac{i2\pi nk}{N}} \quad k = 0, 1, ..., N-1$$

for the other direction.

N is the number of samples in a block, T is the sample time, $\delta\omega = \omega_s/N$, and $F_s = \omega_s/2\pi = 1/T$. The exponential term is still a shortcut to represent a complex cosine wave of either positive or negative frequency, depending on which way you're going. This formulation still assumes complex-in and complex-out data, and each frequency bin is a complex number that can be manipulated to produce the amplitude and phase of that component. If real-only input data is used, the resulting spectrum has complex conjugate symmetry about the point $\mathbf{X}(N/2 + 1)$, which means that there are only $N/2 + 1$ independent spectral points, equally spaced at the reciprocal of the block length in time. In other words, DFTs don't give infinite frequency resolution but divide the Nyquist interval $0 \to F_s/2$ into equal-sized chunks, or "bins".

The way it rolls out, if you transform $1/10$ of a second of data, your frequency bins will be 10Hz apart and wide, starting at zero and winding up at $F_s/2$, and there will be $N/2 + 1$ of them. The "extra" one contains the DC and Nyquist components.

Because DC is purely real and Nyquist is purely imaginary, they are usually combined into one bin and, more often than not, thrown away. The rest of the points correspond to negative frequencies and for real-only input data are identical to the positive frequencies except for complex conjugation. This also implies (correctly) that there's a time–frequency resolution trade-off — the product cannot be less than 1.0, although it can be worse in some cases. A symmetry also allows you to handle real-only data more efficiently by placing alternative samples in the real and imaginary parts of the input data and adding a final "layer" to combine the two resulting spectra into one spectrum of twice the original resolution. This dodge is common for many real-life signals that are real-only, and the math exists to do this in either direction. Because it's not particularly illuminating stuff, but more of an implementation or efficiency detail, I'll save it for a code example of the FFT.

The FFT

Note that the above has a complexity that still goes up proportional to N^2 as the size of the transform increases. You have to do N things to N samples to generate all the frequency bins. This is not good, and the FFT is simply another formulation of this that goes faster by eliminating a lot of basic duplication of work in the original sum formulation. The resulting computation complexity goes up as $N \log_2(N)$ instead, which is a big improvement for N -> large. Going to the FFT formulation adds yet another assumption about the signal: the block length is factorable into a small number of small primes. In virtually all real-life implementations, the FFT is formulated to work on block lengths that are exact powers of two, a further restriction that gives better computational efficiency on many platforms and is easiest to code. Guess what? I'm not going to describe that derivation here and waste our time. All you need to know is that it produces the *exact* same results as the DFT pair, only faster and with less "math noise," and that I'm providing a well-documented, killer-fast Wintel C++/ASM class implementation with all the normal auxiliary routines. In any case, the code itself is clearer and shorter than any words and pictures I've seen so far could achieve — well, the butterfly diagrams are cute.

Various people have simply rephrased the original DFT math as a vector–matrix problem, noticed a lot of identity duplications and multiplies by 0 and ±1.0, and simply rearranged things so that these are eliminated from the computational load involved. It turns out that there are about eight distinct ways to do this reformulation; however, they result in a structure that has virtually no intuitive ease of understanding, as the DFT has, and cannot easily be modified and used sample by sample, so it's not to the point to describe it here. Besides, just about every other DSP book on the planet, including the ones in my bibliography, has the same obfuscatory description of the resulting possible structures, so why waste paper doing it yet again? The trouble is, there are so many ways of formulating this shortcut, which probably means there are even more symmetries to exploit if someone really looks into it. For instance, there are formulations that use "twiddle factors" or probe waves in either natural or bit-reversed order that change fast at either the start or end of the transform and either the inputs or outputs that need bit reversal. Each has a different cute butterfly diagram, and again, unless you're trying to write your own FFT there's no point in covering the half-dozen plus ways here that all result in the exact same answers and computation speed.

The DFT or FFT formulations require unpleasant new assumptions about the signal, and most of the mistakes or bad results achieved by naive practitioners are the result of forgetting these. The assumption is that the signal is assumed to be *periodic* over *precisely* the block length. For the FFT case, there are some further constraints on the block sizes to get the efficiency of computation. This is obviously not the case in nearly all real-world applications, and the failure of this assumption to hold is what causes most DFT artifacts, such as apparent energy leakage from bin to bin. There are ways around most of this, which I'll cover later, but they involve trade-offs, such as doing more overlapped DFTs, which increases the work involved, or windowing the input data, which throws some of it away.

Before I get into that, look at the DSP flow diagram in Figure 3.5 for a single DFT frequency bin because the formulation and slight modifications of it also have other significant applications, and it should help you see what's really going on here intuitively.

Over the block, each input sample is complex-multiplied by a complex frequency signal equal to the center frequency of the bin in question, which is the k term, and the results are summed into the complex output point, which was cleared before beginning. The amplitude output is the square root of the sum of the squares of the two outputs, and the phase of the component is $\tan^{-1}(I/R)$ if you need it in that form. For a complete DFT, repeat this process for each bin separately, going through all the input points for each. This means that a DFT can't be computed in-place, since you need the same input over and over. The usual bin width and spacing are the reciprocal of the record length; for instance, working on 20ms worth of data gives bins that are 50Hz wide, spaced 50Hz apart.

Figure 3.5   Single DFT bin

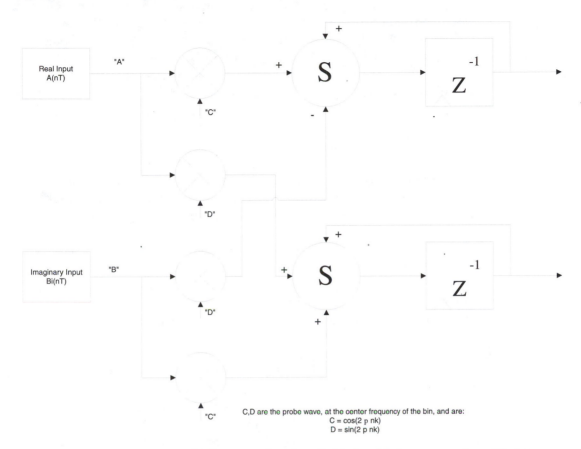

C,D are the probe wave, at the center frequency of the bin, and are:
C = cos(2 p nk)
D = sin(2 p nk)

From a gut feeling stand point, I think of it in terms of multiplying two frequencies and taking their sum and difference frequencies. A handy trig identity given in most math/engineering books is $\sin(x) \times \sin(y) = \frac{1}{2}\cos(x - y) - \frac{1}{2}\cos(x + y)$. There are similar identities for sine × cosine and cosine × cosine as well. You get the sum and difference frequencies in all cases, but the signs and resultant phases differ for the different identities. You can see in this case that the probe frequency is just the center frequency of the bin, so an input component right on the center frequency has a DC output component only. The difference frequency is zero, or DC, and the double-frequency component is averaged out by the summation over the record length. Note that this summation will only average out the high-frequency component perfectly if there are exactly an integer number of cycles summed. You needed the complex probe wave to get phase information independent of amplitude, and vice versa, since with two 90-degree waves, one is always non-zero and the root sum of their squares is always exactly 1.0. This handles the case of the component being exactly in phase or exactly 90 degrees out of phase with the probe wave. If I had just one of them, I could get a zero output for an input component precisely 90 degrees out of phase with the probe, no matter what size it actually was. That can't happen here because I have two orthogonal phases of probe working on the input. Sum and difference frequencies of components near the probe frequency come through but are smoothed by the averaging that takes place over the block length to produce the output. This averaging turns out to be a pretty poor low-pass filter with poor ultimate fall-off and big side lobes, but it's how the DFT is formulated, and it's why the bins "spread out" and have "leakage" between them when the periodicity assumptions are violated.

If you eliminate the imaginary input and its multiplies and adds and substitute *real* low-pass filters (LPFs) for the summing operation, you get what is known as a "synchronous filter," which has some nice properties. Its bandwidth is just twice that of the LPFs, and the LPF shapes are *translated* up to whatever the probe frequency is, and are imaged around it. Synchronous filters (SFs) are usable by-sample, rather than needing blocks of input data to produce a single output.

Its time response is determined by the LPF's time response, so if you use 1Hz LPFs with 12dB per octave slopes, you get a bandwidth of 2Hz, and the time response of the LPFs used. Remember, a difference can be either positive or negative. If the probe frequency is, say, 1KHz, the shapes of the filters are translated (not ratioed), so at 2Hz from the probe frequency (998 or 1,002Hz), you have the response of these filters at 2Hz, or –12dB, at 4Hz, or –24dB, and so on — a *very* sharp band-pass filter, indeed, for using just two second-order low passes and a couple of multiplies by a probe frequency! And this one is easy to make with a flat-top, steep-sided (or any other shape you want) filter, and there's no bin leakage when things aren't precisely periodic. Of course, the result isn't the 1KHz filtered signal, which could be regenerated easily by multiplying the outputs of the LPFs by the complex probe wave again. Rather, it's the energy of that signal, low-pass filtered. The response speed is the time response of these either way. You just can't cheat the old time–frequency resolution trade-off. However, with this structure, you can make a filter bank where each filter has *exactly* the trade-off you want. This is even better than the wavelet transform in this regard, and you can even get sample-by-sample output from it for each filter in a bank. No messy doublings of outputs are required for each octave, as with wavelets, and no restrictions whatsoever are needed in band spacing, width, center frequencies, or coverage. Of course, a sample-by-sample output may be way oversampling a narrow, slow bin and is just barely enough for a wide, fast bin, but you can always decide to throw away any outputs you don't need or want. The point is, with this structure, it's all up to you — nothing is forced on you by the underlying structure. This is why I say wavelets are obsolete and always have been. This structure has been around longer than they have, and is more powerful.

Figure 3.6   Synchronous filter

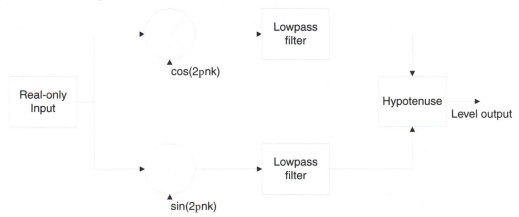

There's another useful slight modification of this structure called a "balanced mixer" in radio circles. You create a complex input from a real input via a 90-degree phase shift, making it "analytic," In other words, it becomes a signal that only has positive or negative frequency components, not both. In this structure, both the input and probe are complex, but the outputs are simply added or subtracted to produce a single real output.

Figure 3.7   Balanced mixer

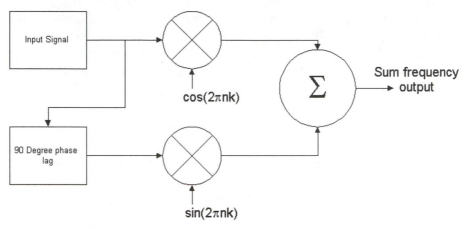

Another trig identity is

$$\cos(x) \times \cos(y) = \tfrac{1}{2}\cos(x-y) + \tfrac{1}{2}\cos(x+y),$$

which, if added to

$$\sin(x) \times \sin(y) = \tfrac{1}{2}\cos(x-y) - \tfrac{1}{2}\cos(x+y)$$

given above, neatly cancels all the sum frequencies or, if subtracted, neatly cancels the difference frequencies. No original or probe signal comes through, and the result is that you can use one frequency to up- or down-shift another — a very handy result for some kinds of special efx. This is one of those places that negative frequencies can be useful for something quite real. If you ever need to frequency shift a signal by, say, 10Hz, this little toy will get it done for you. You use the sum frequencies for output, so if the probe wave has zero frequency, the output equals the input. If a 10Hz probe wave is positive in frequency, you'll shift all the input components up 10Hz. If you negate the sine of the probe to make a negative frequency, you'll downshift the input components 10Hz instead. This cool technique is used for radio station feedback eliminators to handle the caller who has the radio turned up next to their phone. When feedback starts, each repeat is down-shifted, which makes an alien sound that nevertheless is a lot better than a squeal at full power. Because a broadband 90-degree phase shift is difficult to achieve in analog, usually the input signal is split into bands and an easier, narrowband technique is used to get the 90-degree phase shift in each, separately. For octave-wide bands, a couple of second-order all-pass sections designed to have a 90-degree phase difference between their outputs will serve. The results from each band are then combined to produce the output.

In DSP, you can use a Hilbert transform to get a broadband 90-degree phase shift, which can be implemented as a special type of FIR filter. A Hilbert transformer is included in the sample code. I should note that this technique is not very usable for music because it shifts everything by a fixed amount. This means that harmonics no longer are at the right frequencies — adding 10Hz to both a 1KHz fundamental and its 2KHz harmonic means that the second is no longer exactly twice the first. It sounds just like a mistuned single sideband radio instead, and for exactly the same reason. To use this on music, you'd have to split the bands up very finely and use a different shift for each one to make things work out right. One possible good use of this in music would be as part of a chorus effect, where you'd do this for a number of bands and use a pseudorandom control for the frequency shift of each, which could go either positive or negative. This would effectively simulate different vibratos to help the illusion of multiple sources, especially if adjusted to about the right ranges and speeds of shifts for human perception.

The derived DFT and FFT have artifacts because most actual signals are neither periodic at the block length nor zero outside it. Simply by truncating the signal to a block, you've in effect multiplied it by a square pulse that is 1.0 during the block and zero elsewhere. These fast-rising and -falling edges have frequency components of their own, and all the sum and difference frequencies of this and the input signal components are generated by doing this convolution with a square box. The result is called bin leakage, in which energy that belongs in one bin is spread out into several or across the *entire result*. One way to get around this is to use something other than a square window. If you forcibly make the signal tend smoothly toward zero at the ends of the block by multiplying it with, say, a Gaussian bell-shaped curve, the assumption that it is periodic is no longer as badly violated, since the zero at one end matches the zero at the other. This, however, throws away some information, and the result is that each bin is now wider than before, overlapping its immediate neighbors more, but leaks less to and from the bins much farther away. Note that if you can create a situation where the signal is precisely periodic at the block length, there is *no* leakage. Leakage only occurs for components that land off the center frequencies of the bins because summing a noninteger number of cycles of the difference frequency no longer gives a zero-mean result. The time–frequency resolution trade-off has been exercised by windowing. You now see less of the signal because the window removes some at each end, so the frequency resolution is now less than it was with the square window.

You can get all the information back by overlapping DFTs with larger windows, but it's more work. I'd point out that in many practical situations this extra work can make using the DFT or FFT disadvantageous. A very interesting thing here is that if the low-pass filters I described above were translated into the time domain, they'd produce just such a nonsquare window shape. *Everything linear can be flipped between domains at will*! Everyone has their favorite window shapes for various purposes, and some windows are more or less ideal for a given purpose. I'll be giving you a pretty comprehensive set in the code later, with extensive commentary. To make a very long story short, using no window gives the narrowest bins but with the most leakage to bins far away, followed in order, more or less, by Hamming (which gets down to −40dB the fastest, but stops falling off fast after that), Hanning, Blackman, and Blackman–Harris. You can show yourself these effects quickly by generating an FIR filter in WavEd (the waved example program provided), trying different windows, and watching the response plot. In this case, you'll see the filter fall off progressively slower as you go up the window list but fall off farther ultimately. Some purists like the Gaussian or Welch window the best, claiming it has the best time–frequency resolution trade-off. I don't subscribe to the idea that there's one best window. I've found that it depends on what you're trying to do at the moment.

The Convolution Property of the DFT Pair

In the glossary provided in this chapter, I described convolution as a multiplicative transform that combines some impulse response with past signal values. This is generally quite a slow thing to do directly, since you'd need N multiplies and adds per output sample produced, where N is the length of the impulse response you're convolving the signal with. To process a whole signal, this number therefore becomes $N \times M$, where M is the signal length. For some cases, this is OK, but as the impulse response grows longer, as it must to define sharp or low-center-frequency filters, this direct method becomes very slow. Therefore it's nice that there's another, faster way to do this. Although the DFT is not the only transform that could be used — many orthogonal transforms have the convolution property — I'll describe it here because it has the advantage that the time for the FFT implementation of the DFT goes up as $N \log_2(N)$, instead of N^2, as the size goes up. For convolution with FFTs, first forward transform both signals to be convolved to the frequency domain, multiply these two results by each other, then transform the result back to the time domain. Because you usually convolve a signal with a fixed impulse response, you need only transform that one time for any amount of signal you're going to work with. The nice thing here is that you get a whole block of convolution results at one time, rather than having to produce them sample by sample.

There are also some significant disadvantages to using FFTs for convolution. These arise from the assumption that the signal is periodic at the block boundaries that an FFT must make. In other words, you can only do *circular* convolutions with FFTs, in which the ends of both signals to be convolved are assumed to wrap around. This is almost always *not* what you want. You can avoid this by adding zeros to, or padding, the ends of the signals so that what wraps around has no energy and then using a larger FFT to do the job. Unfortunately, this is still not enough. The resulting convolution output now has amplitude loss at the beginning and end due to the convolution with the extra zeros. All you accomplish with zero padding is to avoid the signals wrapping around, but wrapping is still taking place. There are (at least) two ways to cure this problem. One is called select-saving and the other overlap-adding. In these techniques, you do extra, overlapped convolutions, sliding along the data and add the results of these, or you do extra convolutions and just keep the accurate values from the middle of each. Either way is the same amount of work, which means that you're going to have to do twice as many and larger FFTs than you thought at first.

It turns out that this presumably simple technique to speed up the convolution is actually a bit tricky to implement correctly. As usual, the devil is in getting the details precisely correct. An even more complex case occurs if neither the signal nor the impulse response fits into a reasonably sized FFT with its required zero padding. In that case, you wind up doing a lot more FFTs of segments of each signal, shifted appropriately in time, and overlap-adding the results to get the correct output. It is for this reason that I rarely recommend doing convolution with long sequences. It's a dumb, brute force technique that looks attractive until you figure out how much work it's really going to be to do *correctly*. Often, a little more thought will show you another way to get the job done that is much easier than direct convolution or using the FFT method, and when this is possible, that's what I recommend doing. I was able to avoid any need for brute force, long convolutions in WavEd, even for reverb, by doing a little extra thinking.

I provide some sample code to do FFT-based convolutions on continuous data streams, which I intended to add to the FIR filter transform in WavEd in the hopes of speeding up the really long FIR filters; however, a compiler/opsys/heap bug prevented me from actually testing it inside any other nontrivial code, so you're on your own here. The code probably is fine, although it may have a fence-posting or scaling problem here or there. It implements the select-save version of continuous convolution and is in the files `Convolve.cpp` and `Convolve.h`. I've reported the bug to my personal contact at MS/QA, and they are "working on it" but won't have it fixed in time for this book. Take my advice and find another way to do your processing other than with the brute force convolution technique. You'll often find a trick or trapdoor in your problem that will let you accomplish this, and you'll generally be better off as far as both speed and having a better understanding of your problem, which is always a good thing. For WavEd's reverb, where you'd think a long convolution would be required, I (re)invented "sparse convolution," in which only a few delay taps or impulse response coefficients are non-zero By organizing the data differently, I only do those few meaningful multiplies and adds, which is an enormous savings. To simulate the filtering effects of walls and the air, I use IIR filters. The result is both faster and more accurate than any FFT-based convolution would be.

Deconvolution

One of the neatest uses of convolution is to deconvolve. What do I mean by this? Suppose you have a signal that is convolved with a room reverb, and then a microphone's response. You'd like to get the original signal back. One way this might be done is to collect a sample in the room from a known signal, such as the nearly perfect impulse produced by a spark jumping a gap. The resulting impulse response recorded from the microphone includes both the room and microphone responses. Were you to transform this to the frequency domain using an FFT, you'd see that the resulting response is not flat. You could then invert this response by taking 1/(each bin) and using that to convolve with any other signal received from that microphone in that room, as long as the signal came from the same location as the original impulse, so that the room affects it the same. This technique works under many conditions. Analogously, with images, you could find the point spread function as the 2-D analog of the impulse response in 1-D. It's a handy technique for spy satellites. A necessary condition is that the response to be deconvolved contains no zeros (since 1/0 blows up), and to be practical, the response must not contain any nearly-zero values, either, because all real-world processes add some noise to any signal in various places in the chain, and with large gain values at some frequencies, a near-zero condition would simply boost the noise at that frequency, rather than accurately restore the original information. This corresponds to a similar problem in matrix math, where a divide by zero might occur while attempting to invert some matrix. There, the problem is partially solved by a method called singular value decomposition, where zero or near zero values are detected and simply replaced by 1. This is counterintuitive because it involves discarding precisely those numbers that would be largest in the final result! For other reasons, it's known that you can't have much confidence in them anyway, so the procedure works in a practical sense. Other uses of deconvolution might include looking at the original pitch pulse of voiced speech by deconvolving the vocal tract's response.

In general, the trick is to get an accurate copy of the impulse response of the thing you want to deconvolve. There are some automated tricks for doing this, but they go far beyond the scope of this introductory book. The best reference I know of to the entire class of techniques is the paper "Nonlinear Filtering of Multiplied and Convolved Signals" in the first IEEE reprints book listed in the bibliography in . There, the authors show how to take apart and reassemble various kinds of audio and video signals that accomplish some pretty amazing things. Stockham, one of the authors, has used the technique to restore old records, for instance. However, his paper has math that is more difficult to understand than anything in this book, so be warned.

The Cepstrum

One of the more interesting tricks in the DSP bag is the cepstrum, of which there are a couple of flavors. Often, there is some periodicity in the frequency domain, not just in the time domain. An example of this might be a pitched note that consists of a fundamental and some of its harmonics. This shows up in the frequency domain as a series of evenly spaced peaks — a periodic signal in either domain. To further analyze and work with such signals, it is sometimes nice to take yet another step in the analysis: the Fourier transform of a Fourier transform, which would make such a periodic signal finally pile up in one bin. It's hard to define exactly what domain this is now, pedantically, so the whimsical term "cepstrum" was coined to describe it by rearranging the letters in the word "spectrum." There are (at least) two major classes of cepstrums that are commonly used. The first discards the phase information from the first layer of Fourier transform, takes the log of each bin considered as energy, then transforms again.

The log used this way has some interesting properties. For one thing, it tends to equalize the importance of all the wiggles in the original spectrum by making little ones larger and large ones smaller, so each contributes more equally to the result, despite the overall slopes in the original spectrum. It tends to smooth skinny peaks into more sinusoidal shapes, so their energy looks more periodic in the final result. It also puts you into a domain where things that were originally multiplied are now combined *additively*. This implies that you can now separate things that were multiplied together with a linear process, such as filtering, which can be quite a handy property. Unfortunately, when the phase information from the first layer is discarded, you can't unambiguously inverse transform back to the original time domain data.

Thus was born the idea of the complex cepstrum, in which you take the *complex* log of the first transform before transforming again, therefore losing no information. The problem is, taking a complex antilog unambiguously is no small thing. The problem arises because phase information has effectively gone through a modulo process that discards multiples of 2π (or 360 degrees), and you really need *all* the information to do a complex log correctly. This information can be retrieved by tracking each phase bin back to the start of the analysis when doing multiple records and by phase-unwrapping schemes, which simply go through the data and assume that it's a good idea to add π if that would make the phase curve smoother than before. In this case, phases are allowed to become greater than 360 degrees, effectively eliminating the modulo operation. Because this is usually a good assumption, and because things rarely jump more than π between bins or records, you can usually get away with what seems to be a nonrobust operation. For more information on this technique, check out "Nonlinear Filtering of Multiplied and Convolved Signals" in the first IEEE reprints book, where the originators describe it in far more detail than I can afford to do here.

Some engineering and DSP texts insist that a cepstrum is actually the inverse Fourier transform of a logged Fourier transform that puts you back in some version of the time domain. It's a way of looking at it, to be sure. Personally, I've done this either way for just analyzing signals, and you can get useful results no matter which way you do the second transform. I've even skipped doing the log altogether, simply using the complex output of the first transform as the input to the next, and gotten interesting results. It all depends on what you want to see — different variations simply show different aspects of the original signal better than others. This is an area where more research could stand to be made into technology, especially on the reversible versions. These techniques potentially make available information on various corruptions of signals in fairly automated ways, so they could be used to find things like original impulse responses for later deconvolution of signals. For instance, reverb in a "logical" space corresponds to a series of peaks in a cepstrum, which could be detected, removed by setting those bins to zero, then inverse transformed to get the sound back without the reverb. I know I've left you tantalized with this, and would like to give you more on it, but it would stray pretty far from the subject of this work and take a lot more math and time. I give you all the tools and example code you'll need to chase this down further yourself, however. You can re-enable an experimental cepstral view I *almost* got working in time for the book in the `CWavEdView` class to get started.

Chapter 4

Filter Primer

Although this book is not specifically about DSP, it seems like a good idea to cover the most basic and widely used technique in signal processing: the linear filter. Because the technique is so widely used, a lot of jargon and esoterica have sprung up that can make the topic intimidating for beginners and unnecessarily difficult to understand. I'll skip most of that and mention it only as necessary if I think it will help make things clear.

In the field of linear, time-invariant filtering, there are basically two fundamental types of filters. These are designated IIR and FIR: infinite impulse response and finite impulse response. Each has strengths and weaknesses for any particular application. An FIR filter has only numerator coefficients in the S-domain (see the following section, "S-Domain Primer"), whereas an IIR has both a numerator and a denominator in that domain.

In general, all analog filters are of the IIR type, and digital filters can be of either type. FIR filters generally are easier to design for arbitrary amplitude and phase characteristics but tend to be more computer intensive at runtime, whereas IIRs tend to be minimum-phase filters and difficult to design as phaseless or with arbitrary phase or for an arbitrary frequency response. IIRs are, however, usually far more computationally efficient. The common practice has therefore been to use IIRs whenever possible and use FIRs mainly when a quick design to arbitrary parameters is required, at the expense of run-time efficiency. In this book, I introduce a new technique for designing the more computationally efficient IIR filters to an arbitrary frequency response relatively quickly, so as of now, IIRs can be used in a lot more applications.

Most of the big names in the filter business in fact got their fame by inventing optimal design methods for just high-pass, low-pass, and band-pass/-stop filters. These are very simple types that are hard to assemble into some arbitrary complex shape, but with optimal (in some sense) characteristics for the number of IIR sections used, assuming a simple shape (square box) is desired. There are a lot of ways to skin this cat, and one person may want the flattest passband, another may want the closest approximation to constant delay, and another may just want the best rejection in the stopband with the minimum filter complexity. That's how Butterworth, Bessel, and Chebychev, respectively, got their names in lights. They provided simple design rules for cascading second-order sections with those particular trade-offs in mind, but only for the simplest shapes — band-pass, band-stop, high-pass, and low-pass. Hopefully these names need no further explanation.

The entire gist of classical filter design can be summed up with just a few figures and words. One could easily state that the basic problem for, say, low-pass design is coming up with how to get Figure 4.1 by combining Figure 4.2 with Figure 4.3.

The question comes down to how to choose design values for sections such that they combine into what you want given that the only things you can choose for each are the frequency, damping, and simple type (high-pass, low-pass, band-pass, notch), as far as classical design is concerned, and given that all you want is an approximation to some sort of square box, again as far as classical design techniques are concerned.

Figure 4.1   Resulting filter, both sections cascaded

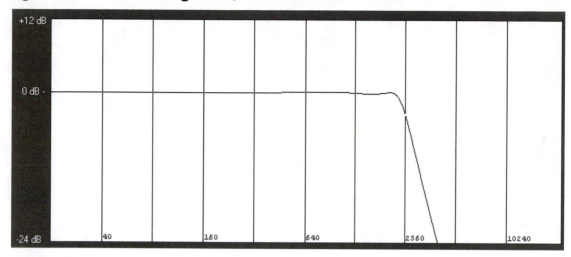

Figure 4.2   Filter section 1, Low Q

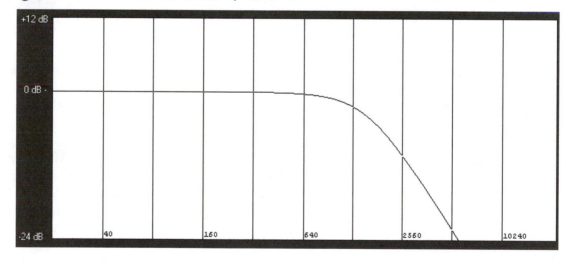

Figure 4.3   Filter section 2, High Q

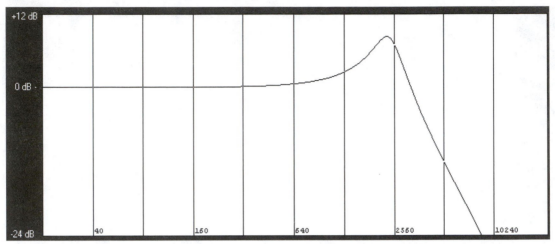

Notice that the result has a flatter passband and a steeper roll-off than either of the individual sections. An individual section was thought to have very limited shape options; basically, you could set its center frequency and Q and whether it was a high-pass, low-pass, band-pass, or notch filter. The fancier stuff I'll describe later just wasn't considered by the older theory, probably because no use for it was anticipated; stereos with equalizers didn't yet exist. In Figures 4.1 through 4.3 I show a slight-dips Chebychev design that combines a low-frequency, low-Q section with a very droopy passband with a higher frequency and Q section, such that the peak of the latter nearly cancels the droop of the former, and the 3dB cutoff frequency that results is my 2,560Hz specification. You may also notice that just after the peak, the higher Q section initially drops off faster than 12dB per octave, which helps the combined response roll off quicker. You can't just cascade identical sections in most cases, because at the 3dB-down point of a section, cascading two sections would cause the result to be 6dB down, so you'd still have to do some figuring to find out what new center frequency you'd need for each to get the 3dB point back where you wanted it again. In general, cascading identical sections turns out to be less optimal than the figures demonstrate; since you'll have to do some math no matter what, you might as well do the right math. For normal filter design, this means you use a plan from Bessel, Butterworth, or Chebychev for nearly all filters.

In this design, *neither* filter section is tuned to the resulting cutoff frequency. It turns out the math that correctly specifies the characteristics of each section to get a nice combined result can get pretty complicated, even for apparently simple cases such as this. It can be tough to draw even a simple square box with wiggly things that only know how to wiggle in certain ways! For band-pass, band-stop, and high-pass filters, the situation is about the same. The name of the game is choosing shapes that can be combined to get what is desired, using a peak here to cancel a dip there, while trying to trade off against complexity, phase, or flatness. What made Butterworth, Bessel, and Chebychev famous was showing the rest of us some comparatively easy formulas for the most common needs so we don't have to work it out over and over for each new filter design. By the way, the source code for the program that created these screen shots is on the CD in the `source\fdraw` directory. You can probably tinker with it and get it to do other interesting things. I just grabbed a couple of standard pieces from elsewhere (WavEd) and slapped them together to get these figures.

One thing I should stress is that *any* filter that can be realized in analog can be realized in digital. But it goes *way* beyond that! Freed from analog constraints, many digital filters can do things that are either very hard or impossible to do in analog. Examples of this are filters with zero (effective) time delay or phase shift, filters that are accurately time varying, filters with precisely zero loss or infinite Q, and several others. In the analog domain, it's hard to remember past values accurately, and you're forced to handle a signal as it comes in. This is not the case in the digital domain.

Even with real-time processing constraints, there is usually an acceptable "pipeline" delay that can be used to hide pretty complex processing in the digital domain. This is a good thing, because even if you are working in an interrupt service routine and handling data sample by sample, the a/d and d/a converters often add significant delay because of their internal resampling filters. Those used in a music product I helped develop, for instance, have a total built-in delay of about 63 samples. In the more general case, where you are working with disk data, much more is possible. In this domain, you're not even constrained to work on the data in any particular order, so you can start at the end of your data and work back toward the beginning if you like, generating a "reversed" time/phase response for the filter in question. You can even do what's called bi-pass filtering, running the same filter over the data twice, once each way. This causes its phase response and delay to cancel out completely. Also, in the analog domain, it's impossible to instantly discharge a capacitor or inductor. In the digital domain, no such limit exists, so a filter can be cleared instantly of any stored energy, retuned accurately on the fly, and so on between samples. Many tricks in DSP are a result of recognizing this and other new possibilities. Not much of this is well documented at this point, since most of the new development is corporate and proprietary. Keep an open mind, though, and you'll rediscover most of it yourself. You might think along the lines of "smart" filters that change on the fly under certain conditions — a technique that finds a lot of uses. One is in the envelope-following "touch wa-wa," which uses signal power as a tuning input to a band-pass filter, which is then passed over the signal to get a wa-wa effect. Most often, you'll use some sort of smart filter when developing a control signal from various inputs.

S-Domain Primer

Virtually all classical filter design is done in what's called the *S*-domain. In this domain, you can plot the roots of the polynomials that describe filter responses, and there is at least a semi-intuitive way to interpret the results. In the *S*-domain, the horizontal (real) axis is usually labeled α or alpha. The vertical axis is imaginary and has units of $j\omega$, often labeled β or beta.

On the S-plane, it turns out that the frequency of a given complex root is the radial distance from the origin, and Q is this distance divided by -2α, so the farther a root is from the origin, the higher frequency it is. The ratio of this to the horizontal distance is the Q or quality of the resonator. Damping, which corresponds to friction, is simply $1/Q$. If a resonator lies on the loss-less $j\omega$-axis, it is a constant-output oscillator. If it lies on the right half-plane, it "blows up," that is, it has an output that increases without limit. Practical filters all have resonators that lie in the left half-plane where some finite loss exists.

The more complex the polynomial is, the more roots it can have. The lowest order, "first," can only have a root directly on the α-axis. All higher orders can have roots anywhere in the S-plane. It turns out that you can simplify your life considerably by factoring all higher order polynomials into second-order sections with possibly one first-order section for odd orders, and this is the assumption I'll make throughout this book. As it turns out, the resulting filters have fewer math precision problems and are analyzed more easily, which is a nice result.

You usually make the assumption that you're going to work with real-only filter elements and signals, which has a further simplifying effect. But first, look at a typical S-plane plot and see what it's good for. Figure 4.4 is a plot of a simple second-order low-pass filter.

Figure 4.4   *S*-domain

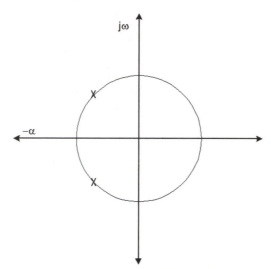

The two X's mark the roots of the filter polynomial. I added the circle to the standard plot. This shows where all roots that correspond to a particular frequency must lie, since the frequency is determined by the distance or radius from the origin. Because these roots are somewhere on the left half-plane, they represent a filter with finite Q. The matching filter equation is

$$\frac{1}{S^2 + bS + 1} \ ,$$

where b specifies the bandwidth or damping (inverse of Q) of the filter, and the filter is normalized to have a frequency of 1 radian. Because the polynomial is in the filter's denominator, its roots, or values of S where it equals zero, are called poles in the S-domain and are plotted as X's. A more general filter might also have a polynomial as the numerator, and its roots would be called zeros and be plotted as circles. This nomenclature may make more sense when you realize that anything divided by zero is infinite or undefined, so it makes sense to call it a pole, whereas zero divided by anything equals zero. Note that this equation has two roots, one that's positive on the $j\omega$-axis, and one that's negative. This is because of the simplification I mentioned above caused by using real-only coefficients and signals. Another way to look at it is by using the old formula for finding roots of quadratic or second-order polynomials, where

$$\text{roots} = \frac{-b \pm \text{sqrt}(b^2 - 4ac)}{2a},$$

which gives two roots for any second-order polynomial phrased as $ax^2 + bx + c = 0$.

When this formula is extended into the complex domain, you get the results pictured Figure 4.4, because your roots will have both a real and an imaginary part, at least for the case of $b \leq 2.0$. Another way to look at this is that the polynomial is factored into two first-order parts: $(S - \text{pole1})(S - \text{pole2})$. So anytime you see a non-zero squared term, you'll have two roots.

How do you interpret this picture in terms of the response of this filter? Imagine yourself traveling up the $j\omega$-axis (e.g., with $\alpha = 0$). The response is the inverse of the distances from all the poles multiplied together; that is,

$$\frac{1}{(|j\omega - \text{pole1}| \; |j\omega - \text{pole2}|)}$$

for this case, with the distances to each pole multiplied together in the denominator. (In the complex number domain the absolute value fences (| |) mean the length, or the root sum of the squares of the real and imaginary parts, also called the hypotenuse or magnitude.) If the filter has zeros as well, the distances to them take the place of the 1 in the numerator. As you move up the $j\omega$-axis, you get farther away from one pole but closer to another, so the response is relatively constant; this is the filter's passband. However, as you go higher than the upper pole, you get farther away from both poles at once, so the response begins to drop rapidly, giving a low-pass filter.

Now look at a high-pass filter. Any low-pass filter can be turned into the equivalent high-pass filter by using the substitution $S \rightarrow 1/s$. This changes the original equation into

$$\frac{1}{\frac{1}{S^2} + \frac{b}{S} + 1},$$

which is sort of ugly and hard to interpret. After multiplying by one in the form of S^2/S^2, you get

$$\frac{S^2}{1 + bS + S^2},$$

which looks a lot like the low-pass filter equation, except that you now have S^2 in the numerator, and the order of terms in the denominator is reversed, which makes no difference. This adds a pair of zeros at $S = 0$ to the plot, since when $S = 0$, the numerator is zero. The resulting S-domain plot is shown in Figure 4.5.

Figure 4.5   High pass in the *S*-domain

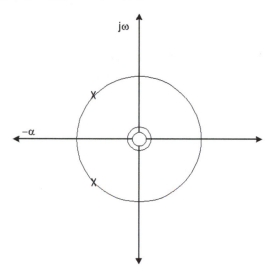

As you can see, it looks like the low-pass plot, except for the pair of zeros plotted at the origin and the original poles. You can think of the two zeros as S^2 factored into two first-order sections, both with roots at zero, or $(S - 0)(S - 0)$, to get S^2. To graphically interpret the result, work your way up the $j\omega$-axis starting on the zeros. This gives a zero response at frequency zero. As you move up the axis, you get farther away from the zeros, but as in the low-pass case, you stay about the same combined distance from the poles, so the response rises. At the limit, far away from the origin, you can see that you're getting farther away from all the poles and zeros about equally, so the response flattens out for high frequencies, just as you would expect from a high-pass filter. The magnitude response equation for this case is

$$\frac{|j\omega - \text{zero1}|\ \ |j\omega - \text{zero2}|}{|j\omega - \text{pole1}|\ \ |j\omega - \text{pole2}|}.$$

For fancier filters, extend the response equation by putting all the zeros in the numerator, multiplied together, and all the poles in the denominator, also multiplied together, where 1 is assumed for any case where there are no poles or zeros. In standard math notation this is

$$\frac{\displaystyle\prod_{N=0}^{N-1} |j\omega - \text{zero}_n|}{\displaystyle\prod_{M=0}^{M-1} |j\omega - \text{pole}_m|},$$

where the $\prod$ denotes the same sort of running operation as the Σ, except multiplications are used instead of summations, and a 1.0 is assumed if the sequence is zero length. N and M are the number of zeros and poles, respectively.

To make a long story shorter, a simple second-order band-pass resonator is given by

$$\frac{bS}{S^2 + bS + 1},$$

whereas that for a simple notch is

$$\frac{S^2 + 1}{S^2 + bS + 1}.$$

Notice that only the order of the elements in the numerator has changed for all substitutions.

The band-pass filter looks similar to the high-pass filter, except now you only have one zero at the origin. This works like the high-pass filter at first, but the slope is only half, since you're only getting farther away from one zero as you move off the origin. Because you're only moving away from one zero, when you get to the place where low-pass action begins, you still have some, but only half as much. The effect is that of splitting the slope-generating ability of this equation into two halves on either side of what would originally have been the low-pass turnover frequency.

The notch filter has two zeros that lie right on the $j\omega$-axis symmetric around 0, which is a little less magnitude than the complex poles — the poles move closer to the zeros as the Q of the filter increases and they come closer to the $j\omega$-axis. This means a notch is a perfect notch. At some point as you travel up the $j\omega$-axis, you will touch a zero, so the response equation must go to zero. On either side of this, you move away from the zeros much faster than from the poles, so the response comes back up again.

By the way, the books give another set of substitutions for bandpass filters that double the order of each section. They basically amount to making a band-pass or band-stop filter out of a pair of crossed low-pass and high-pass filters. Well, that's certainly one way to do it, and may be the best way if you're making a wide band-pass filter that requires a flat top. However, it's perhaps better and certainly easier to understand and analyze such things as they really are — low-pass and high-pass filters.

The technique gives nice-looking filters, but the initial roll-off isn't as good for narrow filters as those constructed by using multiple second-order band-pass poles in what I'll call the Don Lancaster fashion, which is shown in WavEd in the IIR filter transform example. Just for reference, the "book" substitutions for the band-pass are

$$\frac{S^2 + \Omega}{WS} \to S \,,$$

where $\Omega = F_h F_l$, and $W = F_h - F_l$.

Note that because a squared term is substituted for a nonsquared term, the filter order is doubled. For a band-stop filter, you do the high-pass substitution given above first, followed by the band-pass substitution. In the sample code in WavEd, I redid the filter engine, for speed and simplicity, so that it only knows how to eat second-order sections. You'd have to rewrite it to eat these ugly fourth-order sections.

I haven't talked about phase in this domain yet, but it's easy now. The phase is defined as the sum of all the angles from your point on the $j\omega$-axis to each zero, minus the sum of all the pole angles to your evaluation point:

$$\text{phase} = \sum_{0}^{N-1} \phi\, \text{zero}_n - \sum_{0}^{M-1} \phi\, \text{pole}n_m \,.$$

It turns out that you can place zeros on either side of the S-plane and they'll be stable. This means you can have either minimal phase for a filter's amplitude effect by making the center term positive, putting all the zeros on the left half-plane, or maximal phase, as desired, for an all-pass filter, by negating the center term in S. The classic all-pass filter is simply

$$\frac{S^2 - bS + 1}{S^2 + bS + 1} \,.$$

This puts the zeros on the right half-plane and the poles on the left, set up so the distances between them and any point on the $j\omega$-axis is constant, but the angles still change as you move up and down the axis, so you have a frequency-variable phase shift. If you set the damping or b factor low, the poles and zeros are farther out on the α-axis, so the angles don't change as quickly. If the damping (b) is low, or the Q is high, most of the phase change comes just as you go past the pole-zero pair and is less elsewhere. If you don't negate the center term in the equation above, the poles and zeros will lie right on top of one another and completely cancel themselves, resulting in no phase or frequency effect at all. At first, this seems pointless, but it's not. Interesting things happen when you start perturbing the all-pass coefficients so that the numerator and denominator aren't equal anymore. After all, you should be able to manage something more tricky and subtle than just zeroing one or more numerator coefficients, and it turns out that you can.

There is a relatively simple way of viewing perturbations of the all-pass filter by manipulating the numerator or denominator coefficients. The S^2 term controls frequencies above the center frequency (assumed to be 1.0 so far), the middle term controls frequencies around the center frequency, and the simple term controls the DC response. Thus, you can make nonsimple shapes with full generality at least as well as you can with the ratio of two second-order polynomials just by perturbing these coefficients intelligently so that no poles go off the left half-plane. All the numbers for the numerator are safe to play with, but you have to be a little more careful in the denominator.

But now it's time to generalize to frequencies other than 1 radian so that you can see what it is you'll be perturbing. The general all-pass equation is

$$\frac{S^2 - d\omega_c S + \omega_c^2}{S^2 + d\omega_c S + \omega_c^2},$$

where ω_c is the center frequency of the filter and d is its damping factor, $1/Q$. So scaling the entire numerator or denominator isn't a very useful operation since it changes both center frequency and damping. However, interesting things can be done by changing only one term at a time. For instance, if you want a bump surrounded by flatness, first design this filter with a damping factor and center frequency that will determine how wide the bump is and where it is, but before you're done, perturb the center numerator coefficient by making it larger by the bump height you want.

This "widens out" the zero that was canceling a pole in the denominator, allowing the pole to show up. To make a dip, increase the center denominator term. This is safe because you're always moving a pole farther away from the $j\omega$-axis — you can't accidentally create an oscillator this way. In the sample code, look at `setfilt()` for `case 9` in the `CIIRFilt` class.

Suppose that instead, you want a broadband tilt. This is often desirable for equalization, speech processing, and other uses. Design the all-pass filter as before, then perturb diagonally opposing terms in the numerator and denominator and adjust the damping slightly if you want a nice, clean 12dB per octave slope, assuming you started with a damping of 1.414, as for a Butterworth maximum flatness section. You are decreasing the opposing terms, each by half the total tilt, so that the filter gain will pass through 0dB at the center frequency specified. The damping factor originally input to `setfilt()` is adjusted by the inverse of the square root of the reduction factor, adjusting it to track with varying total tilt amounts. I've not yet worked out the math that would allow you to specify this as the tilt endpoints, but it should be fairly easy. Basically, increasing damping reduces the slope and causes the tilt to take more bandwidth to achieve the max or min response on both sides equally. This technique can give a fairly flat, full bandwidth tilt, even at oddball decibels per octave settings. Reducing damping tends to make the tilt overshoot at both ends, so the useful range is ~1.414 and greater. Sample code for the technique can be found in `setfilt()` under `case 12` in the `CIIRFilt` class.

For either of these cases, you can have the minus sign in the numerator second term or not. Replacing it with a plus sign makes a minimum-phase filter that has the absolute minimum phase for the amplitude effect it has. This means that for the parametric EQ, for example, a slider set at zero means that the filter has no effect at all, no matter what Q and center frequency it is set for. If you want a phase shift, change the sign back to negative, in which case you get phase and ringing behavior no matter where the slider is set, just like many analog EQs.

I know that some of you just have to be in agony about the whole idea of 0/0 being 1.0. It sure made this hard for me to understand at first. Even my math-challenged computer agrees that anything divided by 0 is undefined. Well, it turns out that you can get away with this in this domain for the following two reasons.

1. When designing analog filters, exact zeros never occur; no component is perfect. The net result is that the leakage inductance/resistance prevents the filter impedance from ever really getting to zero, and the resulting

divide between the series and parallel branch still works out about right for real cases.

2. When going to the Z-domain for digital filters, the bilinear substitution causes the divide to become a subtract, with the numerator working on past inputs and the denominator working on past outputs. In this domain, things can cancel via the subtraction without causing any math heartburn, so the properties are the same in the Z-domain, in that only denominator functions can create poles or an oscillator, just as in the S-domain. Also, the numerator can only produce zeros, and these can be either on the left or right half-planes in the S-domain or inside or outside the unit circle in the Z-domain.

I'm sure there's an even better, mathematically elegant explanation as well. I admit I don't know it and don't think I need to know it. All you need to know is that this works *every time*.

Although it is clear that the `case 9` arbitrary peak/dip filter has been around for a while, Troy and I seem to have independently rediscovered it. I've not seen a publication that shows how to do this, although clearly, some people already knew how. Many EQ's in waveform editors don't work correctly when tested across a sweep tone. Various ways have been tried that produce cancellations that don't show on the nice user interface display until you actually test them on a signal. The new peak and dip in flatness shapes do not suffer from this. As far as I know, the arbitrary tilt is a new thing, as are a couple of the other less useful options in `setfilt()` that are used to build an arbitrary IIR filter designer in the DirectX example on the CD (the newer types didn't exist then). Check them out and enjoy. You'll see that all the fun stuff is done in the S-domain, then bilinear substitution is used to get practical digital filter coefficients. Messing around directly in the Z-domain is certainly possible, just not very illuminating.

FIR Filters

I'll look at FIR filters first, since they are the easiest to understand. Other nomenclature for these filters includes all-zero or convolution filters, so if you see those terms, think FIR. For FIR, simply convolve the desired filter impulse response with past and current data samples. This is a simple vector dot product, as shown in flow-diagram form for a digital implementation in Figure 4.6.

Figure 4.6   DSP flow diagram of an FIR filter

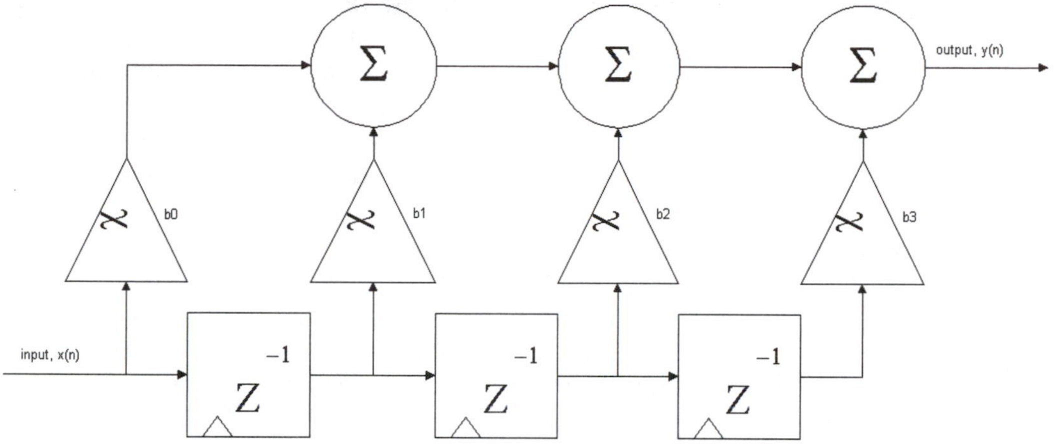

This figure implements the FIR filter equation

$$y(n) = \sum_{k=0}^{M} x(n-k)b(k)$$

where M is the filter order, n is the current sample number, and $b_{(0 \ldots M)}$ are samples of the filter's (obviously finite in duration) impulse response. As you can see, this represents a simple sum of current and past values of the input multiplied by constant coefficients. In a sense, this is a shape-matching operation. In other words, if the signal matches the shape of the filter impulse response, the output is maximal; if it doesn't, the output is minimal. Another way of describing this is as a correlation — it's the same math in another application. This is the direct case of convolving the filter's time response with the signal in the time domain, resulting in multiplying the frequency response of the filter by that of the signal in the frequency domain.

Notice that there is no feedback involved in this filter type. It cannot oscillate, and its output will always die out at most M samples after input goes away. FIR filters in general have all sorts of nice optimality properties in just about every area except how hard they are actually to *do*, sample by sample. You can see from the figure that you need one multiply and one add per filter impulse response point per input-ouput sample. The figure is quite misleading in this regard, since most practical FIR filters aren't of order three, but closer to an order of several hundred or more. This means hundreds or thousands of multiplies and adds *per sample* must be done to generate the filter output. A rough rule of thumb for FIRs is that you should have several cycles worth of the lowest frequency filter feature in its impulse response, or if there are sharp rolloffs, the equivalent rule applies. For instance, if the filter is to go from unity gain to −96dB in, say, 10Hz (at any center frequency), it then needs to be several tenths of a second in length. If you play with the FIR filter transform in WavEd, you can show the effect using the real-time plot of the resulting response. Try making a 200Hz low-pass filter then changing the order until it looks nice. Then try to find a nice order for a 5KHz low-pass filter. Remember, the plot is log frequency and amplitude, so a sharp fall at high frequencies on the plot isn't as sharp per Hertz as one that looks similar at low frequencies.

FIR filters can be made to have zero phase, or in other words, precisely the same delay at all frequencies. Without going into a hairy derivation, this requires that the impulse response be symmetrical about the center point. You can obviously force this for any impulse response that comes out of a design methodology by simply making two copies and splicing them together at the middle, reversing one in time. If the appearance of a true zero phase (instead of constant time delay) filter is desired, you can simply compare the filter output to a delayed version of the input signal, which should be delayed by half the filter impulse length for the symmetric case and odd filter order. I'll note in passing that zero phase is not a normal phenomenon, and the resulting filters don't sound natural for a lot of audio uses, although it may be just what you want in some cases.

Because WavEd works on disk files, I "hid" the half-length delay for you in the FIR transform, so the standard filter types there appear to have no delay at all. I didn't bother to do this for the adaptive type, since I used a style that doesn't force symmetry to reduce its compute time and required order compared to one that does force symmetry. I did this to make it easy on myself and on you.

FIR filters are disgustingly easy to design. For instance, you can simply create a table of desired responses versus frequencies and use an FFT to inverse transform this into the equivalent impulse response. Done! Of course, the impulse response will be chopped off at the length of the FFT used, and that is *not* a good thing. An effect called the Gibbs phenomenon results from suddenly truncating even a small-valued impulse response, and this causes ripple in the filter passband and a failure to fall off as it should in the stopband (assuming you're designing, say, a low-pass filter, but the effects on arbitrary shapes are similar). This is equivalent to the artifacts caused by truncating instead of smooth-windowing the input to an FFT, as described earlier, and the solution is the same.

Use a smooth window to taper the ends of the resulting impulse response so that the filter can have finite length. The results, such as increased main lobe width and decreased falloff rate for a given length, are the same as they are in the forward FFT case.

Simply lengthening the filter impulse is *not* the solution here. All that does is to increase the ripple frequency without reducing its amplitude very much, even for impulse responses that naturally tail off, such as the $\sin(x)/x$ [or `sinc()` function] discussed below. You can see this in the FIR transform in WavEd if you play around with it for a while. I built it to be a better teacher than my words alone ever could be. You can even see the impulse responses generated when the filters are designed by generating some impulses and running the filter over them.

Analytic Design of FIR Impulse Responses

In addition to the arbitrary design method, there are several analytic methods for designing FIR filters for simple shapes, which were found by using substitutions in the continuous Fourier transform to come up with a closed-form solution. For instance, to design a low-pass FIR filter, you can generate the desired impulse response via the equation

$$b(n) = \frac{\sin(\omega_c(n - L/2))}{\pi(n - L/2)}$$

where $\omega_c = 2\pi f_c T$, f_c is the desired cutoff frequency, T is the sample interval, and L is the total filter length. This is just the old `sinc()` function — $\sin(x)/x$ — normalized properly and centered for a symmetric impulse response. Once the b's have been generated this way, they still need to be windowed with a window appropriate to the application to prevent the Gibbs phenomenon from dominating the resulting response.

High-pass filters can be generated similarly via the equation

$$b(n) = \frac{\sin(\pi(n - L/2))}{\pi(n - L/2)} - \frac{\sin(\omega_c(n - L/2))}{\pi(n - L/2)}.$$

Band-pass filters are generated by the equation

$$b(n) = \frac{\sin(\omega_H(n - L/2))}{\pi(n - L/2)} - \frac{\sin(\omega_L(n - L/2))}{\pi(n - L/2)},$$

which could be interpreted as the difference between two low-pass filters. Last but not least, you can define a band-stop filter using

$$b(n) = \frac{\sin(\omega_L(n - L/2))}{\pi(n - L/2)} + \frac{\sin(\pi(n - L/2))}{\pi(n - L/2)} - \frac{\sin(\omega_H(n - L/2))}{\pi(n - L/2)},$$

which is simply a low-pass filter added to a high-pass filter.

All these need to be windowed with a window appropriate to your application. For maximum total attenuation in the stopband(s), something like the Blackman–Harris window serves best at the expense of falloff rate, and where initial sharpness is more important, a Hanning or Hamming window might be the best. You should know that somewhat more optimal design techniques do exist that can produce the same results with a slightly shorter FIR impulse response by tweaking individual impulse response values in an iterative fashion. This is because the necessary windowing to eliminate the Gibbs effect is generally not the optimum way to make use of all the filter taps.

Note that in cases where you are dealing with symmetric impulse responses, you can save half the multiplies normally required in an FIR implementation because of the commutative property of multiplication and addition. Basically what you do is combine input signals at equal distances from the center of the response by adding them before multiplying the result by the coefficient that matches both. Because the impulse response is symmetric, there will be two identical samples at equal distances from the center. On some machines, this would be faster; however, on most floating-point implementations, including the x86, there's little difference; adding `floats` is actually more work than multiplying them because of the extra normalization required. If you're working on a platform where multiplies take longer than adds, remember that taking advantage of the symmetry this way cuts multiplies in half for a given filter length, which might be enough to make this filter type practical.

Hilbert Transform via FIR Filter

In a special case of FIR filter usage, only the phase is shifted, while the amplitude is left alone. To make a real-only signal into an analytic signal that has only positive or negative frequencies, for uses such as frequency shifting and the like, the phase shift needed most often is 90 degrees. A broadband 90-degree phase shift is difficult to achieve with analog gear, usually consisting of a pair of many-sectioned all-pass filters that have a 90-degree phase *difference* between their outputs. In other words, they both corrupt the original signal phase but do so with a constant difference, which is not always what you might desire. In digital-land, however, you can make a single filter that shifts phase by 90 degrees and only adds the usual constant delay, which you can conveniently cancel elsewhere. This is called a Hilbert transformer, and an FIR impulse response can be designed that implements this. Without going too much into the closed-form derivation, the coefficients for a Hilbert transformer can be stated as being equal to zero for even and 2.0/(index $\times \pi$) for odd indexes, with the index going from –order/2 to order/2. Of course, you want to make causal filters, so what I did in the sample code in CFIR was to provide the necessary offset as for the other filter types. The demonstration of this in the FIR transform is not especially useful by itself, but was put there so I could test the code. For real use, just grab a CFIR class. Like any FIR, you need to window it for good results, and a Hamming window seems to work the best for most cases like this.

Adaptive Design of FIR Filters

Because FIR filters are so disgustingly easy to design, partly because it's impossible to design an unstable filter, another significant design technique exists — adaptive design, which can be used in at least two different modes: one where adaptation is used at design time only and the other where adaptation is continuous during use. Both have uses, obviously. I'll look at the basic structure first in Figure 4.7.

Figure 4.7   Adaptive FIR filter

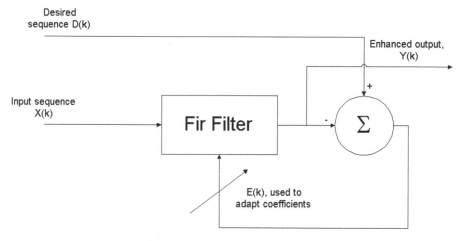

The assumption here is that you have two inputs: the actual data and a "desired response" that you want the filter to produce whenever it sees this data. You may be wondering why, if you already have the desired result, you want to do this at all. Bear with me here because there are important and practical uses of this technology.

Assume for the moment that you do indeed have a way to update the FIR coefficients such that the output of the filter will approach the desired response, as long as you have a sample-by-sample error signal. What can this do for you? For one thing, it gives you another way to design an arbitrary filter with *full* optimization for the given order by generating two signals. One is an input signal with a number of sinusoids at zero phase and equal amplitude — one for each specification frequency. The other is the same sinusoids, but differing in amplitude and phase in the same way that you want the filter to affect its inputs. If you train the adaptive filter on this pair of signals, it will converge on a filter with just this arbitrary frequency and phase response automatically, as close as possible for its order. It will do a better job for a given order than the direct plus windowing method discussed above for a given length at the cost of extra complexity and number crunching at design time. This sort of thing often is completely practical because you can afford the extra computation better at design time than at run time in embedded applications, for example, where low cost of necessary hardware is paramount.

How do you adapt the coefficients to get the desired response? There are several ways, in fact, but the most common way produces the least mean square difference (lowest error power) between the filter output and the desired output. First, run the filter for the current sample $x(k)$ and produce the $e(k)$ error output by subtracting the filter output $y(k)$ from $d(k)$, the desired response. Then, update the **B** vector.

$$\mathbf{B}_{(k+1)} = \mathbf{B}_k + 2\mu e \mathbf{X}_k \ ,$$

where the $\mathbf{X}_k$ vector is the current content of the delay line or the past sample values and μ is the convergence parameter, which controls how fast the filter adapts. Stated another way, once you have an error value for each sample, for each b coefficient, multiply the past data sample (i.e., the one in the same position as the current b coefficient) by $2\mu e$ and add the result to the existing b coefficient. This adds another set of multiplies and adds for each sample for a doubling of the total computation work.

That's all it takes. Most of the complexity involved in practical applications actually has to do with making sure the filter converges at the right rate for the application, or even at all with various nonstationary inputs. You can see that when the signal is high in amplitude the error will probably also be big, so the filter coefficients will move more quickly, that is not usually the desired behavior, so the most common variant on the above scheme is to normalize μ with

$$\mu_n = \mu/(L+1)\sigma^2 \qquad 0 < \mu < 1.0,$$

where σ^2 is the variance (standard deviation squared) of the signal. This allows the user to specify μ in the range 0 to 1.0 and have the adaptation speed be uniform with varying input signal power. A wrinkle in this scheme is that σ^2 is usually computed on a sample-by-sample basis as the square of the current sample, then this computation is passed through a first-order imitation low-pass filter to smooth the estimate a bit but allow it to track changing signal behavior, adding yet another parameter for the designer to specify: the "forgetting" or smoothing factor. This is done by computing each new σ^2 as

$$\sigma^2_{new} = \alpha x^2 + (1-\alpha)\sigma^2_{old} \qquad 0 \le \alpha \ll 1,$$

where α is the low-pass tuning factor, usually in the range of 0.5 (very fast signal power tracking but a lot of ripple) to ~0.001 or so (very slow but smooth), and x is the current input sample. Other schemes are possible and workable. All that's usually required is a fairly gross estimate of current signal power to allow quick adaptation (i.e., a high μ) without tracking "overshoot" on fast-changing signals. In other words, the faster you want the filter to adapt, the more important this estimate of signal power becomes. For the example in WavEd, I used the R-C simulation described below, with fast attack and slow decay, which is best for tracking speech. For many applications, this isn't very important because you want the filter to adapt slowly to ignore sudden signal changes. If you want the filter to track faster on higher amplitude signals, you can just set α to zero and provide a reasonable initial estimate of σ^2. This might be used in some cases where you'd want it to track quickly when a signal is present, but not track in on background.

The possible variations are endless, and most of the useful ones can be described pretty well as heuristic — a "brown" guess as to what would work the best. It often takes a few tries to get a good scheme for some particular type of input signal and problem.

An extremely important implication of this structure is that one might train a filter to anticipate, or predict, its input signal. This would be done with the same structure in the figure, but using the same input signal for both inputs and adding a one-sample (or more) delay between the input and the FIR filter input. This has many applications, such as line enhancement and speech coding. A single sample delay is enough to decorrelate any random noise but allow the signal to look about the same; thus, the filter will learn to ignore the noise, since it can't be predicted, but to pass unaffected except for delay anything that it *can* predict.

Figure 4.8   Predictive adaptive FIR filter

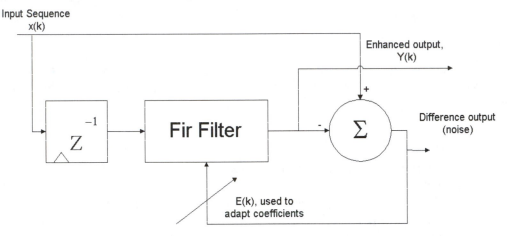

The structure in Figure 4.8 can be used for line enhancement, reverb removal, linear prediction, and so on. Basically, you choose the output and filter adaptation speed parameters depending on what you're doing. If you let the filter track slowly and present it with some sinusoids mixed with white noise, for example, you'll see the sinusoids on the filter output and the noise on the error signal output. In another example, assume the predictable signal is hum, and it's the unpredictable music you're interested in. In this case, you'd use slow tracking as well, since you don't want to track the music, and you'd use the error signal output instead, now with all the hum removed, since it was more stable and predictable than the music.

As noted above, there's more than one way to skin this particular cat. A lot of variations on the above scheme have been worked out to make the filter easier or more stable to compute on integer machines or, indeed, any machine, such as block updating of coefficients after accumulating errors to avoid having to add tiny numbers to huge ones, "leaky" coefficients (which help hide rounding errors in integer implementations), or schemes where just the sign bit of the update is added to the lsb of the coefficient in question. Most of these are roughly equivalent or have advantages not really related to the filter but, rather, to its implementation on a certain platform, and I won't go into them any more deeply here. For more details on this, check the bibliography and the application notes from any of the major DSP chip manufacturers, which go into great detail on the available techniques. After all, they want to sell their chips, and they've discovered that helping you be successful in their application actually helps sales. Both Texas Instruments and Analog Devices, for instance, make a wide range of application notes and practical source code freely available online, and often they are better written, from a standpoint of giving you practical information, than the usual DSP textbooks.

Yet another filter design, which is used in many vocoders and other sound data compression schemes, is used to match an arbitrary input sequence. This usually goes by the name of linear prediction, which is a very important technique, but only for speech professionals. I refer you to the first IEEE reprints book for an excellent tutorial on most of the aspects of this. If a source sound has the impulse-filter model, as voiced human speech does, it is possible to design a filter such that it will reproduce the sound if it is excited by a train of impulses. Why would you bother? Well in this case, you'd only need to send a few filter coefficients across a communications channel along with the pitch rate information to produce an identical sound at the other end, rather than having to send each and every sample, since the sample sequence is so predictable when this sound production model is in use. This can result in enormous data compression, which is the main benefit. In this case, what is designed is an all-zero (FIR) "cancel the input" type of filter, which is then converted into an all-pole (IIR) "reproduce the input" type filter, whose coefficients are then sent across the channel. What you usually do is take a windowed sample of the data, perform an autocorrelation on it, then send this to Durbin's algorithm (named after the inventor), which is a special case of a more generalized n equations in n variables matrix/vector problem solver. This results in an FIR design that has zeros at all the frequencies the input signal has peaks at.

To recreate the signal, the coefficients are simply flipped from the Z-domain numerator into the denominator, after the initial always-1.0 coefficient, which changes it into an all-pole IIR filter that can ring for a lot longer than its order, unlike an FIR design. This means it can stretch an excitation impulse into a lot of useful samples, many more than it takes coefficients to specify the filter. This is how the data compression occurs. Some problems arise with this approach, although they are usually not all that bad in practice. The matrix inversion effectively required here may result in trying to invert a singular matrix if the system is overspecified, and in that case, it blows up mathematically and produces a garbage filter. I should point out that it's almost miraculous good luck that this can work *at all* — designing an all-zero filter to a specification is relatively easy, since without feedback, it can never oscillate; the design problem is one that can be phrased as linear equation-solving. Designing a filter with feedback involves solving a system of non-linear equations, and would take far, far longer to do. As far as I know, no feedforward method of doing this exists. So, our problem here is that sometimes the filter designs that come out of Durbin's Algorithm will oscillate when put into the denominator. This is one reason that this is usually used with the lattice filter types described below, which can be forced into stability at least, by putting some limits on the coefficient values.

This means that you can't simply specify a big filter order to handle all possible situations and get a perfect match all the time. Also, because FIR filters need to be big to have good low-frequency resolution, trying to do this at high sample rates for high-quality speech doesn't work that well, which is why it's usually done at the minimum sample rate possible. This results in a shorter filter that is also less likely to be overspecified for modeling speech. If the filter design blows up, the harm is usually minimized by simply clearing the filter contents at each new pitch interval before re-exciting it, since speech usually has very close to zero energy anyway just before the next pitch pulse — a piece of good luck and *a priori* knowledge you can take advantage of. Because you'll update the filter pretty often — usually every 20ms or so — an unstable filter is unlikely to be out there for very long, which is what makes the approach tolerable at all. Its main advantages are that it can adapt "all at once" on a block of data and that it's computationally efficient to design from a changing input signal. At any rate, the point here is usually to compress a lot of data somewhat lossily into a lower required bit rate. The original motivation for the technique was to get speech into a low-bit-rate digital form so that it could be encrypted easily for military applications, but nowadays it's also used for things like internet telephone coding to reduce the required bit rate and make dialup connections practical. Although many brilliant DSP experts have spent entire careers on this topic, more remains to be done to get the best compression possible and still have a fairly natural-sounding result. I find it fascinating and continue to work on it. I recently patented a robust, real-time pitch tracker for speech that eases certain other design issues, such as when, precisely, to update the filter. This is a big improvement on older pitch trackers, which take several pitch pulses to gain an estimate of pitch rate and which do not tell the user when the pitch pulses occur. The new approach is more natural, because things like pitch pulse timing jitter information are captured, and synchronous filter design updates are now possible. This means the periodic assumptions that most design techniques, including this one, have are not violated, resulting in better filter designs and other advantages. It's a work in progress. When I get around to writing that book about Digital Processing of Speech, I'll have a lot more to say.

A final wrinkle on the design of FIR filters is that it is sometimes possible to design a filter with a high order that gets good resolution when many of its coefficients equal zero. This is a good thing because you don't have to do these multiplies and adds, making the filter computationally efficient while retaining the other nice features. This type of design is used in just about all modern audio a/d and d/a converters for anti-aliasing and resampling purposes. The design techniques, however, are deep, dark secrets I've never had the energy to find out about, and I get the idea that it only works in *very* special cases. However, this introduces the idea of what I'll call "sparse convolution," which turns out to be a very handy concept when simulating reverbs and such, so I mention it here to prime the pump, so to speak.

I can almost hear some of you complaining that since FIRs are convolution-style filters, it ought to be a lot more efficient to use FFTs to perform the convolutions. Indeed, many textbooks say this is the case, citing the $N \log(N)$ nature of the FFT execute time, rather than the $N \times M$ nature of the FIR filter (where N is the total sample size and M is the order). In practice it often doesn't work out that way until the sizes get really big. You have start-up costs (getting the filter response into the frequency domain for convolution), but the real problems are more difficult to solve. (1) You have to do two FFTs per record — one forward and one inverse — and do one FFT-size worth of complex multiplies per record while in the frequency domain to do the convolution, which is now at $3(N \log(N))$. But that's not the worst of it. (2) The FFT does *circular* convolutions — the ends wrap around due to its periodicity assumption — so you must pad the original series with zeros to prevent this, approximately doubling the length of the FFT required — now $3(2N \log(2N))$ — then perhaps round up to the nearest higher power of two in size so the FFT can be used at all. No, I'm not done yet. (3) Because it's a circular convolution, the beginning and end of each record are somewhat attenuated by the effect of the zero padding, so you now have (at least) twice the number of FFTs, half-overlapped, and you need to use either the overlap-add or the select-save method to get around it — now you're up to at least $6(2N \log(2N))$. Get the picture? This works when there's no other way, but you can't claim that it's especially efficient with at least $6(2N \log(2N))$ operations required. Remember that most "O" notation descriptions of FFTs completely forget things like bit reversal (which can be avoided in this case, although almost no one does, by using a DIT (Decimation in Time) FFT one way and a DIF (Decimation in Frequency) FFT the other and by allowing the frequency domain data to remain bit-reversed) and address calculations (which on the x86 platform take about as long as the actual computations). As a practical matter, if you spend a half or a quarter of the time it takes to just code a really efficient dot product for the same platform, the curves (computational effort) cross somewhere in the rather comfortable region of 256 to 1,024 samples or so, not the approximately 64 samples often claimed for the FFT implementation in most texts. Find out for yourself, if you don't believe me. Even on the x86 platform, which isn't exactly optimized for dot products, this is true with a tiny bit of dot product assembly code, and it's a whole lot easier to debug as well.

The upshot of all this is that FIR filters often aren't practical, at least for many real-time digital uses, unless they are very short in length, and they are only used when it's the only way to do the job. There *had* to be a catch, right? To help you learn further and develop a gut feeling for how FIR filters perform, I've added an FIR filter designer and application transform to WavEd. Play around a bit by trying marginal-length filters with and without various windows, and you'll understand this subject a lot better than by reading more about it right now. Because the source code for WavEd is supplied as well, you can look there for the real implementation details, or simply swipe it for your own applications.

Synchronous Filters

I'll close out this section on FIR filters with a hybrid filter that turns out to be very handy when making filter banks and for other uses. Basically, it's a single DFT bin, but instead of summing the real and imaginary results into accumulators over a record interval, you low-pass filter them (with any filter type, but IIR filters are most often used) as you go along, using the hypotenuse (the root of the sum of the squares of the real and imaginary parts) as the amplitude output, at least as most commonly used. You could regenerate the original filtered signal by simply remultiplying the outputs of the low-pass filters by the complex probe wave again and summing those results for the output.

Basically, multiply the input signal by a complex probe wave at the desired center frequency to get the sum and difference frequencies of whatever the input signal and the probe frequencies are. By low-passing the results of the input multiplies, you can get an output that has only the difference frequencies. Where the input signal has frequency components near the probe frequency, they are translated down to nearly DC by the heterodyning action of the multipliers — you only look at the difference frequency component because of the low-pass filters, so the sum frequencies aren't involved.

Low-pass filtering in effect creates a band-pass filter centered on the probe frequency with twice the width of the low-pass filters used, since a difference could be either positive or negative and still produce the same difference frequency. This can make the low-pass filters look very, very sharp at the probe frequency, even though they are minimally complex. Consider the case where the probe frequency is, say, 5KHz, and you want a 2Hz bandwidth. Use a pair of 1Hz low-pass filters; for instance, use second-order filters that have a 12dB per octave falloff, which means that at 2Hz they're 12dB down, at 4Hz they're 24dB down, and so on. Translated up to 5KHz, 4 or 8Hz is not exactly a big fraction of an octave, so to build such a sharp filter up there would take a heck of a lot more complexity (and computation) than this approach does. The price to be paid (always) is that you can't cheat the old time–frequency resolution trade-off here — you'll still have the time response of those 1Hz filters. Remember in the DSP chapter (Chapter 3) when I said that you need a window to "correct for" the assumption of periodicity from the DFT? It's still true, but since you're low-pass filtering instead of just summing here, the effect of the low-pass filters is just such a soft window in the time domain, one that automatically slides along the data stream! (Think of inverse transforming the low-pass filter.) The trade-offs remain the same. You need a low, low-pass frequency to get fine frequency resolution, and its time response means that you have to process a lot of samples to get a valid output — you have to satisfy the settling time of those 1Hz filters, just as with the DFT. Here, though, you can do only the bins you need, and you can make them *any* spacing and *any* width you want, so this technique gives the filterbank designer a lot more flexibility than using DFTs, FFTs, or even Wavelets, not to mention giving a convenient sample-by-sample output as things go along, instead of a batch output per record. You have to have a complex probe wave (both sine and cosine) because you may have an input component precisely on top of the probe frequency, and if its phase is other than 0 or 90 degrees, you won't get the right answer for amplitude from either phase of the probe alone. Combined, however, you get all the information you need. This is the filter of choice when you want very narrow bandwidths at high frequencies or are simply interested in defining each filter in a filterbank separately to precise specifications. If you keep the outputs of the low-pass filters sample by sample, the filter is reversible. You can also regenerate the input (although with some error due to time delay of the low-pass filters) if you want to. In fact, for reversibility, you only need to keep the outputs at some lower sample rate such that you don't violate Nyquist for the difference frequency when sampling.

In general, this filter type provides all the advantages of the wavelet types and more with somewhat less computation. You can have arbitrary time–frequency resolution trade-offs by band, which is better than you can do with wavelets, and it's easy to understand and tune.

I should also point out that since you're doing this a band at a time, a fast, wide band can overlap narrow, slow bands, and sometimes you can combine the information from these usefully to apparently cheat the time–frequency resolution trade-off. You still can't *actually* cheat, but some occasions arise when other *a priori* knowledge of how the signal can behave allows this to work in practice.

IIR Filters

All analog filters are of the IIR type. It turns out that any IIR filter can be expressed as a cascade of second-order sections plus, possibly, a first-order section for odd orders. There are other ways to design analog filters — cascaded, direct, and parallel — but they run into design difficulties, are harder to compute accurately, and so on, so their study is of mostly of theoretical importance, and I won't talk about them too much here. There are many ways to implement a second-order section, some with specific names and various accuracy trade-offs, which is where a lot of the jargon in the business comes from. The major types I'll discuss are the state variable and canonical types, both of which are useful in DSP and are ideal in some way depending on the problem at hand.

State Variable Filters

The analog version of a state variable filter is neat because it generates all the basic second-order simple shapes at once from a single filter, is the easiest to design and tune, and is cheap to make as an active filter with op-amps, although not quite *the* cheapest. That honor would have to go to the biquad filter, which is almost the same, uses one less op-amp, but is much harder to tune. The frequency and damping interact in such a way as to produce a constant bandwidth vs. frequency, rather than one that scales, so it was used mostly by the phone company to make large quantities of filters for various purposes. I won't cover the biquad filter here. The state variable filter has the fewest component tolerance problems. The same characteristics apply to the digital version of this filter, where it has just one serious limitation: you can't tune it above 0.333... of the sample rate. However, for frequencies lower than this, it is more accurate and easier to implement and understand than any other type of filter, especially at low frequencies and high Q's, which is why I'm covering it first. A state variable filter is the direct representation of the good old spring and mass model, with a friction variable to determine how frequency selective it is; that is, how long it takes to decay some amount after an impulse puts energy into it. To make a state variable filter, you need two integrators and some summing. The various outputs correspond to forces and accelerations, or positions and forces of the spring/mass model. The math for the digital version is so simple, I just can't resist giving it here in pseudocode form.

```
lowpass output = oldlowpassoutput + (tunefactor * oldbandpassoutput);
highpass output = input - lowpass output - (damping * oldbandpassoutput);
bandpass output = (tunefactor * highpass output) + oldbandpassoutput;
notch output = lowpass output + highpass output;
```

Figure 4.9 shows how it looks as a DSP flow diagram.

Figure 4.9 DSP flow for a state variable filter

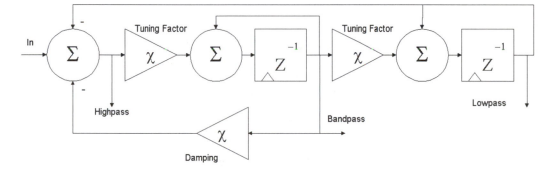

If done in this order, the computation can be done in place, with the new values for each output replacing the old because the old values are used before they are replaced by the new ones with this ordering. You don't need to calculate the notch if you don't need it, or you can make a high- or low-pass filter with a zero in the stopband by using a gain coefficient less than 1.0 for one of the inputs to the notch. How do you generate the magic numbers for tunefactor and damping? The damping variable is the inverse of the

Q covered earlier and corresponds to friction; that is, it's simply 1/Q. The tuning factor corresponds to the stiffness of the spring divided by the mass — it's the bounce speed, so to speak. Because of the inherent one-sample delay, a straight-line fit has errors near the high-frequency limit, so the tune factor equation needs a little more math:

```
tunefactor = 2*sin(pi*fc/fs)
```

where `fc` is the desired center frequency of the section and `fs` is the sample rate.

Knowing this, you can now open any book on analog filter design (e.g., Don Lancaster's *Active Filter Cookbook*), take the F's and Q's from the tables and go straight to digital with it, as long as you don't need to tune above 0.333... of the sample rate. At that point, the tuning factor goes above 1.0, and you can't get a stable filter due to the one-sample delay inherent in this implementation. However, for low frequencies and high Q's, this version has significant advantages over the more general canonical types of filters. In all digital things, there's a precision limit of some kind for number representation. For canonical filters at low frequencies and high Q's, two of the tuning numbers become almost exactly 1.0 and 2.0, and one has to be almost exactly twice the other to set the Q and frequency correctly; therefore, the precision limit comes into play, and the filter may not tune precisely where you want, even with double-precision floating-point numbers. For the state variable filter, there's no problem: `tunefactor` scales more or less linearly with frequency, and `damping` is just 1/Q, so nothing magic happens at the low frequencies, it just keeps working right. If you need frequencies of the order of 1 cycle per month, you should probably be using a slow sample rate anyway. In audio work, I use the state variable implementation when I need low frequencies and high Q's, one application of which is generating semirandom slow-moving control signals for things like choruses by filtering white noise. The result sounds better than using a pure sine wave. You can find source code for this in WavEd.

One thing that's easy to do intuitively with the state variable structure is an elliptic filter. An elliptic filter consists of a normal high- or low-pass filter but adds a notch somewhere in the stopband for the complete suppression of the energy there. These types of building blocks are used in Chebychev type II filters to get the fastest possible falloff outside the passband.

As with many things, however, there's a downside. After the notch, the response must bounce back up to some level instead of falling off forever as it does in a plain high-pass or low-pass filter. The why of this is about to be obvious. To add that notch, you add some of the other filter type's output, which at some point causes a complete cancellation or "zero" when the amplitudes and phases are just right. The basic notch described above adds equal values of high-pass and low-pass filters, so the response jumps right back up to unity on either side. To make an elliptic low-pass filter, you might add, say, 10 percent of the high-pass output to the low-pass output. After the notch, which will now occur at a higher frequency than the filter cutoff, the response will then bounce back up to 10 percent, the contribution from the high pass. Often a combination of techniques works best, using perhaps a single elliptic section followed by several normal sections to allow the stopband to continue to roll off above the notch. Just where the notch occurs depends in a fairly complex way on the amount of the other output that is added and the damping of the original filter. The math is pretty hairy, and the best way to design one of these filters to a specification was defined by Chebychev. Many $100 filter design books contain useful tables with all this stuff precomputed for you. They are worth having if you're going to be doing a lot of filter design.

I can't finish this subsection on "filters that are direct copies of analog implementations" without mentioning the most common and cheesiest digital filter type there is: a digital approximation of the old R-C filter, or first-order section. It is not the same as a canonical first-order section; it's even simpler than that. This is used a lot when just a little smoothing or a sloppy low- or high-pass filter is needed. The equation for the low-pass version is

$$\text{Out} = k \cdot \text{In} + (1 - k) \cdot \text{OldOut} \qquad 0 < k < 1 \,.$$

Figure 4.10 shows the flow diagram.

This takes a single location to hold Out (it can be done in-place as shown) and is tuned by varying k, with small values giving longer time constants and large values giving faster time constants. Notice that the frequency response can never fall below k at high frequencies because there is an intrinsic input bleed-through at a level of k. This can be good or bad, depending, but it *is* a difference between the analog and digital result. It turns out to be not that big a difference in practice. It's as though you'd tuned an analog filter up high by simply reducing the resistor in an R-C network. At some point, the effective series resistance of the capacitor comes into play, causing the same sort of bleed through.

Figure 4.10   Digital "R-C" filter

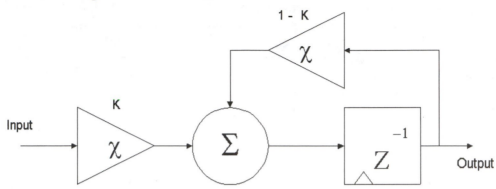

If you want a true analog response, use the canonical form described below and simply set the coefficients related to the S^2 terms to zero to get a first-order section. The CIIRFilt class in WavEd implements this approach in options 10 and 11. For a high-pass filter, just do the low-pass filter and subtract that result from the input sample. Despite its weaknesses, it's often just the ticket when all you need is a little signal smoothing, and it has no ringing or overshoot in the response.

A *very* handy variant of this filter is a peak detector, where Out is set equal to In when In is greater than Out or does the filter normally otherwise. This allows a fast (rather, instant!) attack, but has little ripple on the decay slope if the filter is tuned for a long decay time constant, which is a nice feature in many applications. You can also switch between two different k values, depending on whether the current input is larger than the current output, which is demonstrated in the compression transform in WavEd. Fancier, logic-added versions of this are one of the nice new tricks that digital processing makes easy. You'd be surprised how many tough problems you can solve with this simple construct by adding some logical "smarts" as to how and when the filter is retuned.

Notice that I don't give the tuning relationship here. This is because the filter is so sloppy and prone to bleed-through that the relation is more or less meaningless, and I always wind up fudging it for a particular application anyway. I'll just say that a tuning factor of 0.5 is a really fast (a few samples to track to the input value) or high-frequency filter (with 50 percent bleed through!) and that values below 0.001 or so mean you probably really need something better than this filter for your task. There is sample code in WavEd's compressor class that sets up very approximate time constants, but

I'm not claiming that it's actually *right*. In truth, it's hard to *tell* accurately because of the bleed-through.

Canonical IIR

The most common digital implementations of IIR filters are called canonical. Some texts reuse the word biquad for this type, which I disagree with personally. It *is* biquadratic, but it is *not* the same thing as the biquad filter invented by Ma Bell, which had the name first. This type of filter can have both poles and zeros, so it's really a combination of IIR and FIR filters.

The FIR part of the filter works on past inputs, and the IIR part of the filter works with past outputs. The poles come from the IIR part, which is the denominator in Equation 4.1, and the zeros come from the FIR part, which is the numerator in Equation 4.1. These poles and zeros are simply the complex roots of the polynomial in question, or places where it equals zero.

These filters are usually specified in the Z-domain or are converted to the Z-domain from the S-domain via bilinear substitution. In general, the specification looks like Equation 4.1.

$$4.1 \qquad H(z) = \frac{B(z)}{A(z)} = \frac{b0 + b1z^{-1} + b2z^{-2} + \dots + bnz^{-n}}{1.0 + a1z^{-1} + a2z^{-2} + \dots + amz^{-m}}$$

It can only produce one shape at a time, but the possibilities for shapes are more varied than for the state variable implementation. Like most IIR filters, it's usually applied as factored second-order sections, which are no harder to compute at run time than is the direct form, but are less subject to math precision errors. The canonical type uses all five possible coefficients (really, there are six, but you always normalize so that one coefficient, a0, is always 1.0) for a ratio of second-order polynomials, so it takes a bit more calculation than the state variable on a by-sample basis. For certain special cases, some of the coefficients are zero, so those multiplies and adds sometimes can be eliminated. However, the canonical, while only producing one output shape at a time, can produce more complex shapes than the state variable can since it uses more coefficients. This fact was exploited in the IIR filter designer to create shape classes that are easier to define and minimize shape error with, reducing a hairy multidimensional minimization to a unidimensional gradient descent. A canonical IIR filter implements difference Equation 4.2,

4.2 $\quad y_k = \displaystyle\sum_{n=0}^{L} b_n x_{k-n} - \sum_{n=1}^{L} a_n y_{k-n},$

which is drawn as a flow diagram for $L = 2$, or a second-order section, in Figure 4.11.

Figure 4.11 Canonical second-order filter direct implementation

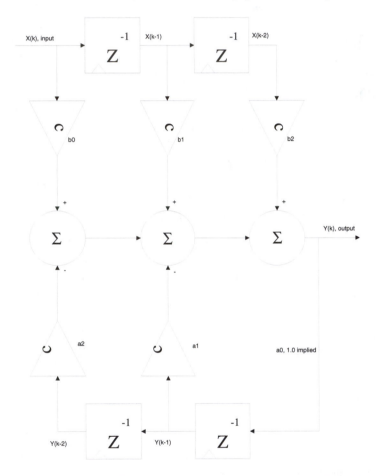

In general, these flow diagrams are implemented in software as follows. Do all the multiplies and adds with each new input sample, then "clock" or update the $z - 1$ registers or memory. These are often implemented as circular buffers, since you will notice that one delay unit feeds the next, and this technique saves having to move a lot of data around for each new sample.

Note that a ratio of polynomials in negative powers of z (the Z-domain transfer function) has somehow changed into a *subtraction* in the actual implementation! This is a *great* outcome for at least two reasons. One is that division is one of the slower math operations on computers, so you like to avoid it, and the other is that zeros in the numerator and denominator now can precisely cancel one another! If it were still a divide, this case would be troublesome. In fact, this example is what suggested that more types of shapes were available than the usual ones discussed in all the filter design books. In this domain, an all-pass filter that only has phase shift can be constructed by allowing a pole and a zero to cancel one another, which in the divide formulation would produce 0/0, or an illegal operation. Once this is realized, there is a whole spectrum of possible shapes that are never mentioned in the filter literature, so you can start to work on the math yourself. By the way, the reason that the divide becomes a subtract is that samples of the output are used in the denominator and samples of the input are used in the numerator. Some hairy math derivation, which I'll spare you, along with the bilinear substitution makes the change.

Canonical filters can be implemented in (at least) two ways. One way uses separate delay lines for the past input and past output samples needed in the direct computation of the difference Equation 4.2, as shown in Figure 4.11, but it turns out that there's a simple way to formulate the math (the *explanation* isn't simple, just the result) to use only a single delay line, which is the most common way these are used, since needing less memory is almost always faster in real-life implementations. It takes up less page space here, too! I'm not going to explain why this works, since that takes a few pages of pretty dense math derivation that is already given in the first IEEE reprints book. Instead, I'm going to show you how to do it.

Figure 4.12 comes about as a realization that both parts of the filter (FIR and IIR) are linear operations and can therefore be done independently and in either order without changing the results.

In the derivation for this formulation, the authors (Gold and Rader) mention having to flip the signs of the "a" coefficients as a result of the built-in subtract in difference Equation 4.2. This generally drives me nutty because some authors just assume you'll use a subtract there after doing the summing with all additions, whereas some authors always use adds everywhere and negate some coefficients to get the effect of subtraction. A word to the wise: If you're trying to use someone's package with some of your own code or some other package and you get a lot of unstable or otherwise weird-acting filters, try flipping the signs of the "a" (denominator) coefficients or changing the add for them to a subtract or vice versa. I can't tell you how many times I've run into this, and there seems to be no actual standard practice — everyone uses the way that suits them best! Murphy's law says it'll be wrong for you. I have one package that has separate parts that don't even agree! I have a lot of lash-ups with such code as

```
for (i=0;i<order;i++) a[i] = -a[i];
```

so that I can use one standard library package with another standard library package. Disgusting, isn't it? And just for grins, the sign of the leading 1.0 doesn't change. That's the implied use of the output feedback at that ubiquitous gain of 1.0. Most filter packages don't even store or pass in this coefficient; just assume it.

Figure 4.12 Alternate formulation of the second-order canonical filter

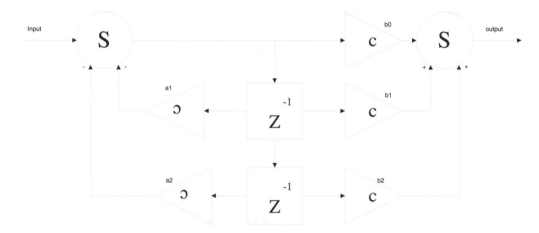

Notice that although I showed the Z-domain transfer function and difference equations for any order, I'm mostly talking about second-order sections here. Although you can design an entire complex filter in "direct" form, instead of factoring it out into second-order sections, there are math precision problems with this approach, and large complex filters will tax even double-precision floating-point math. In general, "it depends"; some design methodologies naturally give one form or the other, and you usually use whatever comes out of that process, unless the need is compelling to do it the other way.

Canonical Shape Options

Along with the so-called standard high-pass, low-pass, band-pass, and band-stop options, filters of more arbitrary shapes are possible with the canonical structure. In general, any shape that can be made by graphing the ratio of two second-order complex polynomials can be created. This seems to be rather a deep, dark secret in the DSP world. The real issue is that simple math to design the standard shapes has been around for centuries but that no simple math exists to design arbitrary shapes — that is, until now. With the help of one of my students, Troy Berg, some new standard types have been implemented, and the design equations are now available. These include shelf, or tilt, bump in flatness, and dip in flatness, with the height of the bump or dip or shelf specified independent of its width, directly, in feed-forward-design math. These make designing arbitrarily shaped IIR filters quite a lot easier than solving a lot of nonlinear equations, as it used to be. It becomes more like doing graphics in the old console text programming days by hooking together little corner shapes and straight lines to get the final shape you want. Because the bumps and dips appear surrounded by flatness (e.g., gain outside the filter's effect area is 1.0, instead of a continued slope of some kind), you don't have to deal as much with interactions between sections, making a design program simple to do. You merely start with a null filter (e.g., a straight line) and compare this with the specification, normalizing the "gain" to minimize error. At the largest error, add a dip or peak of the appropriate height, then adjust its width for minimum total error. Repeat, each time minimizing the total error at each section (by adjusting its height at its center frequency for zero error there, and its width for reduced total error), until the total error is acceptable. That's all there is to it. No complex gradient descent; none of that non-polynomial-complete traveling-salesman stuff.

IIR Filter Design

This section could easily be a book, or even several, on its own. In fact, there are already literally hundreds of books extant on this topic. I had to crack up the other day when a project manager from a company I do work for was here discussing a possible project that involved some very fancy DSP, and he said, "I understand filters and DSP fine, but how do you get those coefficients?" All I did was point to approximately 500 pounds of books on the shelf and say, "well that's a small subset of DSP, and knowing it is what *you* pay *me* for," or I would have had to laugh in his face. It was kind of like saying, "I understand chemistry, but why do some combinations of chemicals get hot when you mix them and other combinations just sit there or produce a precipitate?" So to keep this discussion reasonable, not to mention to avoid plagiarism, I'll quickly gloss over the standard and comparatively limited techniques that are well covered elsewhere. When I need a conventional high- or low-pass filter, I use the techniques and code from the Stearns and David book or from Lancaster's *Active Filter Cookbook*. If it needs to be really fancy, I might go to the pole/zero tables in one of those $100 filter design handbooks. Those books simply list the pole (and sometimes zero) locations for filters of various types that have been normalized to 1.0Hz or radian. It doesn't matter which, as long as you're consistent about your units. These numbers can be converted to frequency and damping values with Equations 4.3 and 4.4.

4.3 $\text{frequency} = \text{sqrt}(\alpha^2 + \beta^2)$

and

4.4 $\text{damping} = \dfrac{2\alpha}{\text{frequency}}$

where α and β are the real and imaginary parts, respectively, of the pole or zero value. Usually, this will give you a filter normalized to 1.0 "something," so you'll want to multiply the frequency result by the desired filter cutoff frequency before using it but after calculating the damping factor for the section.

Filters have existed for quite a long time as math constructs — before even electronics! However, once electronics got going, the development of filter design techniques really took off. A large body of design technique has been developed for the analog electronic world, and it turns out that this is more or less directly reusable in the digital domain, which is what's most often done — it doesn't seem to have been worth the effort for anyone to rephrase the techniques to do most things directly in the digital (Z-)domain since a simple substitution of variables will do the conversion from analog to digital for you. This makes all the pre-existing work for the analog domain immediately available for use. I'll discuss this substitution, called the bilinear transform with frequency prewarping shortly. This is one of the times I'm going to eschew the "standard" explanation in favor of my own take on things, in the hopes that having another viewpoint will help you understand the standard explanation, which was completely opaque to me for quite a while.

Mathematicians have described a construct called the *S*-plane, and anything analyzed this way is considered to be in the *S*-domain. The *S*-plane is simply an *x–y* space where the vertical axis is in units of *j*ω and the horizontal axis is in units of α. You can think of these two as frequency and friction, respectively — sort of (see Equations 4.3 and 4.4 above). On this plane, plot the roots of the filter polynomials for analysis. A second-order filter might be represented as a ratio of polynomials in the complex variable *s*.

4.5 $$H(s) = \frac{s^2 + bs + \omega^2}{s^2 + bs + \omega^2},$$

where Equation 4.5 describes an all-pass filter because the numerator and denominator are equal. When I speak of poles and zeros, I mean the roots of these polynomials (or the equivalent ones in the Z-domain), and a root, or a value of *s* where the polynomial equals zero, becomes a zero when talking about the numerator and a pole when talking about the denominator, which makes sense because zero is zero no matter what you divide it by, and dividing anything by zero gives infinity — a pole. In the form shown here, ω is the denormalized center frequency of the filter and b is its bandwidth, which results from a nonzero α in the *S*-plane plot. Many texts oversimplify the analysis by letting $S = j\omega$ for analysis, instead of its true value, $s = (\alpha + j\omega)$, which led me into a lot of confusion, to say the least. In the *S*-domain, it turns out that the real part of a root, α, is the friction, or loss factor, inside the filter, and without taking this into account, all filters have infinitely narrow bandwidth or are oscillators, so its value is truly important to any sort of understanding and corresponds to the b factor above. As classically phrased, α is negative for friction and positive for antifriction, so a filter's poles have to lie in the left half-plane of the *S*-domain to be stable. The filter's zeros, if any, can lie anywhere, since they are in the numerator, which as you'll see later, implies no feedback is used for them. This means they can't oscillate, but for a minimum-phase filter, any zeros will also lie somewhere on the left half-plane in the *S*-domain.

You can make this all-pass filter into one of the common shapes pretty easily now. All you do is leave some things out of the all-pass filter's numerator. If you leave out the terms on either side of b*s*, you have a band-pass filter. If you leave out b*s* and one or the other of the terms, you have a high-pass or low-pass filter, with the s^2 term controlling the high frequencies and the ω^2 term controlling the DC response. Although this looks considerably oversimplified from the mainstream texts, it isn't, and, in fact, viewing particular filters as perturbations of an all-pass filter leads to a deeper understanding of how it all works. To help you gain that all-important intuitive feel, a program called SDomain generates a simple second-order filter section and then lets you play with the coefficients while watching the response plot. For the all-pass case, zeros in the numerator are used to cancel zeros in the denominator. This is kind of counterintuitive, since zero divided by zero is undefined, not 1.0!

It turns out that by the time you get to the domain where you're working on sampled data, this divide has turned into a subtract because of the bilinear substitution and the use of feedback for the denominator only, so there's no *actual* problem, just a problem understanding how the domain works. To summarize: It's safe to do just about anything in the numerator, and it's normally safe to increase any of the denominator coefficients. The numerator corresponds to the feed-forward or FIR part of the filter, so you can't make it blow up or become unstable by fudging around with the coefficients. The denominator corresponds to the feedback portion of the general case, so decreasing the coefficients that multiply s might cause the filter to blow up or become unstable. In general, it's not safe to mess with the sign of the denominator coefficients — you'd probably move a pole to the right half-plane and make an oscillator. In all cases, the squared term controls the higher frequencies, the middle term controls what happens around the center frequency, and the term that doesn't multiply s controls the low frequencies and DC. Just as you'd expect, increasing a numerator term increases the response in the area it controls and decreasing a denominator coefficient has the same effect. Because there are matching pairs for all three terms, you can always get what you want out of this model, usually without fooling with the "dangerous" denominator coefficients, although a couple of safe changes to these coefficients preserve the overall scaling and center frequency, which is nice.

The "tilt" filter type is accomplished by reducing the right-hand numerator term and decreasing the opposite denominator term, for instance, usually by the same factor. Simply switch which numerator term to decrease (and decrease the opposite side's denominator term) to switch the direction of the tilt. Calculate the amount of decrease by dividing the amount of total tilt you want by 2, since you normally would like the tilt to be symmetrical about the center frequency, meaning the filter gain crosses 1.0 at that frequency and is higher on one side and lower on the other. A wrinkle on the technique adjusts the damping a little (by the square root of the tilt factor) so it continues to track as the other terms are scaled. You can find a code example of this in the SDomain program on the disk, and another in WavEd.

Similarly, if you want a peak or dip surrounded by flatness, all you do is to either increase the middle term of the numerator or the denominator, depending on which you want. The Q of the filter will control the width of this peak or dip just as it does with the old shapes. It seems that as long as you only perturb either the numerator or the denominator center term, and only by leaving one alone and making the other bigger, stable filters always result. The difference between these and a normal band-pass or notch filter is that the slope doesn't continue — after the peak or dip on either side, the response returns to unity. The other difference is that for the old-style notch filter, you couldn't specify a depth. The notch was always pointy and infinitely deep, since it was produced by a complete cancellation. With the new method, the dip just looks like the inverse of the peak — smooth and with controllable depth. This is far better for equalization uses than a notch filter whose depth you can't adjust.

Now I'll move from the semi-intuitive S-domain into a domain that you can actually do some work in, other than analysis. In DSP, this is the Z-domain. There are a couple of ways to get there, but the most popular way is the bilinear transform. The equation to do the mapping is simple (remember that both s and z are complex numbers).

$$s = \frac{z-1}{z+1}$$

This simple substitution maps the entire (infinite length) $j\omega$-axis of the S-domain plot onto a unit circle in the Z-domain plot. Because everything maps, there are some obvious symmetries.

Anything inside the unit circle in the Z-domain corresponds to the left half-plane in the S-domain, or the region of pole stability. Anything outside the unit circle maps onto the right half-plane in the S-domain, or an unstable pole. Because the entire infinite length of the $j\omega$-axis is now mapped onto the perimeter of a unit circle, you'll have to do some scaling. Remember that because of Nyquist, you don't actually *have* an infinite range of possible frequencies representable unambiguously in the Z-domain; rather you are limited to the range of DC to the Nyquist frequency, after which things wrap around, which is one of the reasons you can map between an infinite-length axis and a finite-length circle perimeter successfully: It exactly matches the reality of the case at hand. It turns out that the frequency is "warped" as it is mapped, and you have to "pre-inverse-warp" your analog frequencies before doing the substitution to make things come out right. As you might imagine, this prewarping must take into account both the original center frequency and the digital sample rate to be accomplished correctly. To do this, divide the *S*-domain coefficients by a factor that takes the nonlinear warping into account. This number is $\tan(\pi F_c T)$, where F_c is the desired cutoff frequency and T is the sample interval. This takes the *S*-domain design that was probably normalized to 1 radian or 1Hz frequency and translates it to the desired cutoff frequency, taking the actual sample rate into account. Next, do the bilinear substitution, and out pop the *Z*-domain coefficients, neat and sweet.

With this technique, you can convert from domain to domain at will. The warping and substitution are completely reversible. You can do your thinking in whichever domain makes more sense to you. Normally, you want the lowest order denominator term in the *Z*-domain normalized to 1.0. You don't lose any generality this way, and this term corresponds to the feedback from the filter output to the rest of the math, so making it 1.0 always not only saves a multiply, it makes filter stability somewhat easier to determine.

There is yet another type of IIR filter structure that was developed by Markel and Gray for speech synthesis applications. It needs more multiplies and adds to compute, and to get its coefficients, you need to take *Z*-domain coefficients and put them through a further recursive step-down procedure. The big advantage of this type of filter is that stability is trivial to determine. The stable range for what are called reflection coefficients in this filter type is between −1.0 and +1.0. Also, these coefficients can be put through another simple transform to convert them to a log area ratios form, which will make them correspond directly to the cross-sectional areas of a segmented pipe carrying sound. In other words, each coefficient corresponds to the cross-sectional areas of sections of a pipe whose total length corresponds to the length in samples mapped onto the speed of sound in, say, air. Thus, if you can design a filter that duplicates a human vocal tract, you can then examine the coefficients and interpret them as variations in the area of the tract, going in either direction in your analysis. The effect is such that it's like being able to see into the mouth as it moves, the mapping between a plot of the coefficients and the internal shape of the mouth is one to one. Spooky, eh? Because the vocal resonances and notches are produced by a pipe of variable cross section, it's kind of nice that there is a digital filter type that works so much the same way. I don't provide code for this in the book because it takes longer to do; is mostly only used in direct rather than factored form, which I don't discuss much here; and because there are plenty of other sources for it if you need it. Either the Stearns and David book or the paper in the IEEE reprints book, both listed in the bibliography in Chapter 1, will provide what you need. You can also get the source code for nearly any low-bit-rate speech coding algorithm off the Internet.

The Tools Are Missing, or New Discoveries

I sometimes mentor college kids who want to learn DSP and write code. Troy Berg, who helped with some of the nice UI and DSP code for the work that appears in this book was trying to get a feel for the *S*- and *Z*-domains and was peppering me with questions one day. I pointed out that I'd started backwards myself — I understood filters in some intuitive way long before I really understood the math, and this is what made the math clear to me when I saw it. We were trying to figure out some obscure point at the time, and it seemed we both really lacked that all important gut feeling for the domains. The obvious solution to this (for us) was to program a sample application that would let us noodle various things and see what happened empirically. The result is the program SDomain, which is on the CD in its "research-grade" form. We have a hot interest in being able to design arbitrary-shaped IIR filters quickly, because it has a lot of potential saleable applications, in everything from low-bit-rate coding to speech articulation enhancement. We weren't satisfied with existing theory, which pretty much only describes how to do things like low-pass filters and other basic shapes that don't combine very easily into an arbitrary shape. Also, it seemed that a couple of key insights were missing. For instance, how come we couldn't do *any* shape specifiable by the ratio of two second-order polynomials, and why hadn't the math been developed to go from a user curve directly to this? Maybe there's a curve-fitting transform out there that would translate to the coefficient domain directly, and all we need to do is find it. How could all-pass filters work, since that would seem to imply that $0/0 = 1$, when divide by zero is undefined? We obviously needed to learn more! Some of what we found out is still strictly empirical, and we're working now on making it mathematically describable and robust, although there are some indications that "the tools are broken."

I'll begin with the latter first. One of the things we quickly found out while noodling around was that you can make filters with damping greater than 2.0, and that nothing magic or bad happens when you go above 2.0 — the filter just continues to broaden out as expected. The thing is, this would amount to putting a pole closer to the alpha axis on the *S*-plane than that axis or to having β become imaginary, so that when multiplied by $j\omega$, it becomes real again. In other words, there's no place on the *S*-domain plot to plot such a filter directly. Well, there is, sort of, which I discovered by applying the quadratic root equation to some filters. Once both of the poles hit the α-axis, they split apart, acting like two separate first-order poles, one moving closer to the origin and one farther away. The effect is to change a single second order filter into two differently tuned first order filters. When the *S*-domain coefficients are bilinear transformed into *Z*-domain coefficients and the resulting filter is run, it works! You can see this easily in WavEd's graphic EQ transform, where you can set damping as high as 100 and get a very gentle filter that covers many octaves with a nonmultiple of a 6dB per octave slope. This is not mentioned as a possibility in the literature with a second-order section! It is, however, handy for a lot of things, such as changing white noise to pink, which used to take a mixture of output from many filter sections but can now be done with one. By examining the *S*-domain root plot, you can see that this result is due to a second-order section splitting into two first-order sections. There is no way to simulate a damping factor of more than two with a "real" second-order section, but the symmetries in the math cause the right thing to happen in digital implementations. Encouraged that our lack of understanding seemed to be "not all our fault, the stuff is wrong," we went on to experiment with perturbations of a basic all-pass filter to see what we could see. We discovered the relatively simple mapping that the s^2 term has to frequencies above the filter center frequency, the *s* term has to frequencies around the center, and ω^2 has to low frequencies. I theorized that we could make a bump in flatness with independent control of width by perturbing the center coefficient. It turns out to be true — you can, simply by making the center coefficient in the numerator greater by a factor that is the bump height desired. Because the normal explanation says that this is really the term for Q and the center frequency, this is an interesting result indeed. You can see easily how the pole-zero cancellation might be disturbed here, but it's not obvious, at least not to me, why its height comes out to the unscaled number, independent of the original Q specified. It would seem that as a pole that was canceling a zero was broadened, we'd see something a little different. You can make a dip (not a zero) the same way by increasing the center term of the denominator. This is a nice, smooth dip — no sharp point — and you control the size the same way, independently of width. You can see this demonstrated in WavEd in the parametric EQ.

Another useful shape can be had by perturbing an all-pass filter with a simple tilt around the filter center frequency, boosting either the highs or the lows. To get the filter gain to 1.0 at the center frequency, reduce opposite end terms in the numerator and denominator (which diagonal pair depends on which way you want the tilt). Again, the numbers work out such that to get a 12dB tilt, you halve each term in question. Use half (i.e., 6dB per term) because you're basically doing it twice. To adjust the Q so the shape factor remains as it was before the perturbation, divide the damping factor you'd normally pass in by the square root inverse of the reduction factor you're using. All I can say at this point is that it works like it should. I do not have the full math workup for why this is the magic factor yet, though it's intuitively obvious that some power of two factor makes sense when dealing with a second-order section. At any rate, this is another case where Q's of less than 0.5 are useful for making broadband tilts with a single section.

More work remains to be done on both of these new types of filters so that efficient feed-forward design can be done. For instance, putting a damping of 1.414 into the tilt section results in a tilt of 12dB per octave and flat shelves (after modification by the algorithm). It would be nice to work out the relation so that you'd know just what damping to use for other (smaller) slopes and have a feed-forward way of knowing just how far on each side of center the tilt extends versus damping. It might be nice to have a feed-forward way of knowing where the 1dB points for a given peak or dip are versus Q and height as well. I'm working on it, so stay tuned. I hope you'll play around too.

I know this has been tough going. It took me some years to get a good feel for it myself, and I have about 30 books (some of them with 1,500+ pages) that attempt or purport to cover just this one subject. All fail, mostly by being completely inaccessible, some by being incomplete. I hope that all the code examples I provide, with programs that show in real time what happens as you noodle around with the standard filter types, impart at least some intuition about how filters work. Clearly, linear time-invariant filters alone could be a lifetime study, and they are for some people. Most of us, however, would like to know just enough to get on with the interesting things, and that's what I've been shooting for here. At worst, you can take my word for it that "it works like this," swipe the code, and have a ball with filters.

I haven't said too much about time-varying filters, but obviously they have their uses. Speech itself is created by slowly time varying filters. Changing various parameters quickly compared to the filter center frequency has "interesting" results. Remember, since the energy stored in a digital filter is virtual, you could even change this on the fly by changing the filter Q or the amount of energy it stores for a given input level. Sweeping center frequencies quickly can produce the effect of vibrato in the output of an all-pass filter. Wah-wahs are made with just a single peak swept through the range of interest. The reason I don't mention these much here in the theory section is that the math in some cases is poorly developed and in all cases is obfuscatory; besides, it's more fun to find this out on your own. You can experiment with these and other phenomena with the morphing parametric EQ transform that is included with WavEd on the CD instead. Enjoy!

Chapter 5

Analysis of Sounds

One of the tasks DSP and special efx programmers are often called on to perform is the analysis of sounds. You may need to extract various information about a source signal in order to properly process it or merely to understand it so you can synthesize similar sounds later. It's important to understand sound production models in general for other reasons, as well; for instance to gain insight on what types of processing might sound natural or unnatural. There are quite a lot of parameters you might like to discover about a sound, such as whether it has a pitch; the beat, other types of periodicity, or lack thereof; how fast it builds up and dies away (which may be frequency dependent); and others. If the sound generation model is an impulse-type source modified by a filter, you may want to separate the filter from the excitation and acquire the parameters of each separately, a technique commonly needed in speech analysis and channel bandwidth compression. Because this is such an important area, a host of more or less standard techniques have grown up around it.

I'm sure you know some of the old saws about assumptions, but when simplifying assumptions are possible, it pays to look into them to ease or make possible the task at hand. In the game of sound analysis, more or less standard models are often used to simplify the analysis, called sound production models. In these models, certain things are assumed to be true or false, and as with all assumptions, you must check their validity before blithely proceeding with an analysis based on them! Fortunately, this is often pretty easy to do, you just have to remember to do it.

Sound production models can be phrased in a variety of ways, from equivalent circuits to mechanical models, and you usually can flip from one representation to another usefully. Thus, you can visualize a simple harmonic motion as either a spring–mass model or as a tuned circuit, depending on what is important to the task at hand, for instance, so a sound production model simply gives a vision of how the input energy is used and modified in various stages to make a final sound.

Sound Production Models Overview

The single most common sound production model is a source of some kind that creates impulse-like excitation, which is then filtered and radiated. This would cover human voiced speech (the vowels, basically) and many musical instruments. In some cases the filter's stored energy reacts back on the source and affects its frequency and other characteristics as well, which is the case for the violin and wind and brass instruments. For a lot of these cases, the source is a relaxation oscillator driven by a more or less constant force, be it air pressure or a bow. For other cases the energy is put in as a single impulse by a guitar pick or a piano hammer, for example.

Because the violin model is probably the easiest to visualize, I'll use it as an example. As the player draws the bow across a string, the string initially sticks to the bow. As the bow moves, the string displaces along with it for a while. Finally, the increasing restoring force provided by the spring-like nature of the string causes the string to release from the bow and move back toward (and eventually past, because it has mass and therefore inertia) its original position. This excitation can be thought of as sawtooth; that is, the input force builds linearly until the string releases and suddenly jumps back to zero.

If it's this simple, what factors other than bow pressure and string adhesion cause a particular note? In this case, the string's spring–mass system acts as a filter that reacts back upon the input. The player tunes this very high Q filter by adjusting the effective length of the string according to where it is pressed against the neck of the instrument. Because the string stores energy and "rings" at some frequency in the absence of further excitation, its motion affects the instantaneous force at the spring–bow interface, causing it to break away from the bow at the proper time for the excitation to be in phase with the existing string energy. This type of source–filter interaction is what tunes all of the wind and brass instruments, as well. Once the violin string is set in motion, some of its energy is transferred via the bridge to the soundboard of the instrument. The soundboard acts like an additional fixed filter tuned mainly by the stiffness/mass ratio of the wood. A good instrument will have a number of modes of vibration in the soundboard, each of which boosts some frequencies and dampens others, making a big contribution to the overall sound character. Finally, the sound radiates from the soundboard, which because it has limited size, acts as a high-pass filter. If you were going to simulate a good violin, you'd have to take all these different aspects into account to do a decent job.

In the case of human speech, there is not as much source–filter interaction, and it may be instructive to look at why. The relaxation oscillator in human speech is physically a couple of flaps of skin and tendon that are held together by muscles and forced apart by air pressure from the lungs. The mass of the structure is considerable, so it takes a rather large pressure to operate or affect. The human vocal tract following this structure, however, is fairly big and free flowing for air and of relatively low Q; in other words, the filters created by the air chamber in your mouth don't store very much energy, and a lot would be needed to affect the pitch of the vocal chords, so the back-reaction in this case is minimal. An interesting fact about the vocal tract filter is that when the larynx opens, it reconnects the lower airway to the upper tract. This has the effect of removing additional energy from the vocal tract filter, damping and retuning it slightly. Most linear prediction techniques don't do a very good job of simulating this aspect and try to get away with simply using a lower Q filter for the entire waveform duration. However, the right way is to adjust the filter damping upward near the end of each pitch interval, as nature does when the larynx opens. For unvoiced human speech, there is a noise source followed by a usually less-complex filter. The source is air rushing through some constriction in the mouth, which produces a noise-like source. The reason the filter is usually less complex and lower Q than for voiced sounds is that this restriction is usually nearer the front of the mouth than the larynx is, so there's simply less vocal tract available to do filtering.

For all normal sound production, there is a final postfilter due to radiation efficiency. The long and short of this is simply that until the circumference of a given source is as big as a wavelength, the source isn't very effective at radiating a given frequency. The effect shows up as a 6dB per octave tilt in most sources for frequencies whose wavelengths are less than the source circumference, and it is one reason large speakers are required to effectively radiate low bass. Sound is produced by compressions and rarefactions of air, and if a source is too small, the air around it simply moves back and forth without being compressed — the air's low density means it just gets out of the way rather than compressing if the source has low frequencies and is therefore slow moving. This also is reflected in perceptual models, where things that produce a lot of low-frequency energy, especially in relation to how much high-frequency content they produce, are perceived as big, since no small thing can radiate low frequencies very well.

For large classes of sound sources, the original excitation isn't very impulse-like, so the nature of the excitation comes into play as a separate issue, especially if you want to resynthesize the sound later. Examples of this include striking a string with a soft piano hammer, which imparts mainly lower frequency energy, some types of finger picking of stringed instruments, and the human voice. These differences in excitation allow more expression by the artist than would otherwise be possible in a fixed excitation/filter model. The human voice is especially interesting in this regard. Humans are able to vary source excitation in quite a few ways. We can vary how leaky our larynxes are, to get a more or less breathy sound, we can vary how hard the vocal cords slap together, and perhaps most interestingly, we can control deviations from periodicity. That "gravelly" sound so popular in some singing styles, and in Bill Clinton's voice when he's trying to get out of trouble and sound sincere, is called *diplophonia*. In diplophonia, every other pitch pulse is slightly delayed from a perfectly periodic situation. This isn't perceived as a halving of pitch in speech, which is more or less what it really is. It's interesting that some professional politicians and others have learned to invert their normal speech patterns — normally diplophonia comes when you are relaxed and sincere.

Sometimes the filter in question is neither linear nor time invariant. The human voice is mostly created by a time-varying filter. Another example is a drumhead, in which the tuning of the modes of the head is affected by the amplitude that the head is currently ringing with. You can hear this effect if you listen carefully to a drum that has been struck strongly — the sound starts out with a high "pitch," and the apparent pitch decreases as the sound dies away. In addition, nonlinear effects in the drumhead material create harmonics of the original excitation and create beats between the principle modes. In the case of a snare drum, the snares all add their own little hits as they bounce off and recontact the bottom head. Cymbals are another example of nonlinearity so complex that they're tough to simulate well, with different modes having different travel speeds through the metal because of the nonlinear behavior of the bending metal. Unlike air, metal can support shear waves and other modes that all have different propagation speeds through the medium, and so more modes are possible and their tuning can be very complex.

One of the first things to do when analyzing sounds is to look at their spectra versus time. This can be done with filter banks, sonograph machines, or various simulations of these that use either FFTs or wavelet filters. Many presentation modes for this data have come into fairly common use, the sonograph and waterfall plots being perhaps the most common of these. In a sonograph, time is the horizontal axis, frequency is the vertical axis, and energy at some time/frequency point is coded as the brightness or color of the pixel. This presentation method was developed for and works especially well on speech. A waterfall plot is created by simply graphing amplitude versus frequency and, at each new time step, moving down the screen and perhaps to the left or right for the next plot, producing something that looks like a waterfall. You may have seen an example of this in a movie that shows submarine sonar plots, for instance. It certainly looks cool when done in real time! When the plots reach the bottom of the screen, they either restart at the top or the entire screen scrolls diagonally up and to the right. Either way, the idea is to see the development of a sound's frequency components versus time.

Because a filter bank is easy to understand and, in fact, produces an "ideal" frequency versus time resolution that other methods don't do as well, I'll look at it first, beginning with a definition for "ideal" in this instance. Although there are other useful definitions for other purposes, ideal here will mean a mapping that corresponds closely to human perceptual capacity. It turns out the human ear has a phenomenon associated with it called "critical bands." A critical band is defined as a bandwidth (that varies with the center frequency) within which a loud tone can easily mask a soft one. For instance, a loud 100Hz tone can easily mask a soft 110Hz tone but hardly masks a 1KHz tone at all. It turns out that the critical bandwidth of the ear is approximately 100Hz at the low end, moving to about 5KHz at the high end. The curve on a log/log plot has a deep "belly" in it, meaning that fairly high resolution is retained until around 500Hz, when the curve straightens out and starts to follow the usual exponential curve seen a lot when looking at human perception. This is probably an evolutionary adaptation to speech and animal sounds, since without this extra resolution it would be hard to understand either.

If you were making an ideal filter bank, you might want filters that are spaced about a critical bandwidth apart and are about a critical bandwidth wide to match the ear. This means you'd have a number of approximately 100Hz-wide filters starting at low frequencies and gradually getting wider and farther apart at the higher frequencies. If you "push" your time resolution as hard as you can, you will find that you have better time resolution with the higher frequency, wider bandwidth filters, just as the ear does or wavelet filters almost do — they are a discrete approximation to the continuous way the ear appears to work. The big problem with this is how do you then plot the data for analysis? In other words, for every *meaningful* sample from the 100Hz-centered, 100Hz-wide filter, you have 50 from the 10KHz-centered, 5KHz-wide filter. What to do? Well, you could simply duplicate the output of the 100Hz filter 50 times so you have something to plot below all the samples for the 10KHz filter, and that's what is done in practice most times when this trick is used. However, interesting variations are possible. An ideal filter bank can be constructed easily using the synchronous filters described in Chapter 3 as a DFT bin, followed by low-pass filters instead of simple accumulators, and sampling the outputs of these at the various appropriate rates for their information content. Whatever filter type you use, you're interested in the energy of each output, not the actual output signal. If you're not using a filter type that gives this directly, you'll have to rectify the output and use a low-pass filter on it to remove ripple. Another possible approach would be to sample all the filter outputs at the highest needed frequency for any, which would capture some extra ripple in the lower frequency bands. This may actually be useful information to the skilled examiner when seen in phase with the low-frequency ripple component. When looking at the overall plot, this ripple conveys timing (phase) information on the low-frequency components that may be driving some system nonlinearity to produce rapidly changing high-frequency components. At any rate, it makes the display part much easier to code. For reference, Table 5.1 is a design for a filter bank that has often been used for human speech analysis.

Table 5.1   Filter bank for human speech analysis.

Channel Number	Center Frequency	Bandwidth
1	240	120
2	360	120
3	480	120
4	600	120

Table 5.1   Filter bank for human speech analysis.

Channel Number	Center Frequency	Bandwidth
5	720	120
6	840	150
7	1,000	150
8	1,150	150
9	1,300	150
10	1,450	150
11	1,600	150
12	1,800	200
13	2,000	200
14	2,200	200
15	2,400	200
16	2,700	200
17	3,000	300
18	3,300	300
19	3,750	500

You can see that the bands vary in width but stay fairly narrow and don't quite scale in width with center frequency. This represents one of those hard-fought compromises between mimicking what the ear can hear and how hard the filter bank is to compute. For music, you would probably use a different set of bands and widths, extending higher and lower in frequency.

FFTs, on the other hand, are the most commonly used frequency analysis tool, although they are perhaps the worst suited for this particular application. The trouble here is that the FFT bins are a linear function of frequency, which is not quite the way the ear works, and there is a time–frequency resolution trade-off that applies to the entire analysis frame. You simply need long FFTs to get any decent frequency resolution in the low frequencies, and this smears out any attempt at getting good time resolution in the high frequencies. Actually, there *is* a way to improve this that I've not seen stated anywhere else. Take both long, high-frequency resolution FFTs and short, high-time resolution FFTs of the signal to be analyzed and make two separate images from these. Now convolve them; that is, multiply one by the other. One way to do this is to replicate the big FFTs horizontally and the small ones vertically (i.e., each output goes into more than one vertical bin) so the image x/y sizes match. Usually the FFTs are integer multiples of one another's size since they are powers of two, so the interpolation here is easy. The multiplication of these two images will impress the time resolution of the short FFTs onto the long ones. This amounts to creating "impossible" resolution, and indeed the process has artifacts because of this, but the result can be very useful once you know how to interpret it, and it's fast to compute compared to other methods.

Another possibility is to use the long FFTs with a lot of overlap, then pass a horizontal high-pass (actually, a high-frequency up-tilt) filter across each horizontal line of the resulting image, which improves things somewhat and, again, is relatively fast to compute. Either way, you're better off than using a single FFT-based analysis, although it is still not as ideal as a purpose-designed filter bank. The main attraction of the FFT method is that it's fast, and in some cases it's meaningful to modify the FFT data and inverse transform it to get modified signal data, which is not as easy with other methods.

A variation that's getting a lot of brouhaha these days is the wavelet transform. These are described in great detail in Masters' second book (*Signal and Image Processing with Neural Networks*). What they amount to is doing DFTs with twiddle factors that already have had a window applied to them and which scale easily by octaves, with fewer than the normal amount of bins, but on a per-octave basis. For each octave, double the amount of outputs per bin; in other words, do twice as many transforms for each octave on half as much data apiece. As far as I know, there's no way to do these as fast as an FFT, so they amount to doing a lot of FIR filters, which is the hardest possible way to do a filter bank. I don't see what all the fuss is about, personally. The technique has always been around and is not as good as doing the old DFT bin complex multiply followed by a low-pass filter and sum of squares when an arbitrary time–frequency resolution trade-off is wanted because it *still* has artifacts resulting from the periodicity assumptions involved with the accumulation over a record, which the synchronous filter technique does not suffer from. Wavelet transform is just a buzz phrase that is supposed to impress the uninitiated, I suppose. Just nod knowingly, then do it the right way, instead.

Parameter Extraction

When analyzing sounds, you're often interested in extracting various parameters so that you can synthesize them again, attempt some real understanding about how the sound was produced, or perhaps even discover what it has gone through on the way to the ear as a separate issue. Parameters of sounds that are often looked for include attack rate, decay rate, average spectrum, attack and decay times of individual spectral components, detection of any implicit comb filters which can indicate either multipath (reverb) or part of the normal sound production mechanism, and perhaps components that can be expressed as sums and differences of other components or, in other words, indications of intermodulation distortion. In addition, sometimes it's possible to see the excitation separately from the filter.

For sounds with a single impulse for excitation, the attack and decay rates of various spectral components can reveal the effective Q of the various filters in the impulse filter model. It can at least tell something about the excitation. Comb filter detection might tell about sounds bouncing back and forth in the source or about the acoustic environment in which the sound was captured. Best of all, you can often capture enough information to separate the actual excitation signal from the total of that convolved with any filters present and, thus, know all that's going on. It turns out that you can do this in a variety of common and important cases. In a landmark paper in the first IEEE reprints book, Stockham and others point out that any filter convolves (multiplies) its response with the original excitation function, so if you take the log of the spectra, you're now in a domain where things that were originally multiplied together can now be separated additively, using normal filters before converting back.

The effective Q of any particular filter can be estimated by looking at either its bandwidth or, in the case of an impulsive excitation, its effective decay rate. This can be done simply by looking at some filter bank output or FFT bin and fitting a curve to the decay rate, the parameters of which can be manipulated to determine the Q of the filter, at least if the filter bank output has enough time resolution that it will not swamp the original filter's time response out. Code to do this comes with the Stearns and David book in the bibliography, as does code to estimate rise and fall times generally, so I'll refrain from duplicating their words and code here.

When you want envelopes in a small frequency bandwidth, you're usually hard up against the old time–frequency resolution trade-off — the filter that creates the narrow bandwidth has its own time response that is added to the thing you're trying to measure. Even in the full bandwidth case, you can have troubles. Consider trying to acquire an envelope by full-wave-rectifying the signal (the absolute value function in DSP) then low-pass filtering the result to remove the ripples caused by the individual signal wiggles. Well, you just added the time response of that low-pass filter to whatever you were trying to measure. If you try to make the low-pass corner too high in frequency, low-frequency sounds in the input signal will come right on through. One approach is envelope detection in bands, so at least for the high-frequency (and presumably wide) bands, you can have good time resolution. A cool trick in my opinion is called "Tom Gamble's brown envelope detector" after its inventor, in which another copy of the signal is made that is phase shifted, say, 90 degrees. You then take the sample-by-sample maximum of the two or the root sum of squares to get your envelope, which now needs a lot less filtering since you've removed a lot of the ripple; the 90 degree phase shift fills in the gaps. Where the original signal was crossing zero, the phase shifted one will be at a peak. For signals that have a single sinusoidal component, this works very well and works better than doing just the one-phase version for complex signals. In a digital implementation, one could use a Hilbert transformer to create a wide-band 90 degree phase shift and not have any serious constraints on signal bandwidth for this technique.

Comb filters can be found easily by looking at a cepstrum of the signal. A cepstrum is defined as the FFT of a log of an FFT. Anything that shows up periodically in the first FFT will therefore roll into a single bin in the FFT of that. The log has two functions: one is to emphasize the wiggles in the first FFT, no matter what absolute amplitude they were originally at, and the other is to get you into that magic domain where things that were originally multiplied together can now be treated as additive contributions. This unfortunately involves taking the log of complex numbers (at least if you want to inverse transform later), which is a rather tricky affair, the problem being that there is some uncertainty about the phase of any component that wraps around at 2π. This is solved by tracking the phase from time T_0 and adding 2π whenever it wraps around to keep the total correct before doing the complex log, which then works correctly. Sad to say, the subject of cepstrums and homomorphic filtering really deserves its own book: all the details and tricks just won't fit in this one.

To summarize, most of the sound production models you will encounter while working with audio will be small variations on the basic excitation/filter model. For cases of sounds in metal, more complexity has to be added to handle the nonlinear effects of bending, and the fact that metal (such as in a cymbal) can support more types of waves than merely compression waves.

Using a cepstrum, one can often separate the excitation from the filter model when desired to study or modify these in isolation. It is important to remember that a source's ability to radiate sounds is affected by its effective size, and that this constitutes part of the filtering in the model. Many artists have learned to create fine variations in excitation to create a desired sound, and this is why many synthesis attempts fail — because they fail to take this into account. More information is contained in the signal than is obvious at first glance. Most of the parameters of a sound production model can be extracted via the use of filterbanks, level detectors, and plain old common sense. I urge the reader to develop their own battery of techniques as they progress.

Chapter 6

Synthesis of Sounds

Although it's fun and definitely useful to modify existing sounds, some people have the most fun simply making them up from scratch, as modern-day keyboard synthesizers do. I want to show you not so much how to design a synthesizer, although this should help in that effort, but to give you some good guidelines on the various ways you might accomplish sound synthesis, with a side order of how to realize that particular sound you may have going around in your head. As such, I'll not cover the sometimes popular FM synthesis techniques much at all. Although they are a cheap way to make complex sounds, the controllability and ability to do realistic synthesis of any but bell-like sounds is very limited. The emphasis here is more on making realistic sounds via the sound production models covered in the previous chapter, which can then be modified with the techniques covered elsewhere to get them just the way you want them.

As I stated before, a sound production model is a way to describe how some source of energy winds up producing a sound that you can hear. In general, energy enters a resonator of some sort, which may have complex characteristics that affect how and when the energy couples into it. Then it is coupled into the air, which adds the radiation characteristic, bounces around some, which adds comb filtering and reverb characteristics, and finally enters the pinnae of the ear, which add their own filtering and reflection, after which you finally hear the sound. So there is a pretty good model of how most natural sounds are produced at the outset, and it is one that's fairly easy to translate to the software domain. Most realistic sounds are produced using this common model with only slight variations, such as whether the basic resonator affects the energy input or not and what modes the basic resonator vibrates in. Drumheads and cymbals get into vastly different modes than strings do, for instance. Their two-dimensional natures allow for shear vibrations that do not occur in either strings or the air and for modes that are far from being harmonically related to one another. In addition, drums and cymbals have some nonlinearity as they bend, which causes distortion so that low-frequency modes can give up energy into high-frequency modes long after the original excitation is gone. For more on this, see Morse, *Sound and Vibration*, listed in the bibliography in Chapter 1.

Unrealistic sounds can be created in just about any fashion that will produce numbers you can output through a d/a converter, but some ways seem to work better and be more controllable than others. The sky's the limit here. You can do just about anything, but it's nice to work in a domain that has a correlation between what you specify and what you'll eventually hear. For instance, an experienced practitioner can simply draw a waveform that can then be repeated at any rate to get a sound. I put a sample arbitrary waveform generation transform into WavEd so you can play with it, but you really need some experience in what to draw to get any kind of predictable result, other than a trashy distortion sound. All in all, this tends to be less useful than just using a conventional sound production model, which after all, you can modify to your heart's content via any special efx you choose later. FM synthesis excels in producing complex sounds that nevertheless sound pretty fake for anything but bells, and you more or less have to be able to intuit Bessel functions to make them fly predictably. This is a bit beyond the scope of this book. You can find out more about FM synthesis in Hal Chamberlin's book, *Musical Applications of Microprocessors*, part of which seems to have been the design document for the popular Yamaha DX7 FM synthesizer.

General Synthesis Techniques

Synthesis, as opposed to modification or a special effect, is usually done via the source–filter model. If you are attempting to duplicate a natural sound, you usually analyze a sample or samples of the real thing to obtain the model parameters. If successful, the job is nearly done; you just plug the parameters into a software simulation of the model and get your desired output. More often, you can't get all the details you'd like to have with analysis, or at least simple analysis, alone, and you wind up tuning your simulation later by hand and by ear. The old analog synthesizers were remarkably versatile in allowing this, which partly overcame the limitations of a small selection of excitation waveforms and postfilters. Pianos, however, were notoriously difficult to synthesize this way, and the reason may be instructive.

A piano is constructed of extremely high Q resonators (the strings) and a few poles worth of low-Q resonators (the soundboard and radiation model), all of which is excited by an "almost" impulse — the hammer strike. Although a click to simulate the hammer strike was usually available, most of the analog synthesizers simply lacked a sufficient number of high-enough Q filters to simulate the string fundamentals, their individual harmonics, and the soundboard, so they used continuing excitation and filters instead of the more accurate high-Q filters to represent the string. A string is not simply a single high-Q resonator but actually is simulated somewhat more accurately by a delay with feedback, with peaks at various harmonics of the fundamental. Even a delay with feedback can't do this quite right, because a string has end effects that detune the upper harmonics slightly. In addition, because of radiation coupling and a Q relatively independent of center frequency, the high harmonics of a piano string lose energy at a faster rate than the fundamental frequency does, so each needs its own amplitude envelope function to do an accurate job. This was well beyond the resources of early synthesizers.

Similarly, a violin has such a complex set of resonances in the soundboard that most early synthesizers couldn't do a very good job simulating one instrument, much less simulate a group of stringed instruments. Now, however, these limitations have been overcome, so the important thing is to understand what you want, at which point you can usually synthesize it, if not always in real time on cheap hardware. If you are out to synthesize realistic string or percussion sounds, the first thing you really should do is get a copy of Morse's book, which describes how to calculate end effects in strings and modes of sounds in 2-D and 3-D objects that can transmit shear waves. I can't do too much with that topic here, since it really does take a book or more to adequately describe. Once you know about the harmonic detuning effects of strings, you have a couple of options available to simulate them. You could use a bank of frequency generators, like the table method described below, to create each harmonic component separately, each with its own amplitude decay envelope, or you might get close enough by generating only the fundamental frequency, multiplied by its amplitude envelope then passed through a distortion-causing transform that acts differently versus input level, also described below, to generate the harmonics. Although it is fairly difficult to define the exact distortion table entries needed to generate the right balance of harmonics versus input level, it is an easy technique to experiment with in a feed-forward sense.

I suggest trying various polynomials based on trig functions as a way to start filling the table. I put a distortion transform into WavEd that lets you draw arbitrary curves for each polarity of input. A diligent reader might want to add the ability to put in formulas and to have either polarity of input produce either polarity of output as well. I didn't do these things due to time limitations and because it would have made the normal use of the transform for a fuzz box less obvious.

Fourier Synthesis

Because any sound's spectrum can be reproduced to arbitrary accuracy with some combination of sinusoids, this technique came to be called Fourier synthesis. This can be done by FFT, if discrete bin spacing is good enough, or by summing a number of sinusoids that have been generated some other way to arbitrary frequency resolution, something the FFT cannot quite do.

This accomplishes true continuous Fourier synthesis, as if doing discrete Fourier transforms, but with infinitely small bin spacing. The old time–frequency resolution trade-off prevents perfection when using FFTs, and although you can generate frequencies that fall between bins by incrementing phase between FFT records before inverse transforming, there are some amplitude artifacts due to imperfect joining at the record boundaries, and you still will have trouble simulating an attack much shorter than the FFT record length. That technique is therefore primarily of academic interest. It is much more useful and direct to generate a number of sinusoidal waves and sum them up after applying the appropriate amplitude envelopes to each. You can easily get away with not postfiltering using this method, since you can vary the amplitude of each component separately to emulate the effects of a postfilter.

Nonlinear Synthesis

A number of natural sounds are produced via mechanisms that incorporate significant nonlinearities in the model. An excellent example of this is the sound of a ruler being plucked while hanging over the edge of a table. When the part of the ruler overlapping the table surface is in contact with the table, the resonant frequency of the system is determined by the length hanging over the edge. When the ruler end swings upward and causes the overlapping portion to rise above the table, the resonant frequency changes, since the vibrating portion has suddenly changed in length. This might amount to feeding an impulse into a high-Q digital filter and retuning the filter based on whether its current output was positive or negative. Actually, it's a little more complex than that, since each time the ruler hits the table again, there is another little high-frequency impulse added to the system from the contact. However, a competent coder should be able to figure out how to accommodate this complexity.

Although the above example is extreme and the sound produced may not be the most musically useful (although interesting if you slide the ruler to and fro), it serves to illustrate the idea. More common places in nature where this effect occurs are in cymbals and drumheads, and these are notoriously difficult to synthesize well. One reason is that in media that can transmit shear waves, there are many modes of vibration that can couple nonlinearly into one another. Also, in these media the modes tend not to be harmonically related to one another. Remember, any nonlinearity causes new frequency components to be generated that weren't originally present. In the cases of drumheads and cymbals, the various modes of vibration are not normally harmonically related to each other, or even close as they are in strings, which makes things even more complex. If you add snares to a drum, the model becomes so complex that a direct simulation is probably more or less hopeless. You are saved in this case by the fact that you have passed what human perception can differentiate, so you can simulate the system adequately with the first few major drumhead modes and some additive white noise. You would compute the Q of these modes based on the current amplitude, then suddenly reduce the Q once the amplitude dropped below some level, because below that level, the drumhead no longer throws off the snares on each vibration; the snares simply damp the remaining motion very quickly.

Tables in Sound Synthesis

A very powerful technique in just about any programming is the use of tables to map a table index onto some other sort of data. Signal processing is no different here, and tables are often used to precalculate something that will then be used over and over. Although this may seem like a limiting methodology, I'll show you just how flexible it can be. Table outputs can be modified by using them to index into yet other tables to get some very complex results with very little work.

The function probably needed most often in signal processing is the lowly cos(), used to generate sinusoids and many other things. In most computers, this function turns out to be one of the slowest and also tends to accumulate errors if the input is a huge multiple of pi. Often, you really don't want to scale the input angle by an irrational number (pi) in order to get the cosine of half the circumference of a circle, or 180 degrees. After all, you know $1/2$ to infinite accuracy, and you don't know pi quite that well. Tables can help solve this problem and make arbitrary frequency generation a trivial operation. A downside is that unless the table is very big and contains high-precision numbers, some error can creep in as noise. For most practical applications, however, this is no big deal, and the table sizes needed to get the desired quality are reasonable. I'll look at the simple application of generating an arbitrary frequency using a table of cosines.

First, set up the table. For this exercise, the table will contain the full cycle of a cosine wave with an amplitude of 1.0; that is, the results of cos() evaluated from 0.0 to 2π. Theoretically, you could get by with one quarter of a cycle and a little symmetry logic, but I'll keep this as simple as I can for now. You have to decide on a table length for starters. For reasons that will become clear later, it's a good idea to make the table size a power of two. For example, I'll make this table size 1,024 so that it takes precisely a 10-bit integer for an index. You could use this integer as in index and count it up on successive samples to get a cosine wave with a period of precisely 1,024 samples, but that's not flexible enough for most uses. What you really need in order to generate arbitrary frequencies is a way to increment this index by an amount other than 1.0 per sample.

Here's how. Suppose that these 10 bits are really the top 10 bits of a much longer 32-bit integer, for example, and that all the arithmetic is unsigned. After all, you're using this number, or its top 10 bits, as a table index and sign is not a relevant concept at the moment. You can and should think of this 32-bit number as the instantaneous phase of the cosine wave. A given frequency simply means that this phase is changing at a certain rate, such that it goes through an entire cycle at the desired rate. As you increment this number to get successive samples of the desired wave, it will overflow once per cycle of the output sinusoid. The problems caused by large multiples of 2π are solved right off the bat. The index simply loops around the table and begins again with this technique, keeping the lower phase bits intact, so that errors do not accumulate. To generate an arbitrary frequency to fine accuracy with this method, simply do not use a unity table index, or phase increment. Instead, use a number that is directly proportional to the desired frequency. This number can have bits below the 10 bits you're actually using as a table index, which corresponds to fractions of the minimum table index increment, and bits up in the 10-bit area you use for the table index to generate frequencies that have a shorter period than the table length in samples. You can get as much frequency accuracy as you want by making the phase accumulator as big as you need, although in practice, I haven't yet run into a case where 32 bits wasn't enough and was usually much more than needed. To make this clearer, refer to Figure 6.1.

Figure 6.1   Table-driven arbitrary frequency generator

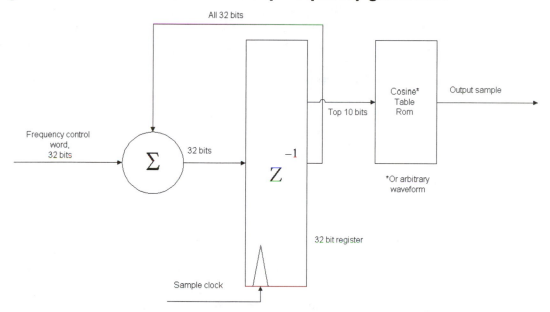

Figure 6.1 is a diagram of a hardware implementation of this concept, but I'm sure you can see how to convert it to software. At each sample time, add the frequency control word to the current phase accumulator and use the top 10 bits to look up the corresponding value in a cosine table. Astute readers will jump ahead and realize that the table need not contain a simple cosine function, and I'll get into that in a bit. First, I should tell you how to calculate the frequency control word for a given sample rate and output frequency and discuss some of the limitations. First, if you use a frequency control word that is greater than half the maximum number that fits in 32 bits (i.e., greater than 2^{31}), you'll have fewer than two samples per cycle. Good old Nyquist then comes into play and reminds you that you can't unambiguously have a frequency that is greater than half the sample rate,

and you will indeed have aliasing under these conditions. Equation 6.1 generates a frequency control word given a sample rate, a register length, and the desired output frequency.

6.1 $FrequencyControlWord = 2^{register\ length} * desired\ frequency/sample\ rate$

You have to truncate the frequency control word (FCW) to an integer to use it in this scheme. The available frequency resolution is therefore determined by the number of bits in the register. One part in 2^{32} is usually good enough! For a Nyquist frequency of 22,050Hz, this amounts to about five-millionths of a cycle per second or so, or about 200,000 seconds to accumulate a single cycle of error compared to "perfect" resolution. Your application might be able to use a smaller register, say 16 bits, and live with one part in 65,536 accuracy.

An occasionally useful wrinkle on this basic scheme is to use another adder to produce a phase offset separately. This adder goes between the register output and the table input and need have only as many bits as the table index requires. This is useful for generating signals that phase shift keyed modems and the like. Slight frequency differences are also producible by varying this phase adder cyclically.

You can see how you might save some memory with this scheme at the cost of only a little more computation by taking advantage of the 4 way symmetry of the cos() function. To do so, you make your table so that it only contains the first quarter cycle of the cosine wave, so it uses a quarter the memory for a given time resolution. Assuming you stay with the 10 bit index example, what is done now is to use the top two bits of the ten bit table index to drive some minimal symmetry logic, and use the lower 8 bits to get the basic cosine sample. For instance, if the top two bits equal zero, you just use the lower 8 bits to provide your index as before. If the top two bits are 0,1 (meaning you are in the second quarter cycle of a cosine), you subtract the lower 8 index bits from 256 to get your table index. This works because the second quarter cycle is simply a mirror image reversal of the first quarter cycle. The second half cycle of a cosine wave is precisely the same as the first, only it is inverted, so when the top two index bits indicate 1,0 or 1,1 — you proceed as before, but also negate the output from the table. This technique works for any very symmetric function to reduce memory requirements. On some machines, it may actually be faster than the full table because of cache issues.

For those readers who jumped ahead, no, you're not limited to a simple cosine function in the table. You can use just about anything that repeats periodically. However, because anything but a cosine function will have higher frequency components associated with it, it will now be easier to generate aliasing, so you are limited to a longer output period by any higher frequency harmonics in the table waveform. This is how a lot of early sampling synthesizers worked. They perform pretty well over a limited range, but they have one serious defect. Remember that most sounds are produced with a model that has both an excitation function and a fixed filtering function. You can see the weakness of using this alone; in this technique everything in the spectrum scales, not just the excitation, so to get a sampler that sounds natural, you have to separate the excitation function from the filtering function and keep the latter constant as the desired pitch changes, just as real instruments do. The soundboard and box of a violin do not change in size with the pitch being played!

I include a nice example of table-driven waveform generation in the WavEd program. In this example, you put a waveform into a table graphically and have the option of smoothing the curve and doing some postfiltering. The code is not for the simplest case, because for that example, the table size isn't necessarily a power of two, but it should serve to help you understand table-driven waveform generation.

Waveform Modification via Table Lookup

One way to modify a waveform is to use its value as an index to a table. Two good examples of this are the distortion transforms in CoolEdit and WavEd. In this transform, you draw the desired response curves, which are put into a table. When the transform is done, each input sample is used as a table index, and the table content is used as the transform output. In this way, you can simulate the way tube amplifiers fail to put out enough current near waveform peaks, or you can simulate a speaker that has some crud in the voice coil gap, some kinds of cone breakup, and other "instantaneous" effects. By definition, this is a nonlinear technique, so you have to watch for aliasing in the result; although, in some cases it might be desirable. You can even simulate fold-back current limiting by using a sine table — as the input increases, so does the output at first, but if the input continues to increase beyond half of the full scale, the output goes down again. Various other nonmonotonic table functions might be used to create an effect that varies with overall input amplitude. My favorite use of this trick is to simulate even-order distortion products, such as those produced by the deliberately unbalanced output stages of such amplifiers as the Marshall brand. This is done by simply having different slopes for a straight line (perhaps flattened at the top to simulate tubes running out of current carrying capability) in the positive and negative half-cycles of the input. Of course, it takes more than this to accurately simulate a complex circuit that has various elements storing energy, but this is one of the more crucial elements. A better simulation would also include various kinds of pre- and postfiltering and would use lines that are not quite straight. You might even copy the tube curves from the tube manuals for an accurate simulation of tube distortion. I have done this with the distortion transform provided in WavEd, and the results are as expected. Adding a small amount of hum and noise to the result creates quite a good simulator for the popular "tube sound".

Time-Varying Aspects

It's important to note that most real sounds have some common characteristics, and if you want to simulate something realistic, you should take these into account. Most natural sounds have an exponential decay, and the high frequencies tend to decay faster than the low frequencies, unless some significant nonlinearity is present that produces new high-frequency content from the low-frequency energy. Things like cymbals actually reorient their radiating surface at a low frequency after being struck, which dramatically changes the actual sound at any remote point. I'm sure you can come up with plenty of other examples on your own.

Summary

In this chapter, I've mentioned the usual approaches to synthesizing sounds without going into too much detail on any specific approach, since each should be a book in its own right. I know I keep saying that, and someday I'll write those books, but for now, it's fun to play and experiment with what you have now, and who knows what you might come up with if I don't tell you what not to try? Instead, I give you WavEd, which is a nice platform to fool around on, complete with the tools to analyze the results. Enjoy!

Chapter 7

Development Tools

You can't do a meaningful project these days without seriously considering what tools to bring to bear on it. Long gone are the days when you could reasonably hand-assemble meaningful code and punch it on a paper tape or do the hardware equivalent of building an opamp out of discrete components, for example. You *can* still do it, but your project would take so long that it could never be profitable compared to other ways of getting it done. Even in these days there are various important trade-offs to consider before diving into a special efx, or any, project, such as what language(s) to use for the programming, how to debug a potentially real-time process without spoiling its real-time properties, what to use as the test platform(s), and so forth. An important consideration is what you want to do with the result. I define two major categories of output here: research grade and production grade. The former describes a research situation, as in code or hardware, that you're just going to lash up once to find some answer. You don't care what it looks like or how long it will live if it gets you that answer. The latter comes into play if anyone else is going to use your product; more so if you're hoping for a lot of people to use it willingly. Sometimes you're not sure at the outset which is really going to be the case, which complicates optimal decision making.

If I'm *sure* I'm doing a research-grade project, I use the following guidelines.

1. Don't reinvent any wheels you don't have to, period. Use an off-the-shelf host as a platform for your effect, even if it's not quite ideal for the purpose at hand. Use any existing code bases, whether optimized for this use or not. The main consideration is not having to write and debug, say, an FFT for each new thing you try. There are obviously cases where existing code has such different requirements from your idea that you won't find ready-made solutions. The main idea is that development speed rules this domain, not code execution speed.

2. Use the easiest and fastest tools and simplest test platform possible, especially at first. If you're thinking beyond "at first" see the guidelines for production-grade effect building.

Usually, I'm not *sure* at the outset of a project whether it is of research or production grade. If whatever I'm working on turns out great, I'll be faced with converting it to a reusable format, redesigning the algorithm for top speed, and dealing with any new user interface issues. If it's complex, I might forget the details of the early parts of the project by the time I decide to port it to a universal format. On the other hand, the idea may fail, and if I've done a lot of work to get to that point, well, it's mostly wasted, although not entirely, since I have a bit of new knowledge. Surely, however, any extra time I spent making it fit some model with tools or platforms that are harder to use but more optimized in some way is just blown. If I'm at all undecided, I usually compromise more or less on how to begin, so I'm not too far off either way (but more on that later). The following are my production-grade guidelines.

1. The platform or platforms the result must be compatible with is paramount. You'll spend a lot of time handling the details of the different platforms correctly, and it's hard to be an expert on all of them at once. Pick one you're going to stick with if at all possible. If you are stuck with having to support more than one, you may want to build yourself an "in-house" model first that's designed to provide the common basic services and will be easy to plug into the common plug-in formats. It's too bad the current standards don't quite agree on what constitutes "common basic services." They usually include preview and preset support and sometimes need support for control changes during processing. Depending on the standard, the data flow may be push or pull.

2. I program in C and C++, initially, that being the language with the most power and flexibility as well as transportability, but I plan on converting some pieces of the code into assembly language for the hardware platform(s) I might encounter. As I've said before, speed can mean more than you may think in several areas. Make sure you know how to do this before you count on being able to do it quickly. It can take a long time to smoke out all the little details of getting the tools to work together. It sure is nice to have a buildable example or two to copy from. For instance, see my article "Mixed-language Programming with MASM," in *Windows Developer's Journal*, Vol. 8, No. 12, December 1997 for information on how to integrate MASM with DevStudio C and C++ code.

3. If you haven't, make a dirt-simple example or null transform for the platform to check out your tools and abilities. Often you can save time by reusing such a project, and most of the available platforms give examples of these with their development kits. It usually isn't a waste of time to add a few of your own improvements to these for use as boilerplate on later projects. I often add things like common stubbed-out functions or functions with default behavior, just to remind myself that this or that function name will be called by the host, and I add comments with important platform defines (structures, constants, and types) so I don't need to look elsewhere for these things as I code. These make for good documentation as well.

If I'm not really sure whether my project is research or production grade, I usually program in my own model, which I'm familiar with and which automatically forces me into most of the desirable design patterns. If it works out well, I start thinking about its portability to other standards as early on as possible. This at least avoids some duplication of work in the cases where the effect turns out well. Because I have done a fair amount of production-grade code, I have the benefit of having created my own examples for every platform I've supported, so the extra cost in case of failure is minimized. It doesn't matter all that much to me what model I use, though some have more overhead and boilerplate components that must be specially modified than others.

Most people don't spend enough effort deciding at what level (research or production) they want their project to work or on the design process as a whole. You are a lot better off doing this up front, especially with experience. Here's some of mine. You have to pay attention to what information your toy is going to need, and when that information becomes available; it may not be the most convenient time or even under your control — your environment may be either push or pull for any part of the interface separately. Initially, you'll want to use the simplest methods to move data to and from user interface controls. Your effect may plug into one of the many standards that has its own control over external presets, and it will have to deal with that, in some cases even at first, since it expects to be able to call you. A *big* flaw in all these standards is that presets are not generally usable *across* applications — even when they support the same standard — since the application itself handles the preset data and stores it in some application-specific structure and location. Oops. Some host platforms allow real-time control modification while another thread is moving data through the transform, and you'll want to provide for this ability early in the design, if it's needed, in order to save a lot of rework later. A lot depends on what the host provides and what you have to provide, and this is why I initially begin with the host that provides the most and needs the least.

I bet the title of this section suckered you into thinking this was going to be about the merits of various compilers and platforms. Well, it is, or will be, as I go along. My point here is that things like documentation and sample source code along with knowledge of the standard plug-in formats are *the most important tools*, as is a basic familiarity with all your other tools, such as knowing how they work when you're not doing something special. For instance, people who aren't good at using a wave editor need to practice with it before using it as a test platform. You need to know how it works and is commonly used by its customer base and how other plug-ins like yours normally work, for comparison.

Cycle Times and Productivity

I'm sure there are a few of you out there who, like Sherlock Holmes, work everything out before testing, then just type in all the code and debug it. Even, or maybe *especially*, if you're that type, you'll be debugging by and by, and you might as well plan for this contingency up front. Personally, I like to make small, incremental changes to a working null transform and test fairly thoroughly as I go along. I feel like I have a better foundation throughout the project that way, and the stabler the foundation, the higher you can build. I get very excited about tools that allow me to do this and do it fast, so I get less worn out trying to find some obscure bug. I use Microsoft's Developer Studio for just about all software development for any platform, since that's where a lot of the best tools are right now. I like that I can use whatever build tool for whatever piece of the project; for instance, I use the Texas Instruments C compiler/linker for project modules written for my TI-based hardware, and DevStudio is customizable enough to use the right tool for the right module when I have mixed host and embedded product modules in the same project. I *like* syntax highlighting: "Don't read code without it," I always say.

A lot of times, conventional debug techniques *do not work* for real-time software, so the ability to quickly make a change and test again is paramount. The usual method is to make and test debugging tools or functions first. Please do not be fooled because I'm not spending a lot of time on this subject. It's *so* important that I don't want to confuse the issue. The faster you can go around the loop, the more you can do, period. It's worth it to have an EPROM emulator and not have to wait for a burn or erase on even a trivial hardware project. It's worth it to have 10 fewer seconds build time, even if it means a faster and more expensive development machine. You will go farther before you tire out, you will be encouraged to test your code more early on and try more things, and you will learn more. The potential benefits are just too great to ignore, and the deficits will be too great if you have only "good enough" tools for the job.

On the other hand, don't get bogged down in tools so complex (of which there are plenty) that getting familiar with them each time takes so long it's not worth it. I prefer simple tools, defined as what I can either remember or relearn in order to work easily and with confidence. You'll have to make your own decisions on this, because the criteria are unique for each programmer.

The Fun Factor

Hey! Using your tools should be fun so that you'll be encouraged to refine them and use them to their greatest advantage. Your fun factor will obviously improve as you learn just what to expect from a tool in the usual cases and as your tools become faster and more suited to the job at hand by your customizations. Other things that improve the fun factor and make life easier are prebuilt examples and debugging hooks and tools. I even make custom appwizards to automate various things and make it easy to get started.

For a lot of testing, you must have some way to stream debugging information as the test proceeds, without disturbing the test itself too much. Not providing this, if it's not already available, is a big mistake. Hitting a breakpoint in a real-time process may cause errors that it's too late to step out of meaningfully and may even leave you with a corrupted file system, depending on what is going on at the time and how the debugger handles the rest of the system. Most likely, you just won't have the information you need any-more, so hitting the breakpoint only tells you if you made it there, but not much else.

Tool Automation

I will spend a few words on the subject of normal tools. Tool automation, whether by built-in functions or a called batch file, is a handy but double-edged sword of its own. When I'm working on something based on an ActiveX, for instance, I can automate my development environment to autoregister the control and bring up the desired test application when I hit the debug button. This saves a lot of time: one keystroke to build and start testing versus one for build, then some typing to copy and register the control, then some more to invoke the test app, then some more to open a file to test; that is, one keystroke instead of navigating a number of File-Open dialogs, resulting in a *major* improvement.

It is a double-edged sword, though, because on a recent project it faked me out and cost half a day of time before I realized what was going on. In this case, an employee was working on an ActiveX control for me, and like me, he had several disk drives. He'd moved the project to where it was convenient for him and set up all the automatic registration and so on. When he gave it to me for a last test/debug session before the product shipped, Dev-Studio silently ported the project files to the new location but failed to fix the automation stuff; it had no way of knowing about what we were doing with the batch files. The result was that the wrong control was registered at each build or was not registered at all and wasn't checked (a lesson in why not to use the "silent" switches) since it wasn't where we told the system it would be. We were working for embarrassingly too long a time on some subtle aspects of a bug and wondering why none of our changes seemed to have any effect. Finally, we put in a "stop right here and throw out a mes-sagebox" line in code we knew had to be getting hit. When no message box appeared, we realized what was wrong. We'd been testing an old version of the control that was still registered in my system, all along. Once we repaired our batch files to register the new version, everything was fine again. So, use the automation. It will usually help you a lot, but beware the oops factor.

Test Methodology

There are two basic aspects of doing a special efx plug-in. One is satisfying whatever requirements the host platform places on a more or less null plug-in. The other is getting the special effect itself to work right. Luckily, you can share almost all of the work involved in the former via prebuilt examples, but in general, you wind up changing these a bit and still have to test that part anyway. Also, not all of the provided samples are going to be bug free. This has been true in my experience for every plug-in development kit from every source so far. I provide some examples and wizards with this book that have been tested, but if my luck is like everyone else's, some fiddling change elsewhere will render them buggy, too — the breaks of the game, I guess.

Depending on threading issues, these two aspects may have some confluence that you'll have to pay attention to, if only to convince yourself it doesn't matter. You really should think it out or you'll get nasty surprises later. Personally, I favor the method of unit testing first and building things up unit by unit, testing the units as I go. As each new unit comes into being, I also then test the combination to make sure that it works as expected under a variety of conditions. At least initially, there's less to test so it's easy. It's important to do the most fundamental or basic things first when they'll be tested the best. I use either debugging facilities made available by my "official" tools or some special bits of code I've written and tested previously to send information from inside my plug-in to a trace window, or a console, or a wave file, for examples. This way, without disturbing the flow of data and events through my plug-in too much, I can see the stream of debug information it generates as it goes along, and I can see where things go bad based on this stream without halting the process uncontrollably, as hitting a breakpoint would.

Plenty of books exist on software testing, and they're probably better than what I'm saying here, where applicable. I haven't had time to read them, myself, but least make a list of things the code should do — you should have done this already or you can't tell me you're doing good design — and test them all each time and in various orders and ways, if possible.

I'm not a complete fan of structured testing for one big reason: No matter how you plan testing, you're still only going to find the types of things you anticipate. Although it's cute to say that you should expect the unexpected, it's not really possible in practice. Most people don't try pulling the plug, or various other things of that nature, while their code is running to see what will happen since, subliminally, the developer wants the code to pass the test. It's a lot easier to be brutal when testing someone else's code.

I remember a large government project I was involved in, in the 1970s. We had worked and slaved to make a multiprocessor (PDP-11s) lash-up do a lot of amazing things and had tried to design a foolproof user interface (UI) for it because the users weren't planned to be computer experts. We made a point of using the word "foolproof" in our reports, which was a big mistake. When the colonel who was the government's representative saw this, it constituted a challenge, so when we went to testing for acceptance, he simply turned around and *sat* on a keyboard. Needless to say, the random combination of keystrokes, control and function keys, and repeats brought the system down more or less instantly. Remember, this was in the early days of working with a lot of hard limits. We couldn't have had code bloat if we'd wanted to in those relatively small machines (by today's standards), so we couldn't really check explicitly for all the stupid cases. Everything had to be done with cleverness instead of brute force. At any rate, the colonel simply said something like, "Gentlemen, when trying to make something foolproof, you'd better assume a *very* ingenious fool!" The moral of this story is to be sure to try some really stupid things when testing — sit on the keyboard, try shutting down the dialog while it's working, try shutting down the application while your plug-in is working, and other things like that, as well as the more obvious functional tests that assume a competent and unpanicked user. All your users won't be both competent and unpanicked.

My last layer of testing is to send the project to someone who's only moderately facile with the subject or procedures in question, with no help whatsoever, and see what they say. This is brutal beta testing, if you will, and sometimes it can be a very fast way to find errors based on bad assumptions in meat-and-potatoes UI issues and user patterns. These are the most important types of problems to find early in the process. After all, your users will be the final testers, like it or not, and they're not nice to you if you let really dumb-appearing things to happen. Not realizing that some things only seem dumb in hindsight, they'll assume you made a conscious decision to make things the way they are.

Debugging

I have trouble thinking of anything tougher to debug than a remote or object-based plug-in for a platform that I can't trace into or see the source code for and that has its own UI, multiple threads whether I want them or not and multiple source languages with real-time requirements and bugs that may be changed or hidden by the small delays or memory footprint changes caused by the inclusion of debugging information or debug helper routines and macros. Can you? I write embedded operating systems, too, and believe me, that's *far* easier. At least there I have full source code and control over everything. This section contains tips and tricks that I've had to learn the hard way so you won't have to.

First, because bugs generally are easier to prevent than to find, it's a very good idea to be sure you *really* understand the plug-in model in question. Start by putting a breakpoint or trace output at every single function entry and try the plug-in on more than one platform, if possible, to see the sequence of calls, which are usually poorly or incorrectly documented and may not be the same for different host platforms. Of course, your code will be expected to work, as well, because, "all the other plug-ins work fine on everything." If the *breakpoints* cause crashes or other problems, it's best to know this right away so you'll know where you'll have to adopt other techniques. You will find that breakpoints often do crash either the platform or the system. Nastier yet, since many platforms are multithreaded, the sequence of events you see when hitting breakpoints may not be the same as when the flow of events isn't interrupted. Knowing the sequencing issues early on will save you a lot of heartburn down the road and prevent you from making incorrect assumptions up-front that are hard to change later. Some platforms may call certain initialization functions many, many times, or in different orders, or in several orders within the same initialization, and you'll have to make your code bulletproof against these kinds of thing.

In this case, when I can't use a macro to send information to a trace window, I create a console window temporarily, or sometimes I open a debug file at startup, write to it, and examine it later. However you get there, you should really be an expert at the plug-in model before writing your specific code. Just going in and modifying examples that may not be bug-free will lead you down the primrose path to disaster. Often the examples just don't cover all the platform functionality you'll wind up using, so an early step is to make sure you add whatever basic platform interface functionality you'll be needing to an example and then test that.

I've never been a big fan of IDE debuggers, as you might have guessed. Maybe it's a macho thing from the days we brought up microprocessor systems with nothing more than a hardware reference book (that is, if we didn't also design and wire-wrap the CPU ourselves, in which case our scribbled notes and prints had to suffice), an EPROM burner, an assembler, and a voltmeter. Scopes, especially logic scopes, were for wimps or getting the clock waveform right. I often find these tools are too difficult to learn the details of sufficiently and quickly enough to be 10 times more confident in the test tool than in what it's testing, which is a long-standing rule of thumb in any testing business. I do use IDE debuggers when they take me directly to the problem (or where the problem was noticed), as in a memory access exception. They are of very little help when the problem is conceptual, other than to alert you to its existence. Sometimes this helps, but usually the "breakpoint" mentality just won't get you where you need to go and may be deleterious to system stability, depending on how the debugger handles other threads going on in the system when a breakpoint is hit. So what to do?

I have a number of schemes that I've found highly useful for debugging and for working with big streams of real-time data. Most of them basically amount to `printf()` debugging, and I can already hear the groans of "Move into the correct century, dude," which I'll be glad to return in kind. Hitting a breakpoint can leave an a/d converter pouring data on the floor, a d/a stuck in some bad state, a file system write incomplete, and other potentially annoying artifacts. By the time you see anything on the screen, a serial port may have overrun because its servicing thread was stopped, giving you a false indication when you check it and leading you in the wrong direction. Also, you only get 1 bit of state information, and that little is usually too late. What you often want and need is the sequence of events and data that led to the problem, not the symptom after the fact. Sometimes you can get this from a debugger unwinding the stack, sometimes not — all too often not — because the previous component running was the opsys or the host platform, in which case, all you get is the final call into your code that blew up because something it needed wasn't initialized correctly yet. So in general, I find other types of tools far more useful than the standard debugging facilities.

Most plug-ins must be lean and mean, which often prohibits the use of things like MFC that have built-in `TRACE()` macros. Often, I avoid the C runtimes and use the Win32 APIs to make them even smaller. You can set up your own debugging output window or use the Windows facility that creates a console, which you can then use `printf()` or `cout` against to display a stream of debug data that usually doesn't disturb the process too much. However, sometimes these aren't fast enough not to disturb the process or you need more history. In that case, the preferred technique is to create a temporary file and write to it. If you do this in text mode, you can open the file in the IDE for examination later. This is my method of choice when I need a lot of information, especially about the order in which things happened over a long run. Don't count on things always happening in the same order! If there are multiple threads present, as is the usual case, you'll have to ensure that the debug file accesses are in a critical section or are otherwise protected. My cheesy and cheap implementation simply flushes the file buffer at each write request, inside a critical section, which is much slower than normal, but sometimes necessary. Another useful technique when working with WavEd is to generate another wave file of a particular channel count and stream transform variables into it for later examination using the platform itself. (Unlike most other wave editors, WavEd can handle any channel count the wave format supports, which is a UINT. I don't know why all the others don't do this, it's not as though it's hard to substitute a `for(all channels)` for an `if(mono) else stereo` construct.) This technique is most handy when you're debugging something that usually works, but sometimes blows it (I've been using it on a pitch detector that I'm developing). Having a stream of the internal variable values synced and displayed with the original signal can help you learn very quickly what works, what doesn't work, and most importantly, why something doesn't work or what type of input confuses it. You can't get this kind of information with a breakpoint debugger, so I say to those fanatics, "Move into the correct century, dudes!"

I strongly suspect that this sort of thing is the main reason that real-time code is considered 10 times harder to write than normal code. Everyone is using the wrong tools and mental models! Maybe it's just 10 times harder to find a programmer who can write real-time multithreaded code as easily as normal code. With a little experience you'll become one of these valuable and highly paid programmers, it just takes a somewhat different mental modeling process and an awareness of various extra gotchas.

The code in Listing 7.1 allocates a console and redirects the common stream I/O functions to it so you don't have to use the crufty console write and read functions. I adapted this from code published by Andrew Tucker ("Adding Console I/O to a Win32 GUI APP" *Windows Developer's Journal*, Vol. 8, No. 12, December 1997) for my own use.

Listing 7.1   A console debugger.

```
// I'll assume you've already included windows.h someplace else


#ifdef _DEBUG // don't want this or its space overhead in shipping version!
#include <io.h>  // need all three of these, probably not for anything else
#include <iostream.h>
#include <fcntl.h>
```

Listing 7.1 A console debugger. (continued)

```
// code modified from a Dec '97 WDJ article by Andrew Tucker
    int hConHandle;
    HANDLE lStdHandle;
    CONSOLE_SCREEN_BUFFER_INFO coninfo;
    FILE *fp;


    AllocConsole(); // create the one console this process can have
// give the thing some size, so it can hold a few messages we can scroll thru
    lStdHandle =  GetStdHandle(STD_OUTPUT_HANDLE);
                                        // need this a couple of times
    GetConsoleScreenBufferInfo(lStdHandle, &coninfo);
    coninfo.dwSize.Y = 500; // set the line count big enough to be useful here
    SetConsoleScreenBufferSize(lStdHandle,coninfo.dwSize);
// redirect unbuffered stdout to console
    hConHandle = _open_osfhandle((long)lStdHandle, _O_TEXT);
    fp = _fdopen(hConHandle,"w");
    *stdout = *fp;
    setvbuf(stdout,NULL,_IONBF, 0);
// redirect unbuffered stdin to console
    lStdHandle = GetStdHandle(STD_INPUT_HANDLE);
    hConHandle = _open_osfhandle((long)lStdHandle, _O_TEXT);
    fp = _fdopen(hConHandle,"r");
    *stdout = *fp;
    setvbuf(stdout,NULL,_IONBF, 0);
// redirect unbuffered STDERR to console
    lStdHandle = GetStdHandle(STD_ERROR_HANDLE);
    hConHandle = _open_osfhandle((long)lStdHandle, _O_TEXT);
    fp = _fdopen(hConHandle,"w");
    *stdout = *fp;
    setvbuf(stdout,NULL,_IONBF, 0);
// sync cout, wcout, cin, wcin, wcerr,cerr, wclog and clog
// point to console as well.  Requires the above to work.
    ios::sync_with_stdio();


// separate example of how  it works
```

Listing 7.1   A console debugger. (continued)

```
    cout << "cout Console test \n"; // test cout
    printf("printf test \n"); // test printf


// alternate debug streaming output method, which does not require the above
    OutputDebugString("output debug string test\n");
                                            // this text goes to debugger window
// The disadvantages are:  a: you have to have the debugger running, and
// b: you have to format a seperate string so the console and printf or cout
    is less 'code-noisy', and
// c:. Your output is all mixed up in other debug-window stuff


#endif // _DEBUG
```

Once you've run this code, presumably during the creation of your plug-in or application, you can use the standard printf() and cout functions to stream information to the DOS console, which I've made bigger than usual here so that it can contain some history of past events. You can use console input too, although it's usually not suited for real-time applications, and I usually don't even include the lines of code that affect stdin. Primitive? You bet. However, it's far more advanced than using a badly suited tool, such as normal debuggers, for certain jobs. If you come up with a better way, I'm all ears.

Existing Platforms

I can't mention every platform that supports special efx plug-ins of various types, so I'm going to describe a couple I know pretty well that are out there in enough quantity that they're profitable to program to. If I missed your product, I'm sorry, and I would appreciate a complimentary copy of it, if it's any good! Shameless, I know.

CoolEdit's XFM Model

Long before there was COM, OLE, or the rest of the new "plug-in" software tools based on them, David Johnston, creator of CoolEdit, was faced with the problem of how to provide an interface to his popular shareware wave editor so that other developers could contribute their efforts without having to have his source code. The model he came up with is, in my opinion, one of the best, and it has survived time and a few operating systems and versions of CoolEdit with only very minor modifications — a darn good track record by anybody's standards. It's also about the simplest platform out there to program to. You can find his plug-in development kit on the supplied CD. His detailed documentation is good, so I'm just going to provide an overview and discuss the main points you'll need to keep in mind to design a plug-in for any of the wave editors produced by Syntrillium Software, David's company. Using CoolEdit as a host platform to test and develop plug-ins has some other advantages as well. It contains a goodly number of analysis tools to help you debug your effect, such as one that gives you the ability to generate complex analytic test signals, and various quantitative analysis tools, such as an accurate FFT of user-selectable size and window, and various other statistical and direct-sample viewing and editing tools. I've made several plug-ins for the CoolEdit model, and the plug-in model itself never got in the way of getting the job done.

CoolEdit was written by David Johnston (and others, including me, but the main credits show only him). There are many versions (cool96, cool2000, coolpro) that all have the features discussed, and all are sold as shareware or bundled with various digital audio products. David owns Syntrillium software, which is the publisher of these products and several others, and they can be reached at `www.syntrillium.com`, where downloads are available (as well as on numerous mirror and bbs sites).

As with all things, there's some good news and some bad news. The bad news is that I had to write this section twice, and the good news is why. When I began to write, I discovered that the current plug-in development kit only supported straight C coding and an early version of DevStudio. I contacted Syntrillium about this, and together we created an even better version that now supports C++ and compiler Revision 6.0. In fact, I went the extra mile on this one, and on the CD you can find a no-kidding DevStudio appwizard that generates both full-example and stripped-down null plug-ins for the 32-bit versions of CoolEditPro. Zowie! This really beats a simple example that needs customization and a lot of set-up before you can even try it.

To use Syntrillium's model, you provide a plug-in DLL with six specific function names exported. Their products, in turn, provide a structure full of pointers to functions, which you can call to get the host to provide certain services, and various useful platform and context data. It's this structure definition that was upgraded to the style of C++ function definitions.

Basically, you must build a DLL that exports the six specific functions, and after building it, rename it to have the .xfm extension rather than .dll so that CoolEdit will know you intend it to be a plugin. CoolEdit simply searches its directory for all files with the .xfm extension, loads them, and queries each .xfm (which is really just a DLL) to see if it really is a CoolEdit plugin or not. You get a COOLINFO* structure pointer that has functions for reading and writing wave data and for providing progress meters, user-drawn graphs, presets, and so on.

About 65 handy functions are provided by the host application, as well as precreated fonts, data on the wave file you'll be expected to process, and various graphics controls to ease the process and allow you to concentrate on the effect you're developing. The rest of the job of parsing wave files, playing, viewing and analyzing them, basic functionality needed to make an effect usable, is provided by the host program. In effect, it's COM without the GUIDs, registration, and IDL hassles, which just aren't needed for a fixed-functionality special efx plug-in for a waveform editor. I'll cover the 32-bit plug-in API here, since practically no one cares about 16-bit stuff anymore, but you should know that the 16-bit model is essentially the same except for a couple of different #defines and you have to compile your plug-in for a 16-bit opsys model using the 16-bit transform API header files. I'll detail my process as I go along using a real-life port of an effect to this model from another.

To make a plug-in for the Syntrillium wave editors, start with the development kit provided on the CD, which is also freely available from Syntrillium at www.syntrillium.com. Alternatively, you can use the appwizard I created for this book to get started. To use the appwizard, unzip it into program files\devstudio\SharedIde\Template, and DevStudio will find it automatically on the next run. With DevStudio v6, put the appwizard in program files\microsoft visual studio\common\msdev98\template. If you use the plug-in development kit, unzip it in a convenient directory on your hard drive. It contains detailed documentation on the model and the functions and data supplied by the host platform, as well as the examples. You'll want the documentation, regardless, because it is far better and more complete than the documentation provided for most other standards, and it's concise. Sometimes less is more.

The kit includes examples for transforms, generation functions, and file filters, which I won't cover here. However, if you want to invent yet another wave format and allow CoolEdit to work with it, you can, via this filter interface or through a DirectX translation object. The generation functions are really a simple subset of a general transform that just doesn't happen to need input, so by covering the transform plug-in model, you'll have covered the important material.

First you have to install a copy of CoolPro because various settings in the project need to reference its directory for copying and debugging operations, and how would you test if you didn't have a host platform? Once you've done this, you can use the appwizard I provide to create a sample plug-in. The most important thing to remember at first is that some names and other data need to be unique or the new plug-in won't be recognized as new by the platform, and you may wipe an existing plug-in or its preset data. The appwizard lists these in a dialog. If you're not using the wizard, check out the defines near the top of the example and redo the text names they define to be unique. I suggest you have the appwizard produce the full example and use static linking for your first try since this shows examples of host-provided functions and controls, whereas the stripped down example does not. It is a minimal null plug-in on purpose. That's handy once you know the ropes because there's less to modify or delete, but it's not really a good example, per se. I included it for those of you who will really be going places with this stuff later.

Once you've produced the example via the appwizard, you'll want to build it, but first, you should really go to the Debug tab under Project Settings and add some nifty project automation that will make your life easier from now on. I couldn't do this for you in the appwizard because Microsoft doesn't quite give access to all of the DevStudio settings. You'll need to change the "executable for debug" to something like `c:\coolpro\cool-pro.exe`, and specify a test file path on the program arguments line. I use a mono frequency sweep or some other simple test signal depending on what I'm testing. The tools to generate this are also on the CD. You'll want to add a pre- or postbuild step to delete `c:\coolpro\xfm.dat` so that CoolPro will search for and recognize all plug-ins on startup, including your new one. I also add a postbuild step that copies the DLL to the `coolpro` directory and renames it with a `.xfm` extension.

Once you've done this, just hit F5 and it will build, copy the plug-in to CoolPro's directory, reset CoolPro's plug-in data cache so the new build will be recognized, start CoolPro, and open the test file. Not bad for one keystroke. This is a big fun factor issue, and it makes it easy to test at each stage. You should see a new name on the Edit menu under the "special" transform group (you can change this too — it's another define up top in the plug-in code) and a new toolbar icon when CoolPro comes up. Now, you can play! Click on the new name in the menu or the new bitmap on the toolbar and away you go!

How does the example actually work? I'll go through it now, from 30, 000 feet or so. The main file you'll be looking at is *yourname*.cpp, which is whatever name you gave the plug-in the appwizard, or distort.cpp if you just copied the example from the Syntrillium development kit. The includes should be pretty obvious, except for xfmspro.c. This module contains wrapper functions for the function pointers exposed by CoolPro in a COOLINFO* pointer you get on just about every function entry. This makes coding simpler and easier, and there is no need to change the .c extension because the compiler sees it as part of your .cpp file, and which defines functions provided by the host platform that you can call as needed. This module pulls in xfmspro.h, which has the definitions for the various structures and such that CoolPro uses and exposes to you. Significantly, the COOLINFO structure exposes a lot of nice functionality and information to your plug-in. Down in the coding part of the yourname.cpp file are some definitions, placed between special double-star-bar comments that delineate and make them easy to find. This is the information that has to be unique from plug-in to plug-in, such as the names, or that has to track with your change, such as matching your UserData structure definition to defines that the presets need to handle your preset information. This is *crucial* to get right, but I've made it easy by grouping it all together in one place and giving you this fairly big example that demonstrates the usage of most things. I've defined a couple of global statics at the top, mainly for the convenience of finding them later. As you go down through the functions, it gets more interesting.

The first is dllmain(), which all DLLs must have, and it does nothing but return true. The first platform-specific function is QueryXfm(), which is required for this model. When it starts up, CoolPro looks in its directory for anything named .xfm, loads it, and calls QueryXfm() if something doesn't match its plug-in data cache (which you delete on each build to force this situation). It's this function that defines your plug-in's name and various other names to CoolPro. QueryXfm() fills an XFMQUERY structure with various data and names so that CoolPro will know how to handle the plug-in. You'll see that this is where many of the defines mentioned previously are utilized. The function has to return a magic cookie so that CoolPro doesn't load something that isn't a plug-in just because someone named it wrong! There's a fair amount of this "belt and suspenders" action in the overall design, and I personally approve of most of it. It makes CoolPro more robust.

If nothing has changed since the last run of CoolPro, this function will not necessarily be called. For improved loading speed, CoolPro keeps a cache of this information from each plug-in in the file xfm.dat to avoid having to load and query them all each time. Therefore, this function is a bad place to do certain types of things because you can't count on it always being called.

The next function is XfmInit(). It is called approximately once each time CoolPro is run. Again, it is not a good place to do many classes of things, but use it to jam "safety-first" defaults into various static variables and into the dynamic UserData struct pointed to by a handle in the COOLINFO structure pointer that's passed in, just in case the real initialization happens elsewhere. Following this is XfmDestroy(), which only seems to be called when CoolPro shuts down. This is a last chance to free any resources you are holding on to. Shame on you! You should have cleaned up earlier than this!

Now for the fun parts! XfmSetup() shows the dialog of your plug-in. The return value tells CoolPro what to do — nothing for ID_CANCEL and something for ID_OK. CoolPro calls the data processing function XfmDo() if ID_OK was returned from XFMSetup(). There aren't many lines of code here because the real work is done elsewhere in the dialog message procedure and in other functions that move data between dialog controls and the UserData structure.

Next is XfmUpdate(). This is called if you've started a preview and the user changes something in your dialog. This function lets you see both the old and new dialog data so you can decide how to handle this case. XfmDo() is running in another thread, and its parameters can be updated during a live preview this way; that is, if you're careful and pay attention to the normal multithreading issues. I'll go into this more deeply a bit later.

At last, the fun DSP happens in XfmDo(), which is called to transform data either during a preview or for the real thing. CoolEdit handles the differences by using different internal file read and write functions when a preview is in progress. At this point, you have a UserData structure with everything from your dialog (or anything else you put there) and a COOLINFO* to access functions and data from the host. Your mission, should you decide to accept it, is to process the requested amount of data in some way that pleases you and, you hope, the user, and to please do it fast. This is the part you'll sweat optimizing the most and is described in most of the rest of this book. This part exemplifies some of the special tricks that experienced DSP programmers use all the time, which, again, are covered elsewhere. As the comments indicate, you should block the request into decent-sized chunks of data, since you may get a request to process a two-gigabyte file and you wouldn't want your users to wait hours for the Cancel button on the progress meter to work! By blocking the data into reasonably sized chunks, you make the UI more lively, update the progress meter more often, and so on. I've found that a pretty good size is around 0.10 second or so worth of data. Smaller, and you start seeing a lot of function call overhead when you call progress meters, data read and write functions, and so on. Larger, and the UI becomes unresponsive. A really sharp programmer might even replace the #define CHUNKSIZE 4096 at the top with a variable that tracks the data sample rate instead.

Continuing along, you get to the crufty parts that involve Windows SDK programming. There are two functions that move data between the dialog controls and your UserData structure, which are more or less equivalent to the DDX functions in MFC, but here you're doing it all manually. David wanted to keep the plug-in sizes minimal and pull in as little run-time system overhead as possible by design. Last but not least is a window procedure that does all the dialog message handling in pure SDK fashion. There's something important to note here, which is that there are some CoolEdit hook functions that are called for each message and temporarily return control back to CoolPro. These are needed so you can avoid seeing mouse hits on graphs and so that presets and other things like that can work. And they simplify your life.

Until I saw this again, I'd forgotten how yucky and error-prone SDK programming was. Perhaps a reader will write and share some DDX-like functions or wndproc-like classes that are ideal for this.

I glossed over something important above, and now it's time to tame that. XfmUpdate() is called from the CoolPro hook functions when a preview is in progress and the user changes something. Presumably, XfmDo() is running in another thread at the same time. How you handle this situation will have a lot to do with how smoothly your plug-in works. The example shows the simplest way, but there are many others. The trick is to do whatever you do thread-safely and in such a way that you do not disrupt the transform in progress. In general, you want to tell XfmDo() to recheck its setup values and rebuild local information that is required when it's convenient for XfmDo(). It's usually not safe to just jam things into XfmDo() variables. Remember, any sudden discontinuity in the data will cause clicking and other artifacts, which most users will consider to be unprofessional. They don't know how hard it is to avoid these artifacts in situations like this, and frankly, the mark of a true professional is to make the difficult seem easy.

As an example, what I did is partly right and partly wrong. When XfmDo() sees the BOOL flag, it starts rebuilding its distortion lookup table, but it does so slowly a sample at a time, rather than all at once, to avoid clicks.

The filter, if used, is set up again, which is the wrong thing to do. As the comments point out, the right thing to do is actually quite difficult. If you retune a filter that has energy in it, you can get clicks and thumps. If you clear its history at any time, you're almost guaranteed to create a discontinuity and more clicks. A correct action would be go back into the input stream, clear the filter history, restart back a bit, and average the output of the old filter and the new for a while, perhaps with a special window that doesn't weight the new output heavily until the new filter gets going smoothly. Or you could retune the filter gradually so that any energy stored in its history doesn't create a big bump or spike, as it would if you retuned it instantly. You have to do one or the other if you want this effect to be truly professional. Run with the big dogs or cower under the porch! It's just this kind of attention to detail that makes the difference between a good and a bad user experience.

DirectX

Chances are that some of this information will be wrong by the time you read it. DirectX is a model promoted by Microsoft, parts of which were formerly called ActiveMovie, which was originally intended to help solve the Babel of different sound and video file formats for which the model is well suited. It has changed far too much and too rapidly for me to use the word "standard" in any but the loosest imaginable sense, but several players in the wave editor business have adopted it despite this, as well as virtually every multimedia web authoring tool. In effect, this adoption has locked down the important audio interface portions, since a change now would break everything. I find a good rule of thumb for this kind of technology output from Microsoft is that once they incorporate it into any of their own major products, it becomes safe to adopt. Before that, you're taking your life into your hands.

I've also done several production-grade plug-ins for this model, and I found it the toughest to get started on. I hope to ease your way by providing more documentation and example information. If you hope to write special efx for sale to the world, you're simply going to have to bite the bullet and learn to conform to this ever-changing "standard."

My company employs a fellow who does the DirectX programming, so Troy helped me document the process from scratch for you. We used DirectX Revision 5.2a. There are some later revisions, but what you will be interested in has stayed the same. Most of the new improvements in DirectX have to do with either the video portions or various enhancements for the container side, not the actual audio plug-in format because changing that would break too much existing code.

I wanted to excerpt just the parts of DirectX that are relevant here and include the code with the book, but I couldn't get anyone at Microsoft to respond to my questions, and the hint was that I should include all of DirectX. In other words, no one over there really even understands the internal dependencies, and of course, they don't trust a third party to make the separation. DirectX is huge and takes an entire CD of its own, so giving it to you with this book is impractical. The CD is available from Microsoft for what amounts to shipping charges, so you can get it from them.

An inordinate portion of the total effort involved in doing DirectX plug-ins is simply getting the basic environment set up and your first example built. At least this means that once done, you don't have to do it over and over, but you could wish for a better installation situation. First I tried to install DirectX on an NT 4.0 sp4 Enterprise server. I begin by reading the semi-helpful readme files on the CD and decided to do the full install from the root directory of the CD. I right-clicked the `autorun.inf` and selected the install option, which failed with an unhelpful error message. I found that `autorun.inf` only sets the CD icon anyway.

Next, I tried to install manually with the DirectX SDK/Foundation directory and `setup.exe`. It also failed with the message

```
Can't run 16 bit windows program

One of the libraries needed to run f:\.....\setup is damaged,
please reinstall the application
```

but no hint as to what application or library was meant. I decided DirectX couldn't run on an NT4 server. If anybody at Microsoft is reading this and doesn't like my commentary, why don't you make this stuff work right? Why don't you even attempt to qualify for your own logo program? Does anyone over there care at all how much developer or customer time you waste with things like this? Evidently not.

Uninstalling DirectX on the NT machine messed up the icon arrangement in Control Panel and who knows what else. Although the uninstall program appeared to work, it didn't. All the directories and their contents remained on the machine, and I had to hand-delete them. By the way, don't try this if you're on your last 100Mb or so of disk space. My recycle bin had 91Mb in it after I deleted all the DirectX files. After this fiasco, I switched to a Windows 95 machine.

Troy and I were unable to determine from the pre-install readme files whether we had to install both the foundation SDK and the Media SDK or not, so we installed them both and upgraded to Internet Explorer 4. An hour or so later, we thought we had everything we needed. While searching through the huge new directories, we found VC5Kit, which purports to have all the DevStudio settings already set up for you. This, combined with one of the sample filter code directories is supposed to give you a jump start in learning how to use DirectX. You have to add some directories and libraries to the DevStudio search path, copy the files from VC5kit and from one of the examples (I used the `Fdfilter` example from the foundation SDK) to the project, add the `.cpp` and `.h` files from the sample to the project directory, and then build.

Well, not quite. The sample had 88 compile errors, evidently related to a missing header file that was not included in the "helpful" directions in VC5Kit. It turns out that the header wasn't actually missing, but perhaps a compiler bug prevented it from compiling in VC5 service pak 3. Something that was clearly defined in a header wasn't seen as having been defined yet in the file that included this header. You tell me.

I should stop right here and mention that the first time Troy and I attempted this it took about a week to get it all to work and build an example. We weren't depending on the Microsoft software alone. We also had the plug-in development kit available free from Sonic Foundry, which has a subset of the Microsoft material (you'll still need the CD from Microsoft) and far better documentation than Microsoft. You will want to avail yourself of this help, and you can get more information at www.sonicfoundry.com (go to support then developers). They appear to update their PIDK constantly, so it will probably help you as long as you have the matching version of the DX CD from Microsoft.

Rather than thrash about and waste time bashing Microsoft, which seems pointless even when very appropriate, I will just give you a complete working project based on DX5 and the Sonic Foundry plug-in development kit Revision 4. You can find it under `sf_pidk\arbfilt` on the CD (you have to mark the files as not read-only after copying them all), and it builds fine once you add the following directories under Tools/Options in DevStudio: `Dxmedia\classes\base`, `Dxmedia\include`, and `sf_pidk\include`, assuming that you install the Dxmedia development kit in that directory and that you don't rename `sf_pidk` when you copy it from the CD. For your convenience, I downloaded the newest version (6) of the Sonic Foundry PIDK and put it on the CD as well. I was a little reluctant to include this material because I didn't write much of the code myself (Troy did), and I don't want to be drawn into explaining it line by line — especially how it designs arbitrary IIR filters.

Rolling Your Own

You could roll your own platform to test processing and efx, but it's a lot of work, and you want your tools to be better debugged than what you're testing. In fact, writing MusiCad was most illuminating for me. Most of the effort was in the UI and getting the user interaction smooth and robust, not the special efx per se. I decided to save you from all this grunt work for several reasons. First of all, I'm a nice guy, but I also didn't want to litter this book with how-tos for riff files, Windows APIs, and MFC programming. Those are already many books of their own, some of which are at least OK. Instead, I assembled a test platform by ripping out pieces of MusiCad and stripping them of the functionality that you won't need. As a result, you get the benefit of work done by others on the basic wave document/view, including a few special speed-up and UI smoothness tricks, and the result is WavEd, a pretty nifty little basic wave editor.

WavEd

All the hard work of implementing basic functions is done for you, such as Edit-Cut/Paste (which is far more complicated than in word processing because you can have different channel counts, you may not want Paste-Insert, and so on ad nauseam), time domain and spectral (sonograph) views, selection, panning, zooming, scaling, and all the rest, not to mention various data read/write routines (in any number format you want), a versatile Windows `mmsystem.dll` interface, and everything you would expect to use for wave editing. You must have Microsoft's DevStudio v6 to compile everything, and you probably need the latest service pack. Sorry about that, but I had to choose something, so I chose what I use myself and what has the market share. There is no way for an author to win this one and make everyone happy when the code is nontrivial. WavEd, is to say the least, highly nontrivial.

It handles all the basic open it–show it–manipulate it–play it–save it issues. You just add your own processing, which is comparatively simple to do. Simply create a dialog with the processing you want and add a command to the view menu so it can be invoked. In the view, you handle the command by instantiating the dialog and calling `DoModal()`. The new dialog you've created does the rest, having full access to all of WavEd's functionality. This is a good bit easier to deal with than any of the other extant plug-in formats, although perhaps not as modular. You have pointers to the doc and view, so you can use all the data access and undo functions and can get to all the globals to find out what is currently selected and so on. It's all just lying there, quivering, for you to play with as you will.

Some of the code was written by my students a long time ago — several revisions of DevStudio at least — and I'll not much appreciate being flamed about its structure. You're not even supposed to get into it all that much, anyway. Call the functions as needed, and it works fine. I let the students write the code their way on purpose, so that as new requirements came along, they found out, indisputably, whether their way was truly a good way or not. This is a good way to teach design while keeping the students motivated. I think the results are pretty good in the main, but if you're expecting perfect encapsulation, for example, you'll likely be disappointed. In fact, private members and encapsulation were avoided at some levels on purpose, either for speed or for flexibility, and it paid off quite a lot as the project evolved. Stick to the header files to find out what to call and what variables are available if you're a purist. The doc is doubly deep in the inheritance chain because, at one time, we thought a separate wave-oriented layer would be cool. I'm not that sure now that it was a good idea, but it works.

We didn't cheat the MFC way of doing things too much. The payoff is that we haven't had to change the code just because there was a new MFC revision since 4.0. Even the workarounds still worked after the bug fixes were added, although I have removed one or two that are no longer needed. The main cheat is that there are no backup copies of open files: WavEd parties in-place on the real data instead. Although copying (which is inherent, but hidden from you, in MFC) is nice for word processing, but in MusiCad or WavEd you might be opening several gigabytes worth of files when you open a single project. I'm sure you can see why we defeated MFC's hidden save. You do have Undo, which is easy to apply in your processing dialogs, and which we did apply in our edit dialogs. It seems to be enough of a safeguard for this kind of work; it sure is faster than copying whole wave files at startup.

WavEd also allows you to create new wave doc/view pairs from within a processing dialog. You can have up to four channels in a file (MAXCHANNELS is defined in stdafx.h). You can change MAXCHANNELS, but your views and the application will use more memory if you make it bigger, since buffers and so on need to be allocated to handle anything they might see.

When I'm working on a complex transform, I often want to see the internal variables versus time *and* the original signal, and having the ability to handle multiple channels lets me do that, which makes WavEd a super-cool debug tool. You just stream the data out like wave samples into the appropriate channel of the output file and they are there after the run to be observed in graphical form. This technique is noninvasive, fast, and doesn't disrupt the process in progress as breakpoints would. If you have something that only goes wacky once in a while, the ability to have a sample-by-sample graphical history trace of its internal variables is pretty nice. I'll walk you through a simple processing effect here that uses many of the WavEd features so you'll have an example with the text description. Most of the examples in Section 2 of this book are based on this platform for pedagogical simplicity and all are provided in the WavEd directory on the CD. Once you've developed something for WavEd, it will be comparatively simple to port it for the other models, and you'll begin with your port tested in an environment where you have source code for the entire shebang.

Creating a WavEd Transform

Let's go through the motions of creating a new transform for WavEd in the same way I built my first one. First, open the WavEd project, compile the baseline version, and play with it for a while. Get familiar with the various obvious functions because you'll want your transform to work like the existing ones and support the same sorts of functionality. Go back to the project, open the resource view, open the WavEdType menu, and add something like TestXfm to the Edit menu. Use a reasonable ID_ for this, such as ID_XFM_TEST so it's easy to remember. Now, open WavEdView in classwizard, and add handlers for the ID and for its command UI update. In most cases, you need to select something in order for a transform to know what to work with, so I usually use global variable gsZoomInfo.SelLen to conditionally enable the menu item, but you may have other needs. I'll return to the command ID function later. For now, you need something generic (Listing 7.2).

Listing 7.2   WavEd plumbing for a new transform dialog.

```
void CWavEdView::OnXfmTest()
{
 CTestXFM dlg(this); // create your efx dialog
 dlg.DoModal(); // do it and return to the program
}
```

In other words, the dialog you create actually does the rest. This is just the plumbing to let the user start it. You may want to do something else in the view before or after this, but it's all you really need. If your transform needs a valid selection before it's run, you'll want to grey the menu item when there isn't a valid selection (Listing 7.3).

Listing 7.3   Command and Update functions.

```
void CWavEdView::OnUpdateXfmTest(CCmdUI* pCmdUI)
{
  pCmdUI->Enable(BOOL(gsZoomInfo.SelLen));
}
```

Now create a dialog and add it to the project. If you're following my practice of putting all headers into stdafx.h, do so. If you put them after the view and doc headers, you can keep view* and doc* and app* pointers as dialog members (for speed and easier syntax) and not have to do a lot of casting later. Just get these in the constructor from the view* (this) that is passed in to that function, and almost all of WavEd is public (although generally not safe to write on). Usually you add some controls to the dialog to let the user specify settings for your transform, which does its work in the usual OnOK(). You will want to set up an OnCancel() handler too, since this may get hit during OnOK() via a message pump contained in the global progress meter. The user may panic and want to stop cleanly and immediately. Instead of terminating the dialog and leaving a big mess, OnCancel() could set a shared boolean so that the OnOk() would shut itself down. This would provide a clean user interface that would need little explanation in Help: Cancel will cancel, plain and simple. This is especially important if the transform takes a very long time. You may want to ask the user if they want the transform to affect the various channels in multi-channel files and act accordingly, for which a tabbed dialog widget is available in the code.

I'll present a very dumb example here so the transform material doesn't overwhelm the plumbing aspects for now (the meaty transforms are in Section 2 of the book). This dumb example creates a DC level and applies it in various ways to the file data. It is only a few lines of code; however, it will support all the desired default transform plumbing (i.e., it can be used as a boilerplate). Later, when I talk about the actual transforms, I won't have to go over this again. When you write your own transforms, you'll find that the boilerplate code can be copied pretty much verbatim into your new dialog to get you started.

So far, I created the test dialog and gave it the ID IDD_XFM_TEST and the title Test Transform. I added a Preview button with ID IDC_Preview and an edit box with ID IDC_Level. I'll create the matching class now and call it CTestXFM so that I know what to enter when I get to the view plumbing.

First, I need to add an item to the view's Edit menu for wave types. I added Test Transform to the WavEdType Edit menu and gave it an ID of ID_XFM_TEST. Notice that the test dialog is IDD_XFM_TEST and Test Transform is ID_XFM_TEST (one versus two D's), which means they're unique IDs, although it doesn't matter in this case. Now I open WavEdView, bring classwizard up, and add the command and update UI options for ID_XFM_TEST (Listing 7.3) for the command and update functions.

In my case, I have decided to deal with the header/dependency situation differently than in the default classwizard "pragma once" way, so I make more changes. I open the dialog class, delete its inclusion of the application header, and cut its inclusion of its own header file, leaving only stdafx.h. Then I put the include for the dialog header into stdafx.h after the view include and before the last include of the global variables. Now it's time to build and check for typing errors. The menu item greys as it should. If I have a waveform file open, and have made a selection on it, the menu item ungreys. Selecting the menu item now causes the new dialog to open. I now add some member variables and the following lines in the dialog constructor to initialize them.

```
pView = (CWavEdView *) pParent;
                       // stash away some pointers for quick calls later
pDoc = (CWavEdDoc *) pView->pDoc;
pApp = (CWavEdApp *) AfxGetApp();
```

The intent is *specifically* to provide the differentiated services and information the base application contains. With these pointers, I can get to just about anything I want quickly. I don't put m_ on the front of my members when I'm going to be typing the names over and over and the meaning has become obvious. I leave it on when Microsoft tools put it there, because I probably won't use those variables, such as variables added to match my dialog controls, often. These variables usually are converted from user-friendly units to DSP-friendly units at the top of OnOk() before they are used, anyway.

Now I have a do-nothing dialog that I can at least call up on the screen. First, I add a Preview button and a Level edit box to the dialog, then I use classwizard again to add OnOk(), OnCancel(), and OnPreview() so I can get to work on the real thing.

Undo

Because I'm working on a system that would make most developers jealous, I built again to make sure it worked as I expected, and it took less time than it took to type this sentence. This is another case of the quantitative speed of the tools leading to qualitative differences in how I can work. Now I add the following lines to `OnOk()` for the undo plumbing.

```
// see wavEdDoc (look at the undo function) for how to do this
    pDoc->AddUndo(pView,0,gsZoomInfo.SelLen); // Add a Paste-replace type undo
```

I'm using the simple type of undo since this won't be a fancy length-changing transform. The document functions will also add the appropriate types of undos for transforms that change the length of the data. You should check the `AddUndo()` function in `WavEdDoc.cpp` for more information on what you can (un)do here. Now it's safe to do just about anything to the selection data, since it can be undone by either the Cancel function or by the user later. The `AddUndo()` function uses the global selection information in `gsZoomInfo`, so if you want it to save something different, you can temporarily save this and change it before calling `AddUndo()`, but you should put it back the way it was again immediately thereafter. I'll add the `OnCancel()` code now, too, and get it out of the way by adding the following lines near the end of `OnCancel()`, just before I call the base class `OnCancel`.

```
if (gbEditWorking)      // eg we got here via progress meter
                        // message pump while onok() was active
    {
    bQuit = TRUE;
    return; // let onok clean *itself* up and quit,
            // simpler and better all around
    }
Cdialog::OnCancel(); // die
```

Cancel

Depending on your `OnOk()` function, you may want to provide more boolean or other variables to allow your `OnCancel()` button to terminate things cleanly. In fact, that's just what I did here. `GbEditWorking` is a global boolean that all transforms are supposed to set before working on a file and to clear after all work is done, which was used originally to work around a bug in multithreaded file access in the run-time system. It makes a lot more sense to cancel and clean up in the function that made the mess in the first place. So, what I did was add a class member variable `bQuit` (set false in the constructor). This is set TRUE if `OnCancel()` is reached while `OnOk()` is active, as indicated by the global variable `gbEditWorking`, which `OnOk()` sets during its run. If `OnOk()` ever sees `bQuit` as true, it cleans up after itself and returns.

This can happen because OnOk() calls the global progress meter update function in the application, which in turn pumps messages. The motivation here is to give users a way to hit the Cancel button if they change their mind and not have to wait for the transform to complete. This also allows control changes to have their messages pumped during previews.

To show the plumbing, I'll add a dialog to ask whether each channel of a multichannel file should be affected. Some transforms should have this, some shouldn't, and frankly, it's often better not to have it for speed reasons if you don't need it, since affecting just one channel out of several takes some extra work at run time. Some transforms or DSP libraries you might acquire may not have the ability to pluck channels out of multiplexed data, which is the way wave files come. In that case, you'd need extra buffer space to do the demultiplexing yourself, so why not go ahead and let the user decide? That's what I'll do here because it's the most general approach to the problem and there are some very fast assembly language functions to help with the task. There are actually two possibilities here. You could either reuse the same Channel-Map dialog set that was used for cutting and pasting, which would allow you to select not only which output channels to affect, but also which channels to use as input for each output, or you could use a simpler "affect this channel" channel sheet, which is what I'll do here. I'm sure any competent programmer can figure out how to use the former technique as easily, but I don't want this exposition to be completely overwhelming; I merely want to point out the possibilities that exist.

I add the following two lines to the top of OnOk().

```
WORD chnls = pDoc->Wv_Header.wnChannels; // get input channel count
CChnlEnableSheet dlg("Channels to affect?",this,0,chnls);
                                    // setup an enable dialog
```

This code sets up a nice dialog that is provided to ask the user what channels to work on. Now I add the following line after the Undo() code to invoke it and get the user's input.

```
dlg.DoModal();
```

Later, I can use this to decide what to do in the main loop by grabbing the booleans contained in this dialog. Finally, it's time to actually do something to the data, but first, I need to allocate a couple of buffers. I'll need one to read the file data into and another to demultiplex each channel of the input into for the dumber DSP functions that can't work directly on multiplexed (i.e., mixed-channel) data. I'll add some pointers to the class members, set them to null in ctor, initialize them in OnOk(), and add a couple of lines of code to let `OnCancel()` know about any cleanup it might have to do. To avoid tediously telling you about this line by line, I'll go ahead to a complete transform and show you what it looks like afterwards.

MultiThreading Issues

WavEd is multithreaded and was extracted from an extensively multi-threaded program. There were some problems with the nonthread safety of `Cfile` and the underlying C run-time system over and above the obvious (e.g., what happens when one thread seeks a file and another reads it) and having to do with the underlying run time failing to properly flush its write buffers with alternate reads and writes by different threads. This is a true bug in the so-called multithreaded run times. The workaround involves having various processes set flags to indicate when they are partying on files so that the file processing can work inside a critical section and conditionally flush buffers on each call. This is slow — you only want to do it when necessary — so a few global booleans are provided to handle this situation. This leads to an amusing line of code in the wave document class in various places where file access occurs.

```
lockit.Lock(); // critical section
if (gbThisWorking + gbThatWorking + gbTheOtherWorking > TRUE) file.flush();
// other stuff
lockit.UnLock();
```

Anyway, I think it's amusing that something might be greater than true. You don't always have to get fancy to make multithreading work. For instance, you don't need to set the booleans inside a critical section. All you have to do is set them before you call the file code and reset them after you're done. You don't need special synchronization, because only one file processing thread is running at a time in WavEd. In fact, the critical section that protects document reads and writes was the only one needed in all of MusiCad, which had a lot more threads. Now you can see why you should use the `windows.h` versions of various types instead of the newer native types, so this cute trick can work. Other cases like this arise often. This code survived the switch from Win16 to Win32 with no changes for variable lengths because of how they're handled. The variables in `windows.h` are at least guaranteed to stay the same sizes regardless of C language standard changes. The global boolean `gbEditWorking` also serves in this situation to flag the case of Cancel being hit while a transform is in progress. If this isn't set when `OnCancel()` is reached, it means the user simply hit Cancel while nothing else was going on, not that any cleanup needs to be done.

Testing the Functions

To test that Undo, Cancel, and Progress Meter really work, I added a line to null the data for any affected channel in Listing 7.4 (a really fancy transform!).

Listing 7.4   Basic WavEd transform boilerplate code.

```
void CTestXFM::OnOK()
{
WORD chnls = pDoc->Wv_Header.wnChannels;
CChnlEnableSheet dlg("Channels to affect?",this,0,chnls);
DWORD amount, length, FileIndex, i, c;

    UpdateData(); // get the user's input
    // see wavEdDoc (look at the undo func) for how to do the below
    pDoc->AddUndo(pView,0,gsZoomInfo.SelLen); // Add a Paste-replace type undo
    dlg.DoModal(); // ask what channels we're supposed to affect
    pMainBuf = new float[MAXCHANNELS * CHUNKSIZE]; // file read/write buffer
    pWorkBuf = new float[CHUNKSIZE]; // for working on one channel at a time
    gbEditWorking = TRUE;  // tell the world we're partying on it's data
    FileIndex = gsZoomInfo.SelStart; // where user wants us to work
    length = gsZoomInfo.SelLen; // and how much user wanted to us to do
    pApp->SetProg(); // defaults to pos == 0, and pumps messages

    while (length > 0 && !bQuit) // big meat loop, tests for user-cancel, too
    {
    amount = length; // try for it all
      if (amount > CHUNKSIZE) amount = CHUNKSIZE;
                             // but don't bite off too much at a time
      pDoc->GetPackedBuffer(pMainBuf,FileIndex,amount,2); // work with floats

// do any work for this buffer here
      for (c=0;c<chnls;c++) // loop over the available channels
      {
        if (dlg.pages[c]->m_Enable) // if this channel is to be worked on...
        {             // else skip the work and go faster too
```

Listing 7.4 Basic WavEd transform boilerplate code. (continued)

```
            GnCpyffmd (pWorkBuf,&pMainBuf[c],amount,(float) 1.0,1,chnls);
                    // demux to local
// null out the channel
        for (i=0;i<amount;i++)
                    // the sample by sample loop - it all goes thru here!
                {
                pWorkBuf[i] = 0.0f; // a VERY complex transform, eh?
                }
                GnCpyffmd (&pMainBuf[c],pWorkBuf,amount,(float) 1.0,chnls,1);
                    // mux from local
        }           // if do this channel
    }               // for each channel


    pDoc->WritePackedBuffer(NULL,pMainBuf,FileIndex,amount,2);
                    // write data back
    length -= amount; // we did this much data, in samples
    FileIndex += amount; // so step this far for next one
    pApp->SetProg(100 - (length * 100)/gsZoomInfo.SelLen);
                    // do progress, check for cancel
    } // end of big meat loop


// cleanup stuff
    pApp->SetProg();
    delete[] pMainBuf;
    delete[] pWorkBuf;
    if (bQuit) pDoc->Undo(pView); // the user has hit cancel to get us here,
                                  // via message pumped to oncancel
    gbEditWorking = FALSE;        // we're done with any document files
    CDialog::OnOK();              // die
}
//////////////////////////////////////////
void CTestXFM::OnCancel()
{
    if (gbEditWorking)    // eg we got here via progress meter
                          // message pump while onok() was active
        {
        bQuit = TRUE;
```

Listing 7.4   Basic WavEd transform boilerplate code. (continued)

```
        return;          // let OnOk() clean *itself* up and quit,
                         // simpler and better all around
    }
    CDialog::OnCancel(); // die
}
```

GnCpyffmd(), which stands for multiply by a gain factor, copy float to float, and multiplex/demultiplex, is one of the assembly functions I provide. See the files convert.h, mixhelp.h, and mathhelp.h for information on what I've bothered to optimize in assembly language. Even though this does a floating-point multiply for each sample, it's pretty quick and is faster than doing it in C. If I have to compensate for a transform's gain or loss, I use this a lot. There is also Cpyffmd(), which simply moves DWORDs around and can be used for floats or integers. When you use the document class' file write functions to write floats onto a 16-bit file, the write function does the required number type conversion for you. It does this using some ASM routines I provide (see convert.h), which also handle the clipping cases when converting floats to SHORTs, and which are about 100 times faster and more correct for this application than a cast would be. After all, a cast would result in a call to the run-time system for each sample, rather than a call for each block as is done here. The call overhead alone kills approaches such as casting by-sample. You can see that each time around the loop, the routine checks to see if OnCancel() has been hit and drops out if it has. This is checked again at cleanup time and an Undo is executed if the user pressed Cancel. You can see that this is by far the easiest platform to work with: you have Undo, File Read/Write, Mux-Demux, Channel Check, Progress Meter with message pumping — the whole nine yards — with less than one page of code before Preview is added.

Preview

The ideas behind providing Preview are several. If the transform takes a long time to complete, the user may want to see if it's set up correctly before the long wait. In the case of transforms that are difficult to adjust, the user may need to hear the results of any adjustment quickly, or for practical purposes, they can never get the right settings. You, as the designer, should avoid this situation whenever possible, but you often can't. A key word above is "quickly," and how you define it has a lot to do with how difficult it is to implement in real life or even whether it's possible with some types of user input. In an ideal world, for instance, you might assume that any transform is faster than real time and that the user can simply turn knobs while listening to a loop of data to get the adjustment correct. In the real world, however, all transforms are not going to be in real time on every computer, and you may want to allow users to put in precise values via edit boxes. What can you do if the user is in the midst of typing a new number, and while typing, an illegal or otherwise bogus value is entered temporarily? I provide some code to help with this situation in the CEditSlider class and as a global function called EditFloat(), but these solutions still aren't perfect for every case.

The assumption is that while a preview is in progress, the current values are read off the UI controls during processing and the user will not want to see message boxes about temporarily invalid values. This is really a UI question, but it translates into a software design problem as well. There is a workaround for the non-real-time problem: simply loop-play a small buffer and transform into it from the original data at whatever speed you want. If the transform is fast, this will happen so that the user will never hear the untransformed data. If it is slow, the play will eventually go past the processing point, and the user will hear untransformed data until the transform completes on the preview buffer. This solution is actually not all that bad; it may even be a "feature" since the user hears what amounts to an A-B comparison when this happens. At any rate, this is the way I prefer to do it. I consider preview functions that "gap" when the transform or your computer is too slow to keep up to be totally bogus and amateurish. Most applications that use DirectX plug-ins suffer from this. Their authors are probably too tired to do it right after having handled the rest of the cruft necessary just to support DirectX.

In the example preview here, I am not going to support real-time changes to the user controls since I am working from a plain-Jane edit box. The extra code that supporting real-time changes would entail would clutter it up to the point of ridiculousness and really falls under the topic of advanced Windows programming, which is more or less outside the scope of this book.

You could either have a completely separate Preview "do it" function, or you might just fake OnOk() into doing the work for you, since you've already added some plumbing to do that kind of thing. That's what I'm going to do here, and even though this saves some code and makes it easier to maintain, you'll find that it's getting somewhat tricky at this point. Because I'm pumping messages from inside the call to the progress meter to allow the Cancel button to work, I now have to prevent various odd user actions from getting me completely tangled up and add some more plumbing so that OnOK() knows whether to call the base class OnOK() (i.e., die) when done or not. The result is slightly ugly, I admit, but this actually makes for less total work. If you copy this stuff, remember that the exact ordering matters here! Note that I added some lines of code to disable some buttons at various times, so as not to have to deal with some of the more ridiculous cases. This is good practice, and good UI, since graying the buttons is better than just ignoring them. To avoid printing almost the same listing over and over again, I'll now put in the actual transform code for the demo before I show the listing again, since it is only a few lines of code. This is still the simplest transform model there is to implement, in my opinion. You can copy it into the message-handling function of your new dialog, add your own code instead of the simple data manipulation here, and be ready to go. There's no need to change the bulk of the programming for most uses. Listing 7.5 is both the preview and another new feature that allows the user to specify whether to add the edit box value to existing data, overwrite existing data, or multiply by it by the edit box value.

Listing 7.5   Boilerplate transform with Preview.

```
// CTestXFM message handlers


void CTestXFM::OnOK()
{
 WORD chnls = pDoc->Wv_Header.wnChannels; // saves typing later
 CChnlEnableSheet dlg("Channels to affect?",this,0,chnls);
                                    // allow user to spec channels
 DWORD amount, length, FileIndex, i, c; // the usual variables
 float * pWorkBuf; // one channel's worth of a data chunk, demuxed
 float * pMainBuf; // one chunk from the file, muxed


    GetDlgItem(IDOK)->EnableWindow(FALSE);
                    // prevent getting here again while busy!
    UpdateData(); // get the user's input
    pDoc->AddUndo(pView,0,gsZoomInfo.SelLen); // Add a Paste-replace type undo
if (IDCANCEL == dlg.DoModal())// ask what channels we're supposed to affect
 {
    GetDlgItem(IDOK)->EnableWindow(TRUE); // re-enable
    return; // go home, they canceled us

 }
```

Listing 7.5 Boilerplate transform with Preview. (continued)

```
    pMainBuf = new float[chnls * CHUNKSIZE]; // file r/w buffer
    pWorkBuf = new float[CHUNKSIZE]; // for working one channel at a time
    gbEditWorking = TRUE; // tell world we're partying on it's data
    FileIndex = gsZoomInfo.SelStart; // where to start
    length = gsZoomInfo.SelLen; // how much to do
    pApp->SetProg(); // defaults to 0, pumps messages


        while (length > 0 && !bQuit)
                    // big meat loop, tests for user-cancel, too
        {
        amount = length; // try for it all, but
        if (amount > CHUNKSIZE) amount = CHUNKSIZE;
                        // don't bite off too much at a time
        pDoc->GetPackedBuffer(pMainBuf,FileIndex,amount,2); // work with floats


//******* do any work for this buffer here
        for (c=0;c<chnls;c++) // loop over the channels
        {
        if (dlg.pages[c]->m_Enable) // if this channel is to be worked on...
        { // else skip the work, and go faster anyway
        Cpyffmd (pWorkBuf,&pMainBuf[c],amount,1,chnls);// demux to local
// operate on the channel
            for (i=0;i<amount;i++) // sample by sample loop
            {
                switch (m_Mode) // which way are we going to affect data?
                {
                case 0: // mix, or add in
                  pWorkBuf[i] += m_Level; // complex transform, eh?
                  break;
                case 1: // overwrite, or replace existing data
                  pWorkBuf[i] = m_Level; // complex transform, eh?
                  break;
                case 2: // multiply with existing data
                  pWorkBuf[i] *= m_Level; // complex transform, eh?
                  break;
                }
```

Listing 7.5   Boilerplate transform with Preview. (continued)

```
                }
                Cpyffmd (&pMainBuf[c],pWorkBuf,amount,chnls,1);// mux from local
        } // if do this channel
        } // for each channel
//******* end of your work for this buffer

        pDoc->WritePackedBuffer(NULL,pMainBuf,FileIndex,amount,2);
                            // write data back
        length -= amount;     // we did this much data, in samples
        FileIndex += amount; // so step this far for next read/write
        pApp->SetProg(100 - (length * 100)/gsZoomInfo.SelLen);
                            // prog meter, message pump
        } // end of big meat loop


// cleanup stuff
        pApp->SetProg();
        delete[] pMainBuf;
        delete[] pWorkBuf;
        if (bQuit && !bPreviewing) pDoc->Undo(pView);
                                // user has hit cancel to get us here
        gbEditWorking = FALSE;    // we're done with any data files
        if (bPreviewing && bQuit)
                        // preview func has canceled us before we were done
        {
            bQuit = FALSE; // we did this, so get ready for another try
            PostMessage(WM_COMMAND,IDC_Preview);
                            // tell preview to try again, we're ready
        } else if (!bPreviewing) CDialog::OnOK(); // die if we're really done
}
/////////////////////////////////////////////
void CTestXFM::OnCancel()
{
    if (gbEditWorking)     // eg we got here via progress meter
                    //   message pump while onok() was active
        {
        bQuit = TRUE;
        return; // let onok clean *itself* up and quit,
```

Listing 7.5 Boilerplate transform with Preview. (continued)

```
                // simpler and better all around
        }
      CDialog::OnCancel();
}
///////////////////////////////////////////////////
void CTestXFM::OnPreview() // *toggle* preview status
{
 CString btn; // for button text
 DWORD tmp;
 PlayInfo *pRequest; // info to pass to playthread via message

 if (gbEditWorking && bPreviewing)
 {// if OnOK is in progress, let it shut down cleanly and message us back
  bQuit = TRUE; // tell it to quit
  return; // OnOK will message us back in this case, tricky, but it works well
 }
 if (bPreviewing) // if already previewing, stop!
 {
  btn = "Preview";
  SetDlgItemText(IDC_Preview,btn); // tell user we're ready to go again
  bPreviewing = FALSE; // we're not anymore
  pApp->pPlayThread->PostThreadMessage(PTMsg,FLUSH,(LPARAM) NULL);
                      // stop the music
  pDoc->Undo(pView); // put it all back the way it was
  memcpy(&gsZoomInfo,&SelSave,sizeof(Zoom_Info)); // back the way it was
  GetDlgItem(IDCANCEL)->EnableWindow(TRUE); // allow cancel again
  GetDlgItem(IDOK)->EnableWindow(TRUE);     // and OK
 } else // start up a preview
/////
 {
  GetDlgItem(IDCANCEL)->EnableWindow(FALSE); // prevent bogus cancels!
  btn = "Stop It!"; // since we have another, better way
  SetDlgItemText(IDC_Preview,btn);
  memcpy(&SelSave,&gsZoomInfo,sizeof(Zoom_Info));
                    // save global var before changing it!!!!!
```

Listing 7.5   Boilerplate transform with Preview. (continued)

```
// limit preview size to something reasonable, since we're locking
   the ui for processing
 tmp = gipresecs * pDoc->Wv_Header.dwSamplesPerSec;
 if (gsZoomInfo.SelLen > tmp) // long selection
 {
  gsZoomInfo.SelLen = tmp;
  gsZoomInfo.SelEnd = gsZoomInfo.SelStart + tmp;
 }

 bPreviewing = TRUE;
          // this also prevents OnOk from calling CDialog OnOk and dying
  pRequest = new PlayInfo; // create a play request packet for playthread
  pRequest->Mode = 4; // loop play
  pRequest->doc = pDoc;
  pRequest->rdoc = NULL; // not recording
  pRequest->StartSample = gsZoomInfo.SelStart;
  pRequest->NSamples = gsZoomInfo.SelLen;
  pRequest->Flags = PIPC; // play cursor
  pApp->pPlayThread->PostThreadMessage(PTMsg,FLUSHADD,(LPARAM)pRequest);
  // playthread deletes this later!  Do not delete a play request yourself!
  PostMessage(WM_COMMAND,IDOK); // tell OnOK to go, now, and return
 } // start preview
} // on preview-pressed
/////////////////////////////////////////////////////////////
```

Some of the plumbing to implement the smooth preview is slightly tricky, but the good deal here is that you'll never have to mess with it unless you want to. I've made a new transform for WavEd in under five minutes by copying this code into the message handler functions and copying some members to the header and their set-up code into the constructor of a new dialog class. Just try that with any other model! When you look through the source code for the other transforms in WavEd, you'll see that about the only thing that changes is the actual "meat" of the transform code. Often I implement that part in some other class, which is designed not to know about the plumbing detail. That makes it very portable to any other transform plug-in style.

Perhaps I should have put most of this into another class from which new transforms could be derived. You'd still have to deal with OnOk(); this isn't that hard to deal with as is. Real-life experience has shown me that I'd still wind up having to override and mostly duplicate the base class functionality from time to time for various special needs (e.g., transforms that don't automatically shorten a long selection for preview or that handle real-time updates differently). I suppose the code would end up a little shorter with common derivation, but to be honest, most of the working set for this or any other application isn't my code, it's MFC, Windows DLLs and the like. All the transforms put together add less than 100Kb the way it is.

A Shortcut

Now that a lot of transforms for WavEd exist, there's a simpler way to create a new one. First, create a dialog resource. If you're going to support preview, add a Preview button with ID IDC_Preview. You don't have to add the rest of your controls at this point. Now, go into the transform you're going to swipe from and copy the first block of member variables from its header (I organized them separately to make this easier since this is how I created most of my transforms), including the variables that point to the view and the document and contain the sample rate or Nyquist frequency and the booleans that control the preview and timer. If you're going to be using the fancy real-time update EditSliders or CGuins and so on, you'll also want to swipe the message definition for the control change message they send. Copy all this into the new dialog's header file. Also, copy the relevant parts of the constructor code from the sample to the target so that these variables are set up at create time. Now, add OnOK(), OnCancel(), OnPreview(), and perhaps OnTimer() to the target and wipe out any content the wizard puts in them. Go to the sample transform, steal the bodies, and put them in. If you're using the real-time preview update, you'll also have to provide a message map entry (copy this from the sample; the wizard can't handle user-defined messages) and the function body. Now go to the body of OnOK() in the target and comment out whatever was special to the transform you took it from to get rid of compile errors caused by referring to transform-specific member variables that you now don't have. If you're doing this the way I do, you'll want to remove all includes except stdafx.h from the dialog .cpp file and add an include for the dialog's .h file to stdafx.h near the end with the rest. At this point, the thing should at least compile. Fix any compile errors now — they should be pretty self-evident at this point — then add a menu entry with a unique ID to the WavEdType menu for your new transform. Now go to classwizard and add a handler to CWavEdView for this and for the UpdateUI functions. When you go to the view to edit code, you'll see a number of examples of what to do. It's pretty much just enabling the menu entry when you have a selection and creating a dialog and calling its `DoModal()` in the command handler. At this point, you have something you can test. One of the first things you'll want to do is add a handler for `Init-Dialog()` so you can use and initialize things like `CEditSliders` and `CMCGuins`. These are designed to be placed on top of an existing control as a button for the Guins and a static group box for EditSliders. For the Guin case, you add a `CButton` member using classwizard, manually change it in the header to be a `C(MC)Guin`, and move it outside the green-highlighted classwizard stuff. For EditSliders you don't have to do this, but you do have to give each group box a unique ID to pass into the `Init()` function of the EditSlider. (A word of warning: If you use the classwizard to add a `C(MC)Guin` directly, it will do so, but it will also "helpfully" add an unnecessary include of that header to your header file, which will cause compile errors.) Once you've completed these steps, you can play around with the code and actually do something in `OnOK()`. This process only takes a couple of minutes after you've done it once, and it's nice not to have to sweat the details of all the plumbing each time.

Creating a Document

Since I don't have a better place to talk about this, I will describe the process of creating a new, blank document (a wave file and a view of it) in WavEd. I use it a lot, both for debugging and for real projects. For instance, sometimes you might want to have more output than input channels (e.g., a stereo reverb from a mono source), and this is one way to do that. Listing 7.6 is a code fragment that I put in `OnOk()` after `UpdateData()`. I save the global default wave header, make any changes I need to for what I want, then instruct the document template to make a new document and return a pointer to it.

Listing 7.6 Creating a document.

```
// Create an output document file
memcpy(&SaveHead,&gsDefaultHeader,sizeof(WaveHeader));
        // save this global variable before changing it!
// make a new doc file for output, add to mix etc, with # chnls = required
gsDefaultHeader.dwSamplesPerSec = pDoc->Wv_Header.dwSamplesPerSec;
        // = = source
gsDefaultHeader.wBitsPerSample = 16; // by fiat, sorry!
```

Listing 7.6 Creating a document. (continued)

```
outchannels = Params.bHfE + Params.bOOrig + Params.bOPT + Params.bOPTSEF
            + Params.DTTrack + Params.bLPA;
        // number of outputs desired, however *you* compute it
if (outchannels < 1)
{
 outchannels = 1;
 Params.bOOrig = TRUE; // force something to happen no matter what!
}
if (outchannels > MAXCHANNELS)
{
 AfxMessageBox ("Too many output channels selected");
 return;
}
gsDefaultHeader.wnChannels = outchannels;
gsDefaultHeader.wBlockAlign = 2 * gsDefaultHeader.wnChannels;
gsDefaultHeader.dwAvgBytesPerSec = gsDefaultHeader.dwSamplesPerSec
                                 * gsDefaultHeader.wBlockAlign;
gsDefaultHeader.dwSixteen = 16;
gbNewWaveDontAsk = TRUE;
        // since we set this up ourselves, don't let user mess it up
pPODoc = (CWavEdDoc *) pApp->AllDocs.pWaveTemplate->OpenDocumentFile(NULL);
        //CreateNewDocument?
if (pPODoc == NULL) return;
gbNewWaveDontAsk = FALSE;// put it back for later
memcpy(&gsDefaultHeader,&SaveHead,sizeof(WaveHeader));
        // don't need this anymore, so restore it now
        // so, don't use gsDefaultHeader after this here.....
///////////////////////////////////////////////////////////////////////////
```

Of course, you'll have to define some of these variables elsewhere; then when processing your input, you simply read from one file and write to the other. This is super-handy when working on complex transforms because you can see internal variables versus time and the intermediate results as well. You may need this in the case of a transform that has a different output channel count than the input channel. If you're only debugging, you can then have MAXCHANNELS worth of debug information for each sample you process and still write your results to the original file; the Tile-Window (Alt-t) function in WavEd still allows you to see everything in time alignment.

Summary

In this chapter, I've described the non-DSP elements of signal processing and stressed that the most important tools are really your understanding of the platform and the debugging process, rather than a particular compiler or editor. I've also explained some of the tricks I use to make processing easier for me. I hope I've given you the background needed to understand the examples provided by the various platform development kits so that you can get going right away with a lot less effort.

Chapter 8

DSP Programming Methods and Tricks

Although supposedly there are a lot of ways to skin a cat, I've never tried it. I'm fond of mine and don't want to get myself hurt. There are a lot of ways to do almost anything you could do with a computer, but when speed and accuracy count, the number of acceptable ways of doing a particular job is drastically reduced. For virtually all production-grade DSP programming, both speed and accuracy really matter. As a result, many of the more common operations now have a more or less standard methodology associated with them, and this chapter covers the most commonly used tricks of the trade, which vary somewhat between platforms. A major consideration in every case is to program into the strengths of the platform that will run the code. However, these days, most platforms are converging because of various technology limits, not to mention market factors, so the same tricks tend to work on most platforms, the exceptions being special-purpose hardware or DSP CPUs that may have their own built-in ways to do certain common operations.

Speed is the only thing. Wow, that's a strong statement! But in general, it's true that once acceptable accuracy is achieved, it's always better to go faster. Although you might think speed is a purely quantitative measure, many times it becomes a qualitative issue. An example should make this clear. Early on, several people writing waveform editors decided it would be nice to add a spectral view; that is, the classic sonograph with time along the horizontal axis and frequency along the vertical axis, with energy displayed as a color or brightness level of pixel at a given time or frequency. This involved many, usually overlapped, frequency analyses or FFTs at many, many start points in the data: basically one for each horizontal pixel that was displayed for each channel. This can take a lot of time, so although users found it a nice feature for analyzing their data, they really couldn't use it for normal editing functions. It was painful to wait 30 seconds or more for a screen redraw each time a scroll or zoom was required while editing. At the time, I was doing some consulting work, and I offered to speed up a customer's spectral view code. They went for the idea. I was able to achieve a speedup on the order of five or six times compared to their original code, which was no small thing since I was starting with pretty good code in the first place. In fact, it got to the point where the Windows drawing functions were the main time eaters. I found that customers simply raved about the new performance because they could now edit while staying in the spectral view domain! In other words, *without any change in basic functionality whatsoever, new uses became possible,* so a quantitative performance change created a qualitative usage change.

Now that I'm aware of the principle, I'm finding more and more places it's applicable. For instance, as development machines and compilers become faster, a new technique of building and testing much more often becomes practical, and this in turn leads to code with fewer bugs, not to mention fewer cases of barking up the wrong tree for any length during the development process. Again, a simple speed change has manifold ramifications, but the principle of "faster is better" seems to continue to hold throughout the software domain. I can think of a couple of major hardware manufacturers that survive and, indeed, grow by improving speed without really changing anything else at all. As they do, people find different qualitative uses for their machines. No one even considered multimedia applications on the original IBM PC. I did do some midi work on a KayPro, but they were actually a good bit faster than the original PC, but it was so difficult with the existing tools, the effort foundered.

Although much verbiage and programming religion these days revolves around the idea of "write once, run everywhere," I'm not a convert when it comes to DSP or, really, much of anything except Web pages. The truth, sad as it may be, is that the software that has the highest performance available for a platform eventually takes the market. This will never be true of an interpreted environment. DOS, Windows, games, CAD, and many other things that come to mind are all highly optimized in some sense for speed, and all use extensive assembly language components to achieve this. In general software that produces the reaction, "Wow, I didn't know my machine could *do* this!" is the stuff that people buy and this is one of the mantras of signal processing. The faster you can go, the more smoothness or slickness you can offer in the user interface, the more features you can reasonably add to your product, and on and on. My contention is that it really does pay to examine the platform for strengths and weaknesses and take them into account when programming.

I can write high-level language (HLL) code that will run okay or even quickly on a wide variety of platforms, and sometimes that's acceptable for research-grade code, but when it comes to production-grade code and products, I'm ripping off the customer if I haven't made their hardware investment pay off maximally in realized performance. Portability is a convenience to developers, not consumers who only work on one platform at a time. Since what goes around comes around, I resolved to do the best for my customers that I can. This involves getting down to the nitty-gritty of whatever platform strengths and weaknesses there are and programming to or around them directly. In the real world, that's in fact what most of us do. The write once, run everywhere idea just doesn't have much place outside of user interface issues, and it could be argued that even that is the last place it's appropriate. Users often select a platform for its UI standards and feel that if you write code that shows them a picture different from what they're used to, you're not helping them learn how to use the tool, not taking advantage of the strengths of the platform, and not giving them a good deal for their money. In these competitive times, that spells disaster, so if the whole idea of platform specificity turns you off, perhaps you have got the wrong book or are in the wrong business — unless you work for a company that makes Java development tools, that is.

Figuring Out the Strengths of a Platform

There is one platform that has such an overwhelming market share that it makes sense to pay special attention to it. This is the Intel-CPU-based PC running Win32. In the embedded world, it's somewhat less clear, but there are emerging favorites for various tasks, again, as seen by market share. I'll mention some major contenders here, since a lot of DSP programming winds up in embedded applications, but I'll begin with the big market winner first.

Intel Pentium

With the advent of the Pentium, real-time audio processing on the desktop became practical on a large scale for the first time. Before this, you could really only do it with specialized and expensive platforms. At last, the desktop became an interesting target platform for the DSP programmer and there was a need for more qualified DSP programmers. This is due in large part to the significant speedup in the floating-point unit (FPU), often making `float` operations on Pentiums faster than integer operations. It also made it easier to program, as you'll see later when I cover integer DSP programming.

For starters, look at Table 8.1, which is what I perceive to be the main strengths and weaknesses of the Intel Pentium for DSP programming.

Table 8.1   Intel Pentium strengths and weaknesses.

Strengths	Weaknesses
Big address space	Low main memory bandwidth/small cache
Good programming tools	Poor interrupt response and few interrupts available
Well-documented operating systems (really!)	CPU used to do I/O instead of DMA in many cases
Fast floating point	Slow bulk storage

There are more, of course, but these suffice to get you thinking along the right track. I find that on this platform, the FPU is considerably faster than main memory. It's very easy to write FPU code that is limited by memory bandwidth, so when dealing with large streams that don't fit in cache, it really pays to phrase your code so that main memory bandwidth doesn't become the limiting factor, and it really pays to be aware of caching issues. Because disks are slow compared to the rest of the system, it pays to avoid disk access as much as possible, which is again a caching issue of sorts, just at another level in the data flow hierarchy. The big address space makes large tables seem attractive at first, and sometimes using tables is much better than recalculating a complex function on the fly, but the small size of really fast L1 cache memory can defeat you sometimes because every table fetch invalidates a whole line of cache, and you may never need the adjacent table entries.

It's easy to see this machine is optimized to jump around between little tasks, rather than work on a single large looping task. It has evolved along with the popular operating systems and software application designs that do this. In fact, having read Intel's Architecture Optimizations manual, it's clear that this machine was optimized more to make poor code run acceptably than it was to make great code scream. Another complication here is that the popular operating systems (I'm talking about Microsoft products) are designed in such a way that cache is invalidated quite frequently. A single event or interrupt causes a great deal of jumping around from place to place in physical memory, which in turn causes most or all of the cache to be invalidated on each and every interrupt or other context switch. I suppose in a perfect world, you could just program for DOS or Unix, which don't have this problem to such a great extent, but this is not acceptable to most users even for the significant performance advantages you might gain. In a really perfect world, some attention would be paid to these issues by the designers of the more popular Windows operating systems. It wouldn't be impossible or even all that hard to do much better, at least for DSP applications, which depending on how broad a definition you use, could include most applications; even text or database information is a signal, after all.

Such thinking leads to some rules or, perhaps more appropriately, guidelines for DSP programming on this platform as well as most other general-purpose CPUs.

- Touch no data you can avoid touching. Pass pointers; don't make copies when not required for other reasons. There are no other good reasons.

- Once you get data from either disk or main memory, do everything you can to each piece and avoid refetching the same information over and over; it probably won't stay in cache for you. Keep it in registers as much as possible while working on it.

- Work in relatively small memory chunks, both to keep everything in cache and so tasks finish before a context switch flushes your data and code loop out of cache.

- Don't do sample-by-sample function calls of any kind (including implied calls such as those created by casting that call back into the run-time system) because the jumping around doesn't help caching issues, and as in all machines, call overhead is extra overhead.

- Carefully consider (test!) whether using tables for functions that are difficult to compute actually buy you anything. Often it just increases the memory footprint and causes cache invalidation and burst line fetches that wind up being mostly wasted. Virtual memory swapping is almost always slower then computing, too, but it depends. Test!

- Generate and check the assembly language files when using an HLL such as C/C++ to make sure you are really following these guidelines: You will get some nasty surprises when you do. Often, however, you can simply rephrase your code and get the results you need rather than having to use assembly. C compilers spoof you but can be spoofed in turn. Of course, each compiler revision will change your code output.

- Program crucial functions in native assembly to maintain control over call overhead and register usage, as well as code stability across compiler revisions. This helped on the WavEd project when I switched to DevStudio v6. The new version wasn't keeping full precision of running floating-point sums in debug mode, whereas the old version did, which in turn broke some old adaptive filter code that was pushing what could be done with `floats` instead of `doubles`. If you want it done right ....

- Design functions to work on blocks of data to amortize call overhead. For instance, block `float` to `short` conversion and clipping is about 50 to 100 times (!) faster with a custom ASM routine than with sample-by-sample calls into the HLL runtimes that casting implies.

- Work with floating-point numbers. These days, it's almost always faster than integer arithmetic, both straight up and because fewer overflow types of checks are needed.

- Never do sample-by-sample divides if you can help it. Usually, just one to create a scaling factor, using the result in a sample-by-sample multiply will do, instead. This is about 20 times faster!

- Avoid conditionals in time-critical code. Find ways to make those decisions before and after it insofar as you can.

- Don't reinvent the wheel. For instance, go ahead and trust the system's disk cache. Adding your own just increases the memory footprint and the number of times memory cache will be invalidated in your program.

There are times when all these rules can't be met at once. For instance, you might not be able to make an algorithm fit in the CPU or FPU registers. In this case, I often find that it's better to break it into two or more parts that do fit, and refetch the data buffer. You may get lucky and still have your data in cache if you choose block sizes wisely. The principle here is to minimize all traffic in to and out of the CPU chip as much as possible. An FFT I wrote does this by breaking the algorithm into small pieces that fit in the

FPU, resulting in a dramatic speed up because I didn't have to load and reload partial results as much.

Other Platforms

Other popular platforms include special-purpose DSP CPUs and RISC-type processors now coming into more widespread use. Many of the issues are the same, and it's only a matter of degree for some aspects. I'll look at a couple of issues from 50,000 feet. I'm not going to go into the tricks and special programming secrets for each here, since the audience would be small. If your company makes great CPUs and I don't mention them here, perhaps you should consider that as an independent consultant, I have to use tools I can afford, get good support for, and actually are a good fit for the application I'm being asked to create. Chances are that I've been at this a lot longer than your company has, since I got the government into the DSP business by designing and wire-wrapping my own CPUs and architectures before there was such a thing as single-chip CPUs. That said, if you make a great product, and want to send me some samples and tools, who knows? You may wind up the as the star of my next book.

On My Soapbox

I'd like to say a few words about manufacturers that expect you to pay several thousands of dollars for the "privilege" of designing their CPU or FPGA into your products. Their tools are so overpriced that you'd have to make and sell millions of your product to get any payback from your investment on something that is very quickly obsolete. I'm not sure what their marketing departments are thinking, but if they price their tools out of the educational market, whose parts do they think new graduates will be designing into products when they get real jobs? Not theirs. They should understand that it's not just about the next fiscal quarter. I suspect the real problem is that they provide these tools free to the big guns and don't care about fleecing the small developers, some of whom will one day become big developers and suddenly be of interest to these same manufacturers. In the end, I won't design their tools into my products even when I can afford it because of their bad attitude.

One of my current personal favorites is the TMS320C3x series from Texas Instruments (TI). I'd call this a super-CISC CPU because rather than needing an instruction for each miniscule operation, it can do a multiply, and add, and a couple of complex pointer updates in one instruction that takes a single cycle ot execure. I've found that this CPU is comparable to the Intel Pentium in DSP performance even at one-fifth the clock rate, due in large part to having several separate integer and floating-point ALUs and the ability to touch up to five or six memory or register locations per cycle, rather than the usual general-purpose CPU architecture that takes many CPU cycles per memory access. This CPU is highly optimized for straight brute-force convolution (vector dot products) and has built-in circular addressing and bit-reversed addressing modes that make DSP buffering tasks and FFTs very easy to program.

Furthermore, it has a smart DMA controller that can respond to interrupts and save context switching overhead if only simple data motion is needed on external or internal interrupts. Because it has 28 or so GP registers, rather than the paltry few in the Pentium, it is very easy for the C compiler to do a good job of optimization; in fact, it often takes less than one line of assembly code to accomplish a line of C code. Because it has lots of registers, its context switching overhead is abominable if the new context needs them all, so you have to write code to avoid this. I use this processor in many embedded applications when I know I'll be writing new code or changing it again and again, since it's one of the easiest CPUs to program for other than opsys-type tasks. Fortunately, I have already written a nice operating system for it and don't have to do that again. Because of its strengths, most of the limitations and guidelines mentioned above for the Intel platform simply disappear; this is an easy machine to code DSP on. Compare the $9.95 unit cost of the 'C32 to the price of a Pentium or most other DSP CPUs and it starts to look really attractive for various embedded uses.

Another example of a popular DSP CPU family is one of the fixed-point CPUs (c1x, 2x, 5x; that's 16-bit integer to the rest of us) from TI. These find wide application in cellular telephony applications, modems, and disk drives, where the code isn't going to change often or at all. These CPUs have many of the same features as the 'C3x but are all 16-bit integer, usually with a 32-bit accumulator to preserve all the results of a multiply–accumulate in progress. As I cover later, integer can be used for DSP work just fine, but it's harder to code; however, in the applications where these find popular use, it isn't that much of an issue: you code once then build lots of product with cheap CPUs. In other words, this pays when the volume for a single product is very high.

I'm not going to cover the very popular PIC line of CPUs from Microchip here, even though I use them a lot for control purposes, as glue logic replacements, for hybrid designs, and things like that. They just aren't DSP material for any but the very lowest bandwidth requirements, even though they are absolutely great for what they are good at otherwise. I guess that's the reason there are probably more of these in existence than all other CPU families combined. This is certainly true in my house, where there's a PIC or two in just about every appliance and remote control unit I own, whereas there are only a few of the big Intel and TI chips around. The tools for these CPUs are cheap and worth getting. I sometimes use them alongside an embedded DSP CPU to handle the things they are good at and the DSP CPU is not so good at, such as serial port I/O, watchdog timing, reliable power-up reset, and things like that.

Although I haven't worked with them much, I understand that the newer RISC CPUs really scream for integer, more so than for floating-point, operations. The recent "shark" advertisements from Apple certainly bear this out, since they compare integer, not floating-point, performance with Intel chips. As usual, it's what they *don't* say that tells it all when the specsmanship game is played. Because integer DSP coding is both more difficult and generally takes more operations than floating-point code to handle things like over- and underflow and scaling, I'm not sure that a mere four times or so speed advantage actually translates to a real advantage for DSP uses. Otherwise, I suspect that most of the same issues apply since they all use the same external memory and disk storage components. I can't see jumping through the hoops necessary to use these parts in embedded systems, since there are better suited parts available for DSP. If you're going to invest so much time in development and money in parts, it doesn't pay to use a relatively obscure CPU. In saying that, I must also say that I'd never embed an Intel CPU either if I couldn't just use an industry commodity motherboard to hold it; otherwise, it's too much work, too high in cost and power consumption, and all the rest. The reason I'm focusing on Intel is simply that we programmers don't have to build them and sell them: they're already out there en masse in the customer's hands. In other words, they've already bought the hardware "razor," and now we can just sell them the software "blades."

At the truly high end of performance and cost, parallel processors are available from a number of sources. In essence, they contain more than one complete CPU and some special internal communication and synchronization features. These are being used mostly in situations where they can replace several more conventional DSPs, such as in banks of modems. Many of them are 16-bit integer, but a few are adding floating-point capabilities. When prices come down enough to make them practical, I expect to see them turn up on add-in coprocessing cards for the desktop PC platform, where they will make possible some very nifty real-time audio and video processing. Often, however, these architectures fail because they have horrible problems with context switching and interrupts, being essentially vector-only machines in the hands of most programmers. In effect, you almost need another CPU just to handle I/O tasks or you throw away 90 percent of the available performance. The tricks of programming these units maximally well are material for another book, not this one. Some processors are so complex that "no human can really program them in native assembly language" (this comment from an unnamed manufacturer of parallel processors). I guess I'm not human, but I must say, it wasn't much fun.

I understand that the Analog Devices series of integer 24-bit CPUs is finding wide favor in various special efx stomp boxes. Unfortunately, I've not worked with them for various reasons, including simple lack of time or need for that particular batch of capabilities. I can say that anything that makes it into the market has to have a very good price/performance trade-off. As of this writing, they are just starting to make a floating-point version of their CPU line that looks very interesting. However, their "take" on ASM substitutes an algebra-like statement syntax for the more usual mnemonics, which I don't like very much. It's a taste issue, to be sure, but I like to see the flow of code in the usual way. An interesting recent development is the on-the-fly reprogrammable FPGAs that are being used as special-purpose DSP computers. This makes a lot of sense in some cost-sensitive applications, since the very same hardware can be fully optimized for each task in turn. It's sort of like being able to design the hardware as well as program the software that runs on top of it and is a very promising technique, limited in generality as it is. Its main use is to get a little bit to act like a lot, with development cost considerations thrown to the wind. In other words, this trades generality and ease of coding for lower unit cost.

Common DSP Programming Tricks

This is the meat of the chapter where I cover or at least mention the most common and useful tricks to get the most out of a DSP platform. Most of these are pretty universal, working on everything from PDP-8s to Z80s to Pentiums to special-purpose DSPs. Before that, however, I'd like to discuss how to hold your mouth, mind, whatever right for this. In many cases, the approach is what leads naturally to the old and new tricks once you master the mentality and model.

Always think of the sample-by-sample data as a flow, not an amount. Even when you block it into chunks for processing, you should think in terms of flow. You want to pour data through a process as smoothly and with as little "friction" as you can. This flow will have some startup and cleanup costs, but the general principle is to minimize the by-sample or even by-block operations as far as possible. This leads to the idea that the sample-by-sample code is the master portion and the rest is there to support it. Even if it's a design–philosophical "jam-fit," you should always be looking for ways to push all work possible outward from by-sample time to user or set-up time. The by-sample code should not allocate or deallocate system resources or make by-sample calls to anything else whatsoever, and it should have every possible parameter computed for it before you get there. Avoid any conditional processing you can, especially if you can make the decision at block level or higher. If()s and the like can take more time than a second-order filter section on some platforms. Often you can find a more closed-form solution than an if() if you look for it. It takes a lot of cycles to equal what that construct costs in pipeline flushing in most modern machines.

Do all sample-by-sample code in a single loop if it's at all practical, to save looping overhead, keep the code localized, and make cache work better. Consider how long a time slice is for context switching in your system, since this will invalidate cache more or less completely each time, and try to phrase things so that this matters less, both because you'll be done with a block before the switching happens and because your design does not need the cache as much. I have actually seen programs speed up dramatically with the addition of Sleep() in the right place so that a full time slice is available once the loop starts. If this happens to you, you're too dependent on cache. I'm not sure who to thank for the "blessing" that CPUs are far faster than memory buses, but it sure makes a lot of good-paying work for those who understand it. That said, here are some of the popular tricks in no particular order.

Circular Addressing

Many times in DSP you will need a buffer of the past n samples or the past n, the current, and the future n samples. In this case, you can know the "future" by waiting for it and accepting a *pipeline filling* delay, after which you keep up in real time or better. If data is moving to and from the disk, there is no penalty, and even in real-time situations, the initial delay needed to get a bit of the "future" may not be noticeable if it's fairly short. If you program it directly as a shift register of delayed samples, as is often shown in the DSP flow diagrams, your code will be slow and inefficient, requiring a shift of all *n* samples for each new input sample. Most often, a circular buffer is implemented. In a circular buffer, you just put each new sample at the next address and wrap around to the beginning when you get to the top. To get a delayed sample, you subtract the delay amount from the pointer, and if this puts you before the beginning, you add the buffer size to the pointer after this to get back into the valid range. This costs a check and `if()` for each operation, but it is far faster than potentially thousands of data moves to implement a straight delay line as a shift register. You might write a C++ class to do this, but there's hidden overhead involved and you might be making sample-by-sample calls out of the loop. Shame on you! This is a case where the pipeline flushing caused by the `if()`s costs far less than a straight-line approach, illustrating that it's important to keep a sense of proportion. A true fanatic will realize that if the buffer size is rounded up to the next power of two, a faster logical AND can be used on the input index instead of the `if()`. Can you figure out a way to do the same for the delay index? Listing 8.1 is a sample solution (also see the delay-based efx examples).

Listing 8.1   Obtaining a delayed sample.

```
#define size max_delay_needed_in_samples
Float *buffer = new float[size]
int index = 0;
int delay_index;
float delayed_sample;
// assumes delay_amount is visible and set to something less than size
for (whatever)
{
 buffer[index++] = getinput(); // put sample in circ buffer
if (index==size) index = 0; // wrap around if above the top
delay_index = index - delay_amount; // point to delayed sample
if (delay_index < 0 ) delay_index += size; // wrap around if below bottom of buffer
delayed_sample = buffer[delay_index];
}
```

Because `size` and `delay_amount` may be huge, this beats shifting samples through a buffer hands down whenever a large delay is required, even with the conditionals needed to implement it. I'd say the cycle-cost curves cross at about a five- to 10-sample delay requirement, which should illustrate why many rules aren't (and can't be) hard and fast.

Use the advanced version of this trick by making the delay buffer an exact power of two in size. The buffer usually will be larger than you need, but you simply don't use all of it. You can then do a logical AND on the pointer (if you can also align the buffer to the same power of two location) or on an index variable to achieve the circularity property. This avoids if()s and is faster yet. It's my technique of choice, and some machines have this ability built in.

Tables

In the days before CPUs became much faster than external memory, the most commonly used single DSP trick was to put constant functions that were hard to compute, such as sines, cosines, logs, and so on, into tables instead of computing them on the fly. This involved decisions on the range and resolution required to meet accuracy requirements as well as code that took advantage of any symmetry in the original function or that interpolated between table entries. For instance, you can store one-quarter of a cycle of a sine or cosine function and, with a little symmetry math, have both sines and cosines for a full cycle, reducing the required table size for a given index resolution by a factor of four. For FIR impulse responses that are symmetric, this reduces the table size by a factor of two. It's a simple trick that can often bring large increments of speed to a solution. However, in these days where cache matters, tables are (and should be) falling somewhat into disfavor since table accesses may take longer than recomputing the function at hand when all things are considered.

After all, you have to have a pointer to the table base, add an index to it, then fetch the result. That's three operations at least if you don't check the index for a legal range or need symmetry code. When integer additions took less than $1/100$ of the time that an FPU transcendental operation took, it made more sense than it does now. On Intel platforms, where registers to hold pointers are in short supply and the FPU is faster than the integer ALU, this makes somewhat less sense if the function only takes one or a few FPU cycles. It's still a useful trick however, and I still use it a lot. It's just that now, rather than automatically using tables for constant functions, you actually have to think about it first and test it to be sure you don't slow things down by increasing the memory footprint, decreasing cache effectiveness, and so forth. This technique is still widely used in DSP processors since most don't have a built-in function for transcendental math but do have fast memory and few cache issues to sweat.

Work in Registers

Whenever possible, try to phrase problems so that they can be done with a minimum amount of memory fetches. You don't want to load two numbers, add them, store the result, then get another number, reload the original sum, add that, then store the result, and so on just for a simple running sum. In practice, you'd keep that sum in a register and not unload it until you were done. You will be surprised how often C/C++ compilers do exactly the wrong thing in this case, because the sum is usually a named variable, and the compilers aren't smart enough to realize the sum won't be needed until after the end of the loop. For example, the lines

```
tmp += a * b;
tmp += c * d;
```

usually run far slower than the line

```
tmp += a * b + c * d;
```

because the compiler "knows" it won't need the temporary variable between the operations, so it doesn't store it in memory until the operation's done. I see this one even in the latest revision of DevStudio.

A Mandelbrot calculation is the ideal easy DSP case, since there is no need for data memory fetching and storing until thousands of computations have been done on the same set of a few numbers. Most real problems are unfortunately in-between — either they don't fit or they don't need enough computation on each piece of data to take good advantage of the difference between CPU and memory speeds. In the "doesn't fit" case, you can look for many optimizations. For instance, can you rephrase the problem to take two passes over the data? You'd be surprised how often this is faster than the HLL compiler left to its own devices. After all, it involves touching each piece of data (preferably in order) only twice, and again, if you look at the compiler-generated code for the straight "doesn't fit" case, you'll often find that each piece of data is loaded and unloaded far more often than twice, often seven or eight times. Also, sometimes you can make the data that's touched over and over a single piece of data, like an FFT twiddle factor, so that it's just about guaranteed to be in L1 cache all the time. I used this technique in my FFT speedup, for instance, and it worked rather well there.

In the case where each piece of data doesn't need enough computation to take advantage of the CPU resources, you have the chance to either add features that don't cost extra compute time or consider how to combine more than one step of what you're doing into a single process. The goal is always to have as many registers in use as you can and to minimize the memory bus traffic. My general rule for the Intel platform is to use the normal CPU registers for pointers, their offsets or constant adders, and loop counters and the FPU for the signal math. Usually it's easy to use all the registers and be a couple short, but in cases where that's not the problem, you can consider combining a simple filter with a conversion to short integer and clipping if it would have been the next step in your processing anyway. Actually, clip/overflow testing is something the Intel FPU is rather poor at, so for that case I've written some assembly routines that use both portions of the CPU for what they do best.

Work in Reasonably Sized Blocks

Most CPUs have a significant amount of call overhead. Parameters have to be pushed, the control flow change causes a pipeline flush (five or six cycles in the Pentium), the called routine has to save and restore registers, someone has to clean up the stack (who, depends on the calling convention, but this doesn't matter much to the speed issue), and the control flow breaks again on the return. In other words, the time it takes to make one call to a null routine could have been used to do quite a lot of useful work instead. This leads me to want to write all DSP routines to work on big blocks of data in order to amortize this overhead. In so far as it's possible, it's a good idea, but you also don't want to go wild here. If a block is bigger than the cache size or takes long enough to process that a context switch becomes likely during the process, you quickly run into diminishing returns.

The big gotcha that most programmers forget is that many high-level language constructs actually call back into the run-time system, and if you're not careful, you can accidentally create many of these calls for every sample you process. It doesn't matter how efficient the run-time routine is here, it will cost you 20 to 50 cycles per call if it does nothing at all! Most HLLs just aren't designed to do block operations, so it's not really an implementation inefficiency per se. Although the call and loop overheads are large, I've found a useful compromise that, whereas neither overhead is minimized perfectly, makes an implementation more efficient and allows for a lot more code reuse.

I've written a number of block-oriented assembly routines that can be called in succession to do most of the common jobs. Because I'm usually working with block sizes in the 4K sample range, the call overhead is still pretty small (a few per block), and the extra looping overhead (each routine has its own loop) is not too bad either. The routines are structured so that they do as much work as possible while remaining reusable in order to amortize their own loop overhead. I find this approach good enough most of the time, but there are always exceptions, and true fanatics will eschew even this. It works for me because I can then do most of my programming in the HLL and call these atomics as needed when speed counts. I've included them in the code that comes with the book. Feel free to use and extend them as needed. If you come up with a really good idea or extension, please return the favor! You'll note that in most of these samples I used the MASM .REPEAT construct, which is a couple of cycles slower than just decrementing ECX and doing a conditional jump. I just didn't have time to change and then retest them all after I found the faster way, but you can. Since the status codes of the integer CPU aren't affected by most FPU instructions, you can interleave their operations and often hide them completely as far as cycle count goes.

Avoid Convolutions

Although very powerful, convolution is a brute-force technique that is often substituted for a true understanding of a problem that would suggest speedier solutions. I don't care how efficient your convolver code is (by the way, many statements made about the speed of FFT-based convolution are just plain wrong). Also, it takes more samples to do a FIR filter, for instance, than it takes to do the equivalent IIR filter, often hundreds to one in both computation and memory resources. As usual, discovering the better algorithm wins much more often than simply tightening the code of a poor one. There are only a few cases where a straight convolution is really needed, and most of those, such as reverb, that you think need it, actually don't if you're clever. When I see convolution being used, I always try to find a better way to solve the problem, and I almost always do. I consider its overuse the mark of an amateur. There are only a few places you really need this technique. Be smart at compile time!

Avoid Divides

You can almost always avoid sample-by-sample divides in DSP by creating a number that's 1/scale and using it in a sample-by-sample multiply instead. Division has numerous problems other than about 20 times less speed than a multiply, such as the possibility of unbounded output, exception generation, and so on. Just don't do it if you can avoid it. The only case I know of where it's really necessary is in deconvolutions, but even there you can divide just once and use multiplies thereafter if what you're deconvolving is time invariant, as it usually is. One of the big secrets of the DSP masters is that they know how to avoid division in various situations. That goes for square roots, exponentiation, and the like, too.

What Language?

Having read this far, you're probably aware that I'm a partisan of C/C++. I got to this point kicking and screaming, but I'm glad I did, and many thanks to the customers who forced me into it and out of pure assembly language programming with a side order of Pascal or BASIC. No other language has the power to do so much with so few characters, the power to easily "drill down" to a very detailed level when required, and the ease of interfacing to native assembly language. In addition, outside of Microsoft documentation at least (doesn't it upset you when you look up something in DevStudio and only find a Visual Basic example that doesn't address your question at all?), there are more DSP examples available for C than for any other language, FORTRAN included, and C is supported on more platforms than any other language. I suppose you could make a good argument for Pascal, but I doubt BASIC, Java, or a HLL other than C/C++ will find significant usage in serious number crunching anytime in the foreseeable future. Someone might invent something better for this than C someday, but I'm not holding my breath, and I would probably have to be dragged kicking and screaming into that as well. I program the UI mostly in C++ because of its cleanness and power, and I use C and the assembly language native to the platform I'm working on for DSP, using each language for its strengths just as you program into the strengths of a given hardware platform. It just makes sense to me to work that way. Therefore, my answer to the question of what language to use in DSP programming is the usual hedge — it depends. No hard and fast rule can ever substitute for good judgement.

In truth, you could use any language if you were willing to work around its quirks. Having learned the quirks of C, they don't bother me too much any more, especially compared to the quirks of various native assembly languages, wherein one machine puts the source first and another the destination first, some things work only on some registers or directions (nonorthogonality), each machine has a different mnemonic for "load" or "jump," and so on. Now that's annoying.

Integer versus Floating-Point Math

The choice of integer versus floating-point math is more often than not forced on you. As I note above, the choice is often tilted toward floating-point math when working on platforms that support it. It saves so much coding trouble that it's generally worth it, at least for concept testing if not for final production. With integer math, you have to deal with numerous scaling issues that take cycles to accomplish, slowing things down even when the basic operations are faster. Integers are often phrased as fractional fixed-point numbers, in which you simply decide that the maximum possible integer represents 1.0 (or actually, almost 1.0 as the case of 32,767/32,768) and smaller numbers are corresponding fractions. This is fairly easy to implement on most integer machines since the major code adaptation is to left shift after a multiply to strip off the extra sign bit that a signed multiply generates and to keep the top word-sized bits of the double word result. Adds and subtracts work the same as ever; however, it then becomes difficult to represent numbers such as 1.414, for instance, which is a pretty common number in signal processing, being the damping factor for maximally flat filters. I've had to resort to dodges such as adding $0.707 \times somevariable$ twice to overcome this problem, which again slows the integer version of the code down compared to a cleaner floating-point version. You can decide that the binary point is anywhere in the word to avoid this, but then you have to track it throughout your code, making it more complex and tougher to maintain. Yet, it's inescapable that integer CPUs will always cost less than floating-point CPUs, so the issue remains live and probably will forever in embedded systems that are sensitive to the price/performance trade-off by nature. I don't want to get bogged down with a long treatise on how to do fixed-point math. If you have to do it, you can find plenty of good expositions of the techniques in the CPU manuals for the CPU you're working with. In general, the manufacturers "feel your pain" and want to show you the most efficient tricks for their products so you will buy more of them. A good place to find this documentation is in their applications manuals and on their Web pages, and you can usually get it free. That said, I'll move on to the special tricks that each numeric format allows or encourages.

Stupid Integer Tricks

There are some cool things about integers that make them truly useful, even in mostly floating-point implementations. For example, integers work best anytime you need to count or index something because you can avoid round-off errors. I always wondered as a kid why a statement like

```
for (somefloat = 0; somefloat < 100; somefloat += .01) cout << somefloat;
```

didn't give me exactly 10,000 steps: The numbers started accumulating errors from the less significant digits. Well, floats just do that — there may be no precise binary representation of 0.01 — integers don't, so sometimes a mixed implementation is best after all.

Another nifty trick is to multiply or divide an integer by any power of two with a simple left or right shift, which is slightly trickier (but still not hard) with floats. By the way, this is faster in Pentiums than using the multiply instruction for integers. I've cleaned up and sped up a lot of code merely by finding things like buffer[i*2] and changing them to buffer[i<<1]. My favorite integer tricks, though, involve various logical or boolean operations. For instance, getting the result of a modulo operation when the modulus is a power of two is a simple and fast boolean AND operation. This can be especially handy to maintain a circular buffer index. In another trick, you can use an integer's lower bits to hold various flags or booleans then implement a very fast switch statement by using the result as an index into an indirect jump table. Modern HLL DSP compilers are starting to do a variation of this as well. The upshot is, integers are quite useful for many things, not as much for the signal in question as for various logical and control operations you always need while processing the signal using floating-point math.

Floating-Point Tricks

There are some special tricks with floating-point numbers as well. As mentioned above, you can often avoid sample-by-sample divides by generating the reciprocal of the denominator number once and using the result as a multiplier for a whole block of samples. This works in floating-point arithmetic because you can represent numbers greater than or less than one with equal facility, which isn't true with integers unless you put the assumed binary point in an odd place that makes other programming more difficult. Floating-point numbers can be scaled by manipulating the exponent directly with the integer CPU. In the case of Intel or IEEE format, you easily can compare the absolute magnitude to some limit, something the FPU doesn't do well or quickly, which is also why those casts need to call back into the runtime system. Table 8.2 shows the IEEE format for 32-bit floating-point numbers.

Table 8.2   IEEE 32-bit floating-point number format.

Bits	1	8	23
Content	sign	exponent	mantissa, top 1 bit implied

The most significant bit (msb) of the 32 bits is the sign of a floating-point number. This is followed by the exponent in offset-127 coding, which means that an exponent of 0 really means 2^{-127}, and the mantissa follows in positive (straight magnitude) format with an implied msb of 1. In other words, you have a 24-bit mantissa, but since the top bit of the mantissa is always 1 for a normalized number, you don't need to store or show it.

An implication of this is that if you AND off the sign bit using the integer CPU, it can compare much faster than can the FPU the result with a constant, also in the integer CPU, to determine whether the number fits in a short integer. It works because the bits are in order of significance in the format, even though the exponent bits don't mean the same thing as the mantissa bits do. I use this trick when block-converting floats to shorts (the most common 16-bit audio format) in assembly, where it is far faster than the run-time system, primarily because I only have to make one call to convert an entire buffer of data. A variation of this trick is many times faster than using fabs(). In C, you have to fake out strong typing to do this, but that is a small price to pay. Simply make a union of a float and a DWORD, put your floating-point data in it, then AND off the sign bit in the DWORD. Next time you access the floating-point number, it will be positive and of the same magnitude as it was before.

Although the FPU has a scaling operation, there are at least two faster ways to scale floating-point numbers (floats) by powers of two. The most general way is to multiply them by the power of two, which is faster than the FPU scale instruction, especially when working in blocks of floats. For this case, you only have to load 2.0**n once, and you needn't be limited to powers of two or even integers when using the multiply. The other way is to use the integer CPU to directly manipulate the exponent bits in the float.

Because floats don't care much about absolute magnitude internally, nothing special needs to be done in the code as a filter coefficient moves from below 1.0 to above 1.0. Floats have considerably more dynamic range than integers do, so most arithmetic noise or dynamic range problems don't come into play as quickly in floating-point math. There are still limits, however, so don't allow yourself to be lulled into the idea that because something is represented in floating-point format, there will never be a precision problem. The popular "canonical" IIR filter structure, for instance, has trouble with low frequencies and high Q's even when using `floats`, and sometimes, even extended precision isn't enough to guarantee an accurate filter response with that particular construct. In some cases, you still need to avoid some operations, such as adding tiny numbers to huge numbers, if you want accurate results. Even the sweep tone generator in WavEd needed to use `doubles` for a couple of variables, since at low frequencies, the phase adder becomes very small and is added to a relatively large number. When this happens, the numbers need to be denormalized for the add to take place, and if the relative magnitudes won't both fit in the mantissa bits, the result is such that the add has no effect.

As of now, the default math libraries with DevStudio set the FPU to use 53-bit precision internally, which is a good bit faster than full 80-bit internal precision. There are some interesting implications of this. One is that for some operations, you can have better than floating-point internal accuracy as long as you keep partial results inside the FPU. The results aren't truncated or rounded off until you store the number into memory. Another is that the FPU has more than one operational mode, and you can change it if you like. I find the default 53-bit mode just fine for almost everything, and it does have a significant speed advantage over full-precision mode. However, if you like to twiddle things like this, you can. Look up `_control87`, `_controlfp` in DevStudio Help for information on how to do this in C. You can change internal accuracy, rounding method, and whether exceptions are masked or not, among other things. Note that if you change anything for a routine, you should change it back or risk disturbing other system functions and routines that expect it to be set at the defaults.

An Important Caveat

Although the Intel FPU has eight internal data registers, which can be used either as a stack or directly indexed, it's not a real good idea to use them all, especially in in-line assembly. The run-time system and compiler seem to assume there will always be a couple of registers free to be used in stack mode, and your data may get pushed off the FPU stack if an interrupt occurs. Actually, this is such a snaky problem that I'm not sure anybody really knows what's going on. Microsoft claims to have fixed this problem, but I'm not so sure. I do seem to be getting away with using all the registers in pure assembly functions, but I've had some trouble with in-line assembly and even with pure C++ when in-lining a lot of class functions in code that also uses the FPU. If your function suddenly starts acting wacky, you may save a lot of time by looking here for the problem. Check and see if the bottom numbers on the stack are causing your troubles. It's a dead giveaway.

Intel Assembly Tricks

I'm not going into Intel assembly tricks in great depth because Intel makes so many slightly different flavors of the Pentium; a lot of tricks will only work correctly on one flavor or another. Most of us, however, have to write code that will work everywhere, even on non-Intel CPUs. Fewer tricks work on all PC's and they are pretty universal in nature, in that the same guiding principles usually work on any modern CPU. Mostly I'm going to tell you what to avoid, rather than what to do.

Avoid the .REPEAT–UNTIL **construct** in MASM, which uses the loop instruction and is far slower than the following.

```
dec ecx
jnz loop
```

Note that because the FPU doesn't affect the flags register unless you ask it to, which by the way is a slow operation, you can intersperse these instructions with FPU instructions.

Don't bother with the USES **directive** if you're going to use a pushad/popad to save and restore registers; it will generate useless extra pushes and pops. If you're following my advice of doing things in blocks and using the registers as much as possible, you'll be using the bulk pushad to save all registers, and popad to restore them.

Pass pointers in and out of assembly language routines. This way, no matter how large the data really is, you're only passing a few DWORDs. Usually, I work on data in-place or move it from one buffer to another, which only takes a pair of pointers, a DWOR for the amount, and perhaps another DWORD or float for parameter data.

Of course, the previous two recommendations only affect the calling overhead, not the execution speed of the inner routine. If you're following my advice to work in blocks, this is not as important as it otherwise would be, but efficiency is efficiency: any is good; more is better.

Unroll loops when you can but don't take it to extremes. I usually find that working in chunks of about four works well because I don't run out of FPU registers but do get rid of three-quarters of the looping overhead. It's also easy to divide a loop count by four by shifting it right a couple of times but keeping the remainder so you can perform any odd operations after unrolling the loop. If you try to unroll a loop into 16s, then eights, then fours, then the rest, it is easy to waste time checking at each stage, especially if someone calls your fancy routine with only a small amount of work to do. In some situations, I provide myself with both unrolled and straight loop versions to handle both cases well. For instance, I use the unrolled dot product routines for things like FIR filters or autocorrelations, but I also do some neural net work, which needs a lot of dot products that tend to be short in length; therefore, I have one version optimized to do long ones and another optimized for low call overhead.

Schedule one FPU instruction per cycle. FPU instructions often take more than one cycle to finish but often can be scheduled one per cycle (this is what makes pipelining cool) if you don't depend on the results of the previous instruction. There are several tricks to accomplish better performance around this idea, and a good example of this is on the CD in `mathelp.asm` under `adotu()`. Multiplies and adds take three cycles apiece, for instance. In `adotu()`, I've unrolled a loop to do chunks of four multiply–accumulates at a time. The first three multiplies are scheduled by using the `fxch` instruction before adding anything. This often takes no cycles because simple instructions can pair in the Pentium instruction decoder. It also makes programming the FPU a lot easier. This means that the add doesn't try to take place until its operands are ready. Another trick is to intersperse integer pointer updates and loop control instructions with FPU instructions so that work is being done all the time, even when the FPU would have stalled. I found the best arrangement by testing, testing, testing. The *Intel Architecture Optimizations* manual (updated frequently but irregularly) provides some clues on how to do this, but the fastest way on my machine was a little different from what this source suggested, so the moral of the story is *test*. By unrolling the loop to amortize the loop control instructions over more than one multiply–accumulate and interleaving instructions more or less optimally, I've made this approach the theoretical maximum speed for multiply–accumulates on my machine: one per three cycles. My code gets about one per 3.2 cycles, which is pretty darn good (about 110MHz) on a P-II 400.

Interleave integer and FPU instructions. Some of the same issues apply to the integer ALU. There is a penalty for writing to a register (or adding something to it) and then using it on the next instruction. In general, it's a good idea to avoid this when possible by interleaving integer and FPU instructions or just paying attention to what's going on. Often you can avoid the problem by, for example, bumping a pointer at any time after the last use and before the next use.

Use simpler and shorter instructions as much as possible to make your loop tighter and increase the chances that instructions can be scheduled in parallel. Although the Intel CPU doesn't have any sort of autoincrement for registers (a big lack in my opinion), it does have the ability to add some sort of scaled offset with no penalty other than the number of bytes in the instruction. I often find this is the fastest way when unrolling some loops, although it can be even faster to keep an increment in another register and add it to the pointer each round. It is a very short and fast instruction and can often be paired with others.

Group your stores away from your loads. A store to a cache miss on MMX and P-II CPUs requires that the entire cache be fetched before proceeding. It's not exactly the same kind of write-back cache that some of the other CPU flavors have, so if you can do a bunch of stores to the same line together, you only pay the penalty once. Obviously, if cache is tied up doing this, your loads will have to wait. Again, you can usually plan in such a way that these events are separated somewhat. Remember that the top of a loop follows the bottom, so loading at the top and storing at the bottom puts them right together.

Align doubles **correctly.** Many compilers don't. If you must work with doubles, you can use the trick of allocating somewhat more memory than you really need then ANDing off the bottom bits of a pointer to it to ensure alignment. You still have to keep the original pointer around to pass back to delete() or free(), remember. Misaligned fetches take tons and tons of extra cycles during which no other work is done.

Use the integer ALU when it's faster than the FPU. You can do things like absolute value of floating-point numbers far quicker with the integer ALU than any other way. It takes only a single AND instruction to take the sign bit off. You can scale floating-point numbers by messing with the exponent bits as well, although you have to make sure carries don't propagate into the sign bit, for example.

Caching Issues

I've heard some awfully silly talk about cache. It *is* a tricky subject, especially on the Intel platform, because it doesn't load a whole line of data at a time as the better designs do (the main memory data path width is less than a cache line width). Intel CPUs "burst," which basically means that if you need just 1 byte, you still have to exercise main memory for a full cache line's burst and wait for that to finish before accessing any other main memory.

Whether you can access any other cache during the wait depends on what flavor of Intel CPU you have. I've seen claims in supposedly serious magazine articles, and not in the April fools issue either, that prefilling ("warming") cache is faster than using data as you need it. What in the world could they have been thinking? How can grabbing data you don't need yet, and that may be wiped out of cache again by a context switch, ever be faster than just using it as needed, allowing line fills to take place while doing other useful work? Why is it not always faster to do at least some computation while waiting to fill a cache line with the next main memory fetch, rather than waiting for it to happen in the warming method? More likely than not, the difference in timings the authors of these articles think they measured was due to an interrupt or context switch that occurred in one method and not the other. True benchmarking is tough. Trust me, if it looks faster, someone's not timing the entire process accurately, is using some assumptions that don't hold in real life with big streams of data and real operating systems, or does not have well-written code.

According to the Intel, caching is one feature that differs substantially from Pentium to Pentium, so it's probably not worth it to program in a way that only speeds up one type of chip. If you can use it, Intel provides a performance library that checks which CPU type you have at runtime and loads the appropriate optimized DLL for each possible case.

Fortunately, in DSP you usually work with floating-point numbers in order, and they are less than a cache line or even a main memory data path width in size. This guarantees that at least every other line will be in cache by the time you need it, and it's usually possible to do much better than that. When you can, do things in place to avoid making extra copies, which also means that you will take best advantage of write-back cache, since you only invalidate data that's no longer needed. Of course, the real trouble with any cache scheme is that it's a Band-Aid on a basically poor design to hide the fact that main memory isn't as fast or wide (same thing in another guise) as it ought to be to keep up with the CPU and peripheral access needs. However, you're stuck with the platform that your potential customers already own, so there's nothing you can do about it.

Some techniques are obvious. One is that whenever possible, you should use data in ascending or descending order, without skipping, and work in-place. On Pentium II's, a write miss actually causes a burst fetch of the entire line before the processor can proceed. By doing things in order, at least the time-consuming cache burst line-fill won't usually go to waste, which is why using tables (for random table accesses, at least) is no longer as advantageous as it used to be. Think about the cache size as well so that you don't a priori exceed it with data block lengths that cannot possibly fit. This is especially true if you are going to make more than one pass over the data because that will always force misses. I am talking about L1 cache, of course. In most current designs, L2 cache is, at best, only a little faster than main memory and is mainly used either as a specsmanship ploy or to reduce main memory accesses by the CPU to leave more memory bandwidth available for peripheral bus-master accesses. It's not benefit-free, it just doesn't perform as advertised or hoped. Lewis C. Eggebrect (*Interfacing to the IBM PC*, Sams, ISBN 0-672-22722-3) mentioned that this was one of the early fights in the original IBM PC design. The incremental-cost-conscious bureaucrats wanted to make sure the cost of adding to its too-small initial memory would be minimal for each increment, or more truthfully, that the initial cost to them could be completely minimized even though end-user price/performance suffered. Well, it didn't save anyone money, and that design attitude is still costing everyone who uses the platform extra money for a given performance level. A narrow main memory path is only cheaper for the guy fleecing you on a new system, not for you or anybody else. It's a rip-off. Nothing that I can think of has ever been faster and cheaper than wires and wider paths.

The best and safest path is to program as though cache won't help you. If it indeed manages to do some good, it'll just be gravy. This is why I preach my religion of keeping things in registers, touching data only once if possible, and so forth. Cache is by nature too small and volatile (otherwise it'd be main memory), so its effect is unreliable at best, and you shouldn't count on it. In effect, all the tricks that work without cache simply work a little better with it. It's not very productive to spend a lot of time optimizing for something like this, other than to pay attention to what might or might not fit or what would make cache invalid.

MMX

You may have noticed that I've been pretty silent about MMX and related technologies so far. Well, it's on purpose, and not necessarily a cop-out. First, many of your customers just don't have it yet, as I write this, so writing code into the strengths of things like this might be pointless from the customer's perspective. Also, there are many different flavors of proprietary speedup technologies these days, and currently, there is no way for a single set of code to address them all without some gyrations that should be the subject of at least one additional book. What's a developer to do when faced with this mix of not-quite-identical platforms that all have utterly different requirements for optimization? My approach is to find ways to write code that performs well on all of them, rather than writing different code for each and trying to find an efficient runtime way to select which code is actually used. Even within Intel's line of CPUs, this is not all that easy to do, but it is possible.

In Intel's case, they've provided what they call a "performance library" for DSP that, in essence, consists of a separate DLL for each CPU flavor they make, along with some other code that decides at runtime which DLL to actually load. That approach has its own weaknesses, since doing the DLL entry point fixups at runtime means extra calling overhead into each one of these supposedly optimized functions — somewhat defeating the original purpose of optimized code. And, of course, this performance library does not handle AMD or Cyrix CPU's at all, each of which has their own proprietary speedup hardware. Recent developments are promising, but that's about the best that can be said at present. The developer will either have to write separately optimized routines for each technology and solve the problems of selecting which to use at run or install time, or punt, as I have been doing mostly. I prefer the install time approach, myself.

The MMX extensions are not very useful for quality audio processing these days. There was a time when being able to go much faster with integer code was attractive. However, users are now used to having the extra quality of floating point computation, and are unlikely to return to noisy 16-bit processing by choice. I'm sure MMX is great for video uses, but that's not my subject here. It would, of course, be nice if some of the C/C++ compilers in common use would support MMX, but DevStudio does not, and it has nearly the entire market.

Other MMX-like extensions include various flavors of SIMD, or "Single Instruction Multiple Data," produced by Intel and AMD. These new instructions and hardware operate on more than one floating-point value at a time, similar to what MMX does for 16-bit integers. Theoretically, these extensions could double or quadruple the speed of various calculations, that is, if main memory could keep up. The big problem is that they aren't the same on the various CPUs, and not all customer CPUs have them at all.

For most cases, I'm just waiting for something to be standardized. Very rarely, when doing some research-grade code for myself, I will use the extensions, and since I already know what machine I have, I don't have to do anything special. You're probably in a different boat, however. If you want to use this sort of thing, here's a plan that should work for you.

Make as many versions of your release project as there are different types of CPU extensions you want to support. Make your installer code smart enough to install the correct one, with the extension support code already statically linked into it. This will avoid the need for extra runtime entry point fixups, and for reliably detecting the CPU flavor at every run; presumably, your installer can simply ask the user what they have if you don't want to try to write robust CPU detection code. It will also help keep people from making illegitimate copies of your product, because it will usually fail to run at all on a different machine than it was installed on. You should be able to do this without too much effort, because by nature, the streaming extensions work roughly the same on all CPUs; they are all trying to solve approximately the same problem. They only seem to differ in the small details, which nonetheless must be correct or you have broken code on your hands.

Chapter 9

User Interface

I have to start this chapter out by recommending the book *Front Panel* (Niall D. Murphy, R&D Books, Lawrence, Kansas, 1998). In this book, he covers the subject of user interface (UI) a lot more thoroughly than I can in a short chapter. He is an expert in this field; I am not, so I'll keep it short and sweet and only bring up a few topics that are related to DSP or signal editing. I divide my discussion into two parts — one on DSP and Windows and the other on philosophical matters.

DSP User Interface

I did *not* want this to turn into a book about Windows programming. However, some mention of the topic is unavoidable, from both programming and user points of view. After all, you have to fit an application into the platform it will run on such that common user expectations about how things work will be fulfilled.

It turns out that mentality behind the "form" or text-based dialog leads to problems when designing a nice user interface for complex DSP transforms, which have different types of debugging problems than most code and require different sorts of user input than most other applications. At first, I tried to avoid "wasting time" on this sort of thing, but it became clear early in the game that not only would developing nice user interface tools not be a waste of time, it would actually save time and result in a better set of examples. These examples would be better because they'd be easier to set up and experiment with and thus better to learn from and use. Although I also mention some of these reusable classes in Section 2 (Chapters 10 through 16), I thought it would be nice to describe them in one place here.

DrawDat to Guin to MCGuin

You almost always need to graph when you play with DSP data streams. This is handy for both normal use (say, showing a frequency response from a filter) and debugging (say, showing an internal transform variable versus time during a debug run). A lot of times you'd like to have more than one plot. You'd also like it to be quick, and you want it to update automatically as the user changes various controls. This makes transforms better learning tools because you can twiddle the parameters and see the effects right away. Often, you'd like to be able to specify a parameter versus time as an input issue — not an option with edit boxes! Of course, any code that provides these features should be reusable and easy to drop into a new transform. My solution to the first part is a class called CDrawDat. I derived it from CButton, so you can use the visual dialog editor to specify its location, rather than having to create it dynamically.

To use a CDrawDat, just put a button on your dialog, size it the way you want, attach a CButton class member, then change the declaration in the header to CDrawDat. You now have a multichannel oscilloscope that is easy to work with. It can show up to four traces at a time in different colors. You set up scaling and whether it draws a frequency scale or not with a call to its Init() at OnInitDialog() time. To draw a trace, call DrawResp() after preparing a buffer of floats that has length specified in the m_cHoriz member of the class. It's up to you to provide the buffer and take care of getting rid of it after drawing, which takes place during the call to Draw-Resp(), so you can delete[] it when the call returns. Although you don't have to use it this way, DrawDat also contains an array pointed to by m_pFreqs with an entry for each horizontal pixel, which is why it wants to know the sample rate at Init() time. I created this part of the software while devising the filter designers so that it would be easy to have an exponential frequency scale and assign responses versus frequency without computing it over and over. The array pointed to by m_pFreqs contains frequency values in Hz that are exponentially spaced, starting at 20 Hz and extending to the Nyquist frequency. Thus, if you want to plot something versus frequency, you can use this array to find the frequency represented by each horizontal pixel location on the graph.

This is pretty nice by itself, but what if you want graphical *input* from the user so they can specify input versus time, space, or whatever you want to label your horizontal axis?

To tackle that one, Troy derived a class from CDrawDat called CGuin, for "graphical user input." This class has all the DrawDat properties but adds up to another four user-drawable traces. It's basically a simplification of an OLE custom control (OCX) that my company sells. Its Init() is pretty similar but now includes a mask input for setting up which user-drawn traces will be visible and allows their scales to be set separately from the display-only traces. It creates arrays of user points, which are updated as the user manipulates the curves, and sends messages back to the owner window every time a change is made so the owner window can update whatever is desired dynamically.

The UserTraces class member is a doubly subscripted array accessed by

```
Guin.m_UserTraces[whichtrace][whichpoint]
```

where whichtrace is a number from 0 to 3, corresponding to the traces in use, and whichpoint is a number between 0 and m_cHoriz - 1. In other words, there is one array of user points per trace, and each array has one entry for each horizontal pixel in the drawing. These are floating-point numbers that are smoothly interpolated between the points the user inserts on the traces by clicking on them. This is nicer than a simple plot, but what if you want to do something by-channel in the input file or each channel has more than one parameter to set or both? Although you can have up to four user-drawn traces, and four computer-drawn traces on a Cguin, the screen quickly becomes cluttered and hard to interpret and work with if you use them all.

Thus was born CMCGuin, a multichannel version of CGuin. It creates up to four Guins. You still have your mask word, but now each nybble has to have something in it to specify what traces show up in what channels. MCGuin will create the proper number of CGuins tiled horizontally in the original button space. It gets a little complex at this point to access the user points, because you now have to use

```
CMCGuin.m_Guins[whichguin].m_UserTraces[whichtrace][whichpoint]
```

to get to one, but those are the breaks.

These all use the XOR mode for drawing to avoid flicker when you're redrawing a lot. It's a lot faster to redraw 100 pixels than 200 by 400 pixels. The main negative ramification is that when drawing a lot of near-vertical lines, the part of the line going down might wipe out the part going up, leaving a blank spot with a dot at the tip. This is because of the way Windows line drawing works, and without halving the possible horizontal resolution, there appears to be no cure for it. If you erase a curve by redrawing with a black pen, you might put black spots in other curves (which you could avoid by erasing all curves first) or on the scale, which can then only be redrawn in toto to repair the damage. If you use this method to draw amplitude envelopes to show the user relative timing, for example, you should use a little extra low-pass filter to minimize the problem.

I like these tools, and thanks to Troy Berg for helping to write them. They allow you to do more interesting and complex transforms than would otherwise be practical, and they give the user a lot of nice detailed control over things that would not otherwise be possible. If you do much practical coding, I'm sure you'll appreciate the debugging tool potential of these as well.

CEditSlider

A lot of times you'd like to have both coarse and fine control. Coarse control control (such as with a slider) gets you close-enough most of the time and gets you there quickly, but you'd still like to be able to type in exact values once in awhile. You'd also like to protect the underlying process from bogus inputs without flooding the screen with ugly message boxes as the user updates an edit box keystroke by keystroke. To solve this sort of problem, we developed CEditSlider. I say "we" because it started out a long time ago as a component in MusiCad, written by Alan Robinson for me. I then rewrote it to add key-by-key validation and some other nice features. Again, I wanted it to be easy to use, so I made it so that you can either dynamically create these controls or give them the ID of a group box and let the dialog editor take care of placement and the like. Also, many controls in DSP require an exponential control range so that octaves are equally spaced, for example, and of course, you'd like to be able to set the range you want.

CEditSlider does all this. You just place them where you want them at the size you want them by putting group boxes on your dialog template, then you call Init() in OnInitDialog() in your dialog, passing in the groupbox ID you want the control to overlay. There is also SetScales(), with which you can set up ranges and offsets, specify whether the control is exponentially scaled, and indicate which way the maximum is. The group boxes determine whether the controls will be horizontal or vertical based on which way the rectangle is larger. They don't hide the group box static text label, so you can still use that to label your control. EditSliders also send a message to the parent to let it know when the underlying data has been changed, allowing dynamic redrawing or realtime changes during previews.

There are plenty of examples of using these controls in the WavED application, so you should go look there for more details. They are not hard to use and have one feature I really like. When the user types in the edit box, validation is done key by key. There's no way to know when a user is done typing, so this has to done on the fly. If an entry results in a non-number or an out of range number, you don't want to pass it to the underlying process, which might blow up as a result, so this is prevented. However to provide good UI design, you have to let the user know that what is displayed in the edit box is not what the underlying process sees. The solution to this was to change the colors used in the edit box dynamically to obvious error colors, which lets you know right away what the status of the data is without being annoying or corrupting the next entry. The instant you return to the invalid data and enter a number that is in range, the edit box color reverts to normal, and both the slider and underlying process are updated with the new, valid data.

EditFloat()

I wanted a subset of EditSlider-type functionality available for controls that you don't exactly own or create dynamically or have to initialize in `OnInit-Dialog()`. Thus, I wrote the global function `EditFloat()`, which can be called from inside `UpdateData()` instead of using the default MFC DDX/DDV calls for edit boxes. You can add a message handler for `EN_CHANGE` for the edit box in question and have this call `UpDateData()` to get real-time updates, but you can't guarantee messages will always be put through when needed, so you dare not mess with things like edit box colors. Instead, the class puts the text OOR (out of range) or NAN (non-number) after the user input to indicate bogus input. These text tags disappear when the user enters something valid. A separate minor advantage of this technique is used in cases where the valid number range might be dynamic. The classwizard gets unhappy when you put member variables and such into DDV statements as ranges, but `EditFloat()` doesn't have this limitation and neither does EditSlider, although if you want the EditSlider range to change, you have to explicitly call it each time and tell it so.

These routines represent a peculiar set of trade-offs that match my tastes and usages, but I hope you'll find them as useful as I do. Indeed, they represent some of the more important pieces of reusable code this project produced. I already had most of the DSP code lying around from earlier projects, but for this book I wanted to avoid things like ActiveX and COM, install routines, DLL dependencies, and so on, so I wrote these routines. They turned out to be very handy, and I'll be reusing them a lot.

Philosophy of User Interface

When we first built MusiCad, we had some important decisions to make, and I'm still not sure if we made all of them correctly. However, it's instructive to examine the motivations behind some of them. We wanted MusiCad to be a desktop publishing application for music only, not words and pictures, with CDs, not printing, as the output. We felt this would also capitalize on skills that already existed in the user base.

I'm not sure it turned out that way. For one thing, most customers were professional musicians who were first-time computer users, so there wasn't a strong existing skill set. They only "got it" much later, after they'd gained experience using their new machines and discovered that MusiCad worked just like a lot of other desktop applications.

When designing an editor for other than text or pictures, there is a fundamental assumption that is no longer safe to make. In a word processing program, what you see on the screen is sufficient to know the precise content of a file: WYSIWYG. This is not true when looking at a waveform or even a spectral view of a waveform, even for a very experienced user. Hence, you have to have an extra layer letting the user hear what's on the screen or in a file to help them find things fast. There is no such thing as a Find/Replace for sounds (yet)! The big problem is that, for visual or text content, you can find things quickly, and you don't have to look at anything if you don't want to. With sound, this just isn't possible. If a sound is out there, you'll hear it, period. (I first learned this while working on stereos in a TV repair shop. The TV technician could simply turn on a line of TVs and scan the burn-in bench occasionally until they replicated their problems. I was always jealous because there was no practical way I could do the same thing with my stereos.)

It's hard to take in groups of sounds at a time (e.g., from different songs or a whole tune at once), so we had to make it as easy and as fast as possible to play a view, a file, or a selection, and scan files or selections at high speed. We *still* get questions about why we didn't put the standard set of tape transport controls on the toolbar. The reason, of course, is that things like Pause have no meaning since there isn't a unique place "where the tape is now." Fast Forward and Rewind are similarly meaningless. Similar functions are accomplished by the Page Up and Page Down (to scroll the view by view-sized chunks), the Home and End (beginning and end of file), and arrow keys, or you can simply drag the scroll bar on the view. These actions are intuitive and save a lot of screen real estate. What could be simpler? The point is that you can instantly move anywhere, and you are not limited to the linear format of tape media. That is why this is sometimes called nonlinear editing — a term I don't like, but a concept that is very important to the application. It's the whole point or main advantage of processing sound on hard disk instead of tape.

The user also has the decision of *which* open file to play. Normally, standard Windows UI practice would be to play the currently active file, indicated by the color of the view's title bar, but this is not always what is desired here. In MusiCad, you have an output file that is a mix of the project track files. You can view one file while listening to another (the output file) while the Play cursor moves over an input file. This way, you can find a note in an input track that sounds fine by itself but is wrong only *in context*.

To handle all these cases, you'd need a ton of similar looking play buttons on the toolbar. This would be unaesthetic and confusing, and it would waste precious pixels. Therefore, MusiCad and WavEd have features like Play-When-Scroll, which sounds a bit like the Fast Forward button on a CD player. A host of shortcut keys are available for the various playing modes. For example, the Spacebar toggles a view play, "S" plays the selection, "L" loops a selection while allowing you to adjust it to loop just so, and so forth. It's not perfect, but plenty of keys are available because you don't normally type in sound editing outside of efx dialogs. Experienced users do learn to read the waveform or spectral view somewhat, at least at the level of picking out the major sections of music or speech quickly. The Play cursor and automatic playing, where appropriate, help the user learn this technique. You can turn these settings off when you don't need them anymore and they become annoying. The ears just can't look away the way the eyes can, and good UI practice must take this into account.

Another difference between the sound editing model and the word processing model is how basic editing operations like Cut and Paste must be handled. This is easy for the word processing programmer to address: Paste inserts in the format of the copied data. Pasting an OLE object might be a little complex, but there is never any doubt about what the user actually intends. In MusiCad, where individual tracks are mixed together to produce the output, inserting is almost never correct because the rest of the track would be pushed out of time synchronization with the other tracks. Most of the time, you probably want a Paste that overwrites the data. But do you want to replace the original sound or simply add in the new one? In what ratio of old and new? What if the sample rate is different or there are a different number of source and destination channels? Now you can see the problem: even something trivial in desktop publishing takes on a whole new dimension in desktop sound editing. I'm not sure that any program at this point handles this problem as well as it could. You either have what Niall Murphy calls "poorly defined modes" or you must ask the user each time how to Paste, setting the defaults to a best guess about what the user wants to do, which is what MusiCad and WavEd do for both Cutting and Pasting. No program I'm aware of resamples on the fly to handle the case of different sample rates, although a few will warn the user about this.

Yet another problem with cutting and pasting waves and similar data is their sheer size. It's pretty easy to crash Windows by cutting one-half gigabyte or more to the clipboard! To use the Windows clipboard, you pass in a handle to some global memory you've allocated that contains the Cut or Copied data. In the Windows API, global memory allocations are not allowed to fail but may not be able to succeed. Yet, for various types of interoperability between applications, Windows clipboard use has to be possible, so the upshot is that you have two choices: the user can specify whether to use the Windows clipboard or a special file created just for this purpose. Because many other applications can read or insert from a file, you still have interoperability with most if the user is willing to learn which can do what and adjust the settings accordingly — but what a pain!

Even MusiCad and WavEd only have a channel map if the source and destination have different channel counts. They aren't yet smart enough to paste stereo into mono by mixing the stereo channels (although you can do this with two operations), since they don't make the assumption that stereo and mono are the only extant modes. For example, you could spend quite a while handling the case of channel 4 moving to channel 3; therefore, we made any Paste possible, such as a multiple mix-in Paste. However, the user has to have a plan, which many think is something we should've saved them from. As you can see, it is virtually impossible to do all things well on all levels, and you have to do one heck of a lot of work before getting to the fun parts, which might be thought of as the musical equivalent of adding font or table or embedded object support to a word processor.

I call this the "atomic" versus the "complex" approach. What I mean here is that the more atomic a thing is in some sense, the more reusable it's likely to be. For instance, a hammer can pound nails, but it can also form metal, crush glass, and be a weapon, a lever, or a counterweight. Some but not all of its uses take advantage of the main "feature" of a hammer: It can store energy during a long swing and release it quickly when it strikes. A can crusher, however, is useful mainly for crushing cans. By differentiating the function too much, you take away some reusability. The same is true of much software work. When writing helper routines or making user interface decisions, you often strive for atomicity to make them maximally reusable. However, if you make things too atomic, you give up some ease of use, as in the hammer example. Although either a hammer or a can crusher can crush cans, the latter does it more quickly, easily, and safely if one is available in your jumble of specialized tools. You usually end up with a hierarchy — in whatever domain — where there are low-level atomic tools and various combinations or specializations of these tools which promote reusability, ease of use, or, with luck, both.

You can never manage all the trade-offs perfectly in real life. For instance, if we had handled every possible case of Cut and Paste in MusiCad, the resulting dialog boxes would have been several pages in length, which would not have been very usable at all. I bet we would have had even more complaints if we had done that! Instead, we strived to provide as many general hammer- and pliers-like tools as possible and hoped that, like an accomplished carpenter or machinist, the user would eventually learn to use them together or in sequence to get a desired result when a specialized tool wasn't available. Also, finding a particular tool in a large toolbox is a lot harder than finding one in a small toolbox that doesn't have as much search space to fish through.

Another huge issue is UI smoothness, for lack of a better name, with attributes such as speed and responsiveness. A big chunk of screen should not go temporarily blank during scrolling, for instance, and you should not have to wait a long time for a command to appear to take effect. The computer is supposed to be my servant, not the other way around! If I want to scroll through a file, I should be able to do it either quickly and coarsely or slowly and accurately, and I should not see or hear goofy artifacts while it is happening. If I change my mind during an operation that is going to take a long time to complete or I start the operation by accident, I should be able to break out right away. If an operation completes too fast to allow this, I should be able to undo it. Most software gets this right, with some notable exceptions. Try canceling a print operation on a short file in Word on a fast machine, for instance. The dialog box blinks on my screen for about $1/_{20}$ of a second, and if I want to stop it, I have to leave the program, go to the Windows Print Manager, and stop it there. This is bad design as a result of machines becoming faster than programmers expected, to put it nicely. Much more likely, they didn't think about it at all.

Another place that many multimedia programs fail is in special efx previews. If an effect is too slow to run in real time on your machine, it breaks up and you hear little chunks that don't convey the resulting sound well enough to allow proper adjustment of the effect. Most of them don't allow you to easily a/b compare the effect with no effect, either. Again, this is something that is far easier to implement well on a word processor than for general data processing. It isn't handled it all that well (yet!) in WavEd or MusiCad either, but I'd submit that it is handled better than in most other programs. At preview will loop-play a selection while writing the effect data over it. This allows you to hear the a/b comparison as the processing catches up with the playing, and it never glitches or gaps. Sometimes you are allowed to move sliders and such while loop-transforming and playing the effects, but the problems of continuous updates from things like edit boxes that allow precise parameter input are pretty daunting, to say the least, since while you're changing the edit box contents, the number in it may be wildly wrong or simply invalid. Although the problems weren't insurmountable, it took a lot of effort and even fudging to make it work.

Unlike a word processor, a sound processor works with enormous amounts of data per file. In MusiCad or WavEd, you may have 10 or 20 files of 10Mb or more open at a time, and they all need to scroll together. Consider how much data you have to fetch each time you scroll and you'll see what a programming challenge it can be to do this fast. Things like being able to scroll quickly and smoothly suddenly become paramount design issues, rather than the usual afterthoughts. This goes way beyond whether users just find the software slick to use. A poor job on UI makes the editing software either impossible or unrewarding to work with. It's one of the reasons I emphasize speed so much elsewhere — it really, really matters here.

The attempt to make MusiCad and WavEd work like the familiar word processing applications was partly a success and partly a failure for a variety of reasons, not all of which were the programmers' fault. At least word processing users already know how to read! The main advantage of making them work like other editing platforms is the ease with which the users can build an internal model of how they work and what to expect from them in a given circumstance.

I hope that, among other things, this book will stimulate some better thinking about user interface as a whole. In other words, what is the best way to deal with the fundamental differences between the way we use our ears and eyes for this sort of work, and what allows the best use of the new computer-based capabilities by artists who don't want to become DSP experts? If I've convinced you that it's a tougher question than it first appears, I've succeeded.

Section II

Section II Introduction

This section of the book describes transforms, special efx, and signal generation in practical form. Most of the code examples are for the wave–form editing program WavEd, which is supplied on the accompanying CD. I wrote most descriptions at the same time as the code so they'd match accurately. Only later did I realize that the organization of this book would put some of the documentation out of order. Also, as I went along, I got smarter and added fancy UI tricks, helper classes, and the like. These are only described once in the description of the transform that first uses each. In most cases, even when I found a new trick that might save some code or effort, I did not go back and modify each transform that could have taken advantage of it. The upshot is that some basic concepts aren't necessarily first in the text, which is not wonderful. You may want to read this section out of order anyway, depending on what you're interested in, which is one thing that makes books cool. If you want to know the order in which I actually wrote the code and the descriptions, check out the WavEd `stdafx.h` file. All the includes for the dialog headers are at the end of this file, and they're in the order I wrote them, so let it be your guide if you want to follow my progression.

Most of the examples are built into the WavEd project, which is included on the CD. After you copy this onto your disk, you'll have to mark all the files so they're not read-only. This can be done in Windows Explorer by selecting every file in a directory, (ctrl-A) then right clicking to get the "properties" menu, and then unchecking the read-only attribute. You will

237

have to do this for both the main project directory and the res directory under it to be able to edit the code. (Windows can't preserve the original settings of the files when the CD was burned and can't figure out that the files are only read-only after your copy because of the medium on which they're delivered.) In general, the transforms are in two parts: one has the transforms specific to UI, MFC, and WavEd; the other is pure DSP meat code. I followed this rule often enough that virtually all or the nifty DSP code is in its own independent classes, so you can easily reuse them. Where something was really simple, I often violated this rule and put everything into the dialog code — in order to reuse that sort of thing, it's usually a simple matter of cutting and pasting a couple of lines of code, so it's not worth having the class overhead.

I *strongly* suggest that you put the prebuilt version (WavEd.exe) on your hard drive in its own directory, play around with it some, and read the User's Guide and Programmer's Guide in the appendices to get the lay of the land, so to speak, before continuing. Additionally, you should go ahead and open the WavEd project in DevStudio, and use its helpful navigation features to read the code and comments along with the text in this section. In general, the text descriptions assume you are doing this, and do not cover things that are, after all, clearer in the code than any mere description could ever be; C++ is more strongly typed and certainly more *definate* than English. To bring up the entire project, simply open WavEd.dsw, and DevStudio will take care of the rest. You can then use the class browser, or simply the project source file window, to easily jump to any file or function as you read the English descriptions.

I hope you'll have little trouble understanding the naming and other conventions used in the program. In a passing nod to Hungarian notation, I generally prefix pointers with a p, globals with a g, structs with an s, and so forth. I often removed the leading m_ that the tools put on auto-generated class members, because it doesn't help me, and removing it saves lots of typing. It is easier to type a g in front of a global that isn't a class member, than to mark the other 99% of variables that are members with the normal notation. I tried to name functions so that the name describes what they do, which was usually possible. All class names begin with a C in the normal naming convention. Often, my classes have a separate Init() function rather than doing all initialization in the constructor. I did things this way because often a class is used as a static member of another class, and this other class provides setup information only later in the game, long after construction takes place. This happens most often with the DSP classes, which are usually used as members of a transform dialog class that is itself created on the stack with only a single construction parameter that points back to the view that created it. After the dialog is drawn, the dialog class will initialize the DSP helper class(es) using information it has pulled from the view and document, and this technique makes that delayed initialization possible.

WavEd's a nice little program as is, but of course I'm hoping that it will get you started on the path to making something much nicer. There are many places this software can be improved, and I mention some of them as I go along. I didn't do the improvements myself for several reasons. One is so that you can learn by doing. Another is that some of the UI features would be overwhelming to explain and would obfuscate the intention of showing you the DSP goodies, leading to yet another book on Windows programming rather than what I want this one to be about. Another is that if it was a complete commercial product, I'd probably feel like selling it separately, not giving it away with the book. The last reason is simple: time. You have to stop somewhere to get a book to press, and I didn't want to leave anything important out completely, so this meant that some niceties and details that take a lot of programming time had to be forgone. After all, I'm giving you source code for a nice Windows application, and I'll only get a couple dollars a copy after quite a few books have been sold. Other people in the multimedia business who know what I'm up to think I'm nuts, and most aren't very happy about it because it will surely cut into their sales somewhat. After all, some of them have become millionaires or at least get their daily bread from knowing and selling this information precisely because the techniques aren't widely known; they'd prefer to keep it that way, themselves.

Chapter 10

Filtering

As you may have guessed from the theory presented in Chapter 4, there are a lot of possibilities for the use of filters in special efx and general signal processing. Filters may be used on sounds directly or to help create control signals that in turn affect another type of transform. Adaptive filters might be used to code speech, enhance articulation, or remove noise.

In this chapter, I cover filtering effects when the filter is used directly on a signal, and I discuss the sample code that is part of the WavEd program on the CD so that you can understand it well enough to swipe it and use it in your own efx or applications.

FIR Transform

The code for the FIR transform is split into two classes: one that implements the dialog and user interface and another that encapsulates the actual FIR filtering. This separates the plumbing and creates the reusable FIR filter class Cfir, which is used by the CFirXfm dialog class to provide WavEd with a FIR transform.

The FIR filter class has the designers for the "normal" FIR filter types and also includes an adaptive FIR filter. The designer code implements the equations in Chapter 4 ("FIR Transform," pages 69–71), except that the divide by zero cases are handled explicitly, something I didn't mention there. The class also provides a batch of the more commonly used windows instead of truncating the filter impulse response at "order" length. For a quick, practical education on why I do this, run a 1-kiloHertz low-pass filter at order 127 and try the different windows — it's a real eye opener. A little more testing will show you that as corner frequency goes down, filter order must go up to get a good sharp filter. At any rate, the window is put into the impulse response array first, then the sinc-type $(\sin(x)/x)$ impulse response functions are multiplied into the pImp array. It's a cute way to get the job done without modifying window-generating routines. You'll see the same equations as in Chapter 4, only offset by half the impulse response length to get a zero-phase filter and with the divide by zero problem taken care of explicitly.

I recently added a designer function for a Hilbert transformer. This is a special filter type that doesn't affect amplitude, at least if the order is high enough, but does affect phase. The Hilbert transformer shifts phase to lag 90 degrees from the original signal, and while not strictly a "filter," it has a lot of uses, so I've included some example code for you. It needs to be of large order to work well in the bass region, just as with normal filters, and has to be odd order to give a precisely 90-degree phase shift at all frequencies. Although not very useful as shown in this demonstration transform, you now have some tested code to generate Hilbert transformers for the other types of special processing described in the first section, such as the frequency translation or shifting described in , and other uses of balanced mixers. Cfir (the FIR filter class mentioned previously) makes use of a couple of assembly language routines for speed and accuracy so I could control the code regardless of the C compiler. I provided these routines in .obj form, too, in case you don't have a copy of MASM. Before I did this, I was having accuracy troubles with the adaptive filter, since my compiler (MS DevStudio v6) was loading and unloading the FPU a lot and doing it differently in release and debug builds. Any practical DSP coder needs to watch for stuff like this constantly, no matter what compiler you use. With assembly, I can force the behavior I want, which is to keep all sums inside the FPU to avoid truncating them to 32 bits until the last possible instant. The upshot of this is that, even with floating-point numbers, I still have the default 53-bit internal precision, so I don't have to use the much slower double-precision numbers nearly as often. I didn't need them at all in this class, for instance, once I got full control over how floats were used.

If you are going to use this class in your applications, you'll want to instantiate one for each channel you're going to filter because FIRs must keep a record of past inputs to perform the filter, and this class will only do that for one channel, in the pZ array. If you add code to keep several of these arrays, you'd have to add a parameter to the filter calls to choose which one to use each time, as I did in the IIR filter class. I have no idea which way the purists would prefer. Either way, you wind up with either extra memory allocated or extra parameters.

Notice that there is an adaptive filter in this class, too. It doesn't care about windows and such, but you do need to clear the pImp array or at least set it to reasonable values before running it. If you don't, it may take a while to track in from the near-infinite or non-number values that may be in the array at startup. Using Init() with a rectangular window beforehand, or just initializing the pImp array to zeros, will suffice here. This filter operates according to the math in Chapter 4, with one slight enhancement: I made the sigma (signal energy)-estimating low-pass filter into a fast-attack, slow-decay filter so that the FIR filter tracking isn't as likely to overshoot with high values for mu on the peaky signals you see in human speech, for example. To see this in action, make a test file with gaussian noise at about a 3,000-count amplitude and add a 5KHz sine wave with about a 2,000-count amplitude. Run the LMS filter with mu and alpha set to about 0.001 and with order at about 511 or so. Do this while in spectral view and try both the error sequence and the filter sequence (the check box either way). You'll see that it plucks that nice predictable sine wave right out of the noise better and more easily than you could with a standard band-pass filter of the same size and without having to know the frequency first! The error sequence contains the rejected noise without the sine wave. This might be just what you want for some applications, such as removing hum from less predictable music. If you set the order to about 512 again but set mu high (e.g., 0.7), this becomes a speech articulation enhancer. In this mode, it can track fast enough to keep up with speech changes and makes the formants narrower and more well defined with the energy in between (the dips) filtered out. Consonants are also enhanced somewhat. The FIR filter engine needed speed and precision, so I used assembly language. This is pretty simple code, so I don't need to go into a lot of depth.

The CFirXfm class is slightly trickier. You want all the nice user interface touches, such as Undo and Preview, and you want to handle things like the intrinsic zero-phase FIR filter delay, which is equal to half its order. I take care of this in the OnOK() function of the class by taking some data from before the selection and putting it through a "fake filter" before processing the real data. The transform reads data ahead of where it writes thus canceling the delay. WavEd's Document Read function correctly handles the case where you read beyond the end of file by returning zeros, but it doesn't know what to do if you try to read before the beginning of the file, so I handle that as well. This is pretty straightforward, so the code probably is a better place to look for details.

Somewhat different tricks are used if the user selects the LMS adaptive filter, in which the "desired" sequence is one sample ahead of the input sequence; that is, the filter is asked to predict its input one sample ahead. Generally, this has the effect of decorrelating random noise without affecting the signal too much and is the most common use of the LMS adaptive structure. Also, most of the energy of many signals is at low frequencies, where FIRs of limited length work the worst. With speech, for example, you'd like the LMS algorithm to concentrate more on the higher frequencies where the formants are. To make this happen, I pre-emphasize the input signal with a "nearly differentiator," send it to the filter, and "nearly integrate" it when I get it back. This causes the LMS algorithm to pay more attention to the higher frequencies where it can do some good. Feel free to experiment; this is one of the fun filters to play with.

Although I don't want to turn this part of the book into a treatise on Windows programming, there are a couple of UI tricks in CFirXfm that are worth mentioning. One nice trick is to only enable edit boxes and other controls that are logically usable based on which filter type is selected via the radio buttons. This helps the user avoid mistakes, since it is now obvious which filter type is going to run when the OK button is pressed. I did this simply by adding message handlers to each radio button in the group and using a batch of EnableWindow(True); or EnableWindow(FALSE); statements to get the effect. It generates a lot of repetitive code, but it's worth it. In an alternative approach, you could somehow ensure that a group of controls have contiguous IDs, then use the ON_COMMAND_RANGE message to update the whole group in one message handler. I rejected this approach because the platform will be used by a lot of people and I wanted to make tinkering with it as easy as possible. The problem is that DevStudio autogenerates a new control ID number when you add something new to the dialog, and you can just about guarantee that it won't be contiguous with the other controls to support the single-handler approach. This would require you to know how and when to hand-edit resource.h. You win some, you lose some.

Another interesting wrinkle that appears in a lot of the transforms is the DrawDat class. This is just an owner-drawn CButton derivative that I use to put little graphs into dialogs. To use it, first put a button in your dialog of the size and at the location you want the graph to appear. When editing the dialog code with classwizard, add a member variable to match, which will be a Cbutton. Change that definition to a CDrawDat, and you now have graphing capabilities. CDrawDat must be initialized at some point. Usually this is done in an override of OnInitDialog() in your dialog class. Because this particular class is specialized to draw frequency responses with an exponential scale, it needs to know what the current sample rate is so it can draw the scale correctly and, most importantly, create an array of exponentially distributed frequency values that are mapped to the "x" pixel locations in the graph. It also needs to know whether to draw the scale or not and what vertical range to use, so the InitResp() member of CDrawDat sets this all up.

One problem with showing a graph of a filter response is updating it as the user enters numbers, changes sliders, or modifies anything else. To manage this problem, I simply added handlers for the various change notifications from the dialog controls, and in each, I call UpdateData() to make sure all user input is captured, then I recompute and redraw the graph. Again, some may prefer to guarantee that control IDs are contiguous to cut down on the number of handlers. One type of control presents special problems: the lowly edit box, which only gives notifications keystroke by keystroke and, during a change, may contain values that are out of range or invalid, such as when backspace is used to wipe out an old value before a new is entered. To handle this sort of thing, I added a global function to WavEd called EditFloat(), which attempts to do the "smart" thing for such cases, and I substituted the normal MFC DDX/DDV calls in OnUpdate() for edit boxes with calls to this function instead.

Resampling and Interpolation

Another use of FIR filters is as interpolators. Suppose for the FIR low-pass case you want to know the value between two output samples. This is more or less equivalent to shifting the Z's by an amount less than one sample. If you knew how to do that, you'd already have the answer. However, you can shift the filter coefficients the other way instead because you have an analytic way of generating them at any effective time shift. They don't have to be centered exactly since you are plucking these samples out of an infinite-resolution mathematical series. After this, resampling boils down to bookkeeping. You can resample by any ratio without having to up-sample by a big integer number first then decimate by another integer, as some books would have you believe. The penalty is that you either have to constantly recompute new filter coefficients all the time or precompute all those you'll need and use logic to compute which set to use to produce each output sample, which is what the example code does. When you resample downward, you might need to low-pass filter the input signal first to prevent aliasing. You get that free in this technique by simply designing the interpolation filter to the desired response. Otherwise, simply set it at the Nyquist frequency. The transform I provide "knows" what the input frequency is by looking at the input file, and it asks you for the output sample rate, producing a new file with the resampled data, so it serves as an example of that technique as well.

Most of the muck is in keeping track of time so as not to build up cumulative errors and in figuring out how to phrase the code into something you can reuse. I think Troy Berg did a nice job of this.

IIR Transform

The IIR transform also is divided into two classes: one very general IIR filter class called CIIRFilt and the dialog class CIIrXfm, which has the WavEd-specific material in it. Much of the CIIRFilt class will be familiar to those of you who took my advice and bought the Stearns and David book listed in Chapter 1. I simply grabbed some of their code, rewrote it to make it faster and handier, and put it all into a class for IIR filters. In particular, I rewrote the filter engine so that it is about five times faster than the original, which is pretty important for this kind of work. One thing I ditched in the process, however, was the ability to do weird-length filter sections. This code only handles cascaded second-order sections, although you can do first-order sections by setting certain variables to zero. I felt the five times speedup was worth this, but if you don't, buy their book or figure out how to change it back.

CIIRFilt

The CIIRFilt class is a lot more complex than the simple FIR class, so I'll provide more documentation here and walk you through it. Because this class was originally developed to support the arbitrary-response IIR designer, it has a lot of interesting things in it that you won't find elsewhere; I invented them, and this book represents the first disclosure of some of the new techniques. It also has various handy, debugged utility helper routines for arbitrary filter designers and so on. Some are for practical use and some are just for dorking around and finding things out, such as the aptly named dorkfilt(), which I'll get to later.

I want this class to be reusable, but first CIIRFilt must be initialized. I did this with an explicit Init() function instead of in the constructor, since sometimes you might want to instantiate this class before you know the information Init() needs, which is the sample rate in Hertz. Digital filters "think" in terms of the ratio of the frequency to the sample rate, so by passing this in once, I could make it accept all other frequency inputs in units of Hertz, which is a good bit handier and easier to remember and debug. At Init() time, the filter takes care of acquiring memory and setting up various other things for you. It expects to see MAXCHANNELS and MAXIIRSEX defined someplace so it can allocate the right amount of memory for worst-case situations. I defined these in stdafx.h in WavEd before the CIIRFilt.h #include. Some users may want to modify this approach and pass these numbers in to Init() instead.

Once the filter class is initialized, arrays and variables are available. The variable iNS tells the filter how many second-order sections it will have. You must put data into four arrays — pfl[] (pointer to flags), pcf[] (pointer to center frequency), pdf[] (pointer to damping factor), and pamp[] (pointer to amplitude, used only for type 9 sections at present) — to design various types of filters using setfilt(). The numbers in the flags array tell setfilt() what type each section should be, and as I mentioned before, you have more than the normal number of options here. The function setfilt() takes the information in these arrays, designs prewarped S-domain second-order filter sections, then uses bilinear substitutions to create the coefficient arrays, which land in pA[] for the denominator and pB[] for the numerator Z-coefficients. Note that there are two pA and three pB coefficients per second-order section because the leading 1.0 for the denominator is assumed and not stored. I'm sure any clever programmer can figure out how to print these coefficients or output them to a file to design filters for embedded applications, for instance.

The other two input arrays are floats and contain the center frequency in Hertz and the damping factor (= 1/Q). The new options are interesting, and most of them were discovered by using dorkfilt() and a little rederivation of the second-order math. Look at the flags options. Options 0 through 3 are the old standard low-pass, high-pass, peak, and notch filters, respectively. Option 4 is the dip filter. The classic notch filter can only be made with infinite depth and fairly steep sides. If you want something usable in a graphic EQ, for example, you need dip instead. Its depth is set as a function of its width in this particular implementation, but you can also control them independently. Option 5 is the peak filter, which follows the same idea. It works a lot better than using a band-pass pole for an EQ, since it only affects the area right around its center frequency, instead of rolling off on each side forever as a normal band-pass filter does. Options 6 and 7 are the lp with shelf and hp with shelf filters. These operate like the normal lp or hp sections but, again, don't roll off forever — they "shelve out" at some point. You can see the simple fudge numbers I added if you look at setfilt(). There's nothing special going on here; it's just part of the ongoing quest to obtain more complex and useful shapes than square boxes from a ratio of second-order polynomials. Option 8 is the conventional all-pass filter that has only phase shift. Option 10 and 11 are the first-order low- and high-pass sections, just in case you want a filter with odd order. This can be handy for speaker crossovers and such, since odd orders have a 45-degree, instead of 90-degree, phase shift at cutoff. Therefore, you can make filters that don't have a deep notch when added back together as in the crossover cases. And it is just cool to be able to do it with second-order sections. I'm sure you understand that motivation. At any rate, the included *S*-domain application has advanced past even these options somewhat with the function dorkfilt(), which simply allows you to perturb an all-pass filter in various interesting ways, with access to all six of its coefficients in the *S*-domain. There's a pretty simple and intuitive mapping between the coefficients and what happens when you perturb them. You can get results like peaks and dips with independently adjustable widths and heights that only affect the area around their center frequency, broadband tilts of odd numbers of decibels per octave, and all kinds of shapes that are highly useful in audio processing. The most useful new shape is the peak or dip filter with independently controllable width and amplitude. I added this shape as option 9, and it's used in the parametric as well as the graphic EQ transform. Once you call setfilt(), your filter is ready to run using fl() or one of the other engine functions.

Someday I'll write an assembly version of the filter engine, which is why I included a couple of different engine function names in the code as placekeepers. I usually keep a copy of the C function around after I do this, for reference and portability, so having a couple of names to keep them separate is convienient later. It's pretty fast as it is. If you just want to look at the filter shape, you can use response(frequency) to find its amplitude response at some frequency. This works by calling spgain() for each section of the filter, which is the e^{jw} substitution for Z that you see all over signal processing. If you want, you can add phase information to this by taking the arctangent of the ratio of imaginary to real output from spgain() in the response routine and returning a pair of numbers. For the arbitrary filter designer, I was only interested in mapping amplitude response at a number of discrete frequencies onto screen pixels, for which I use `response()`, which returns the value of the combined filter's response at some input frequency in decibels. This is why I had `CDrawDat` precompute an array of frequency values that map to the available pixels. These are used here so that I only call `response()` with the plottable frequencies.

The rest of the routines in this class just support the main actors mentioned above. They were excerpted from the Stearns and David library and rewritten to be faster and cleaner than the original port from FORTRAN. `Spbiln()` does the bilinear substitution needed to go from prewarped S-domain to Z-domain filter coefficients, and some of their complex number functions are dragged along as well, mainly to maintain control over how they're implemented. This is another area where it's often not wise to trust the compiler vendor.

You now have a pretty powerful and general IIR filter designer and engine, so now it's time to talk about how I used it to implement a standard IIR filter transform. I decided to use the filter designs in Don Lancaster's *Active Filter CookBook*. I must say that I've used that book for a couple of decades at least, and I never noticed before that some of the numbers in his tables are simply flat wrong, which I discovered while writing the code for this book. I guess I never built those particular filters when I worked in analog, or I would have known earlier. To be specific, his numbers for slight dips filters are way off for orders four and greater; they produce a dip filter of more than 3dB. To make a long story short, I went to other sources and came up with correct slight dips numbers. Don's numbers for his band-pass filters also were wrong, both in insertion loss and q — some by quite a bit. I still like his method for narrow bandwidth filters over the more standard (now!) method of doing a substitution in the S-domain that results more or less in crossed low-pass and high-pass filters. For one thing, this method is more computationally efficient, and for another, I've already provided low- and high-pass filter designers here, so you don't need both. The problem is that no other source I could find has this derivation, and since Don's numbers are wrong, I had to suspect his derivation as well. Being the good guy that I am, I spent a couple of laborious, boring days tuning test files until I got the right numbers. You can thank me with all the $20 bills you want to send. These band-pass filters only need about half the order of the standard method to get very good initial rejection for things like DTMF detectors in embedded applications, so you'll save some money by using them when you can for similar narrow bandwidth applications.

The transform dialog allows you to design and run low-pass, high-pass, band-pass, and notch filters in any combination. The shape options for the first three are the same as in Don's book, with low- and high-pass having Bessel (what he calls best delay) compromise, Butterworth (which he calls flattest amplitude), and various Chebychev filters (which he calls dips filters). Band-pass filters have his options of maximum peaked-ness, maximal flatness, and various dips levels. The notches are just zeros with infinite depth and a 3dB width that is set by the user as percent bandwidth. I divided the code into helper routines for each type of filter. Some of you are going to look at this code and go "arrrrrrgh," I just know it. I did it this way to maximize cut and paste reusability and to avoid dealing with a big multiply subscripted table. You'll like it when you're able to steal just the part you need for your application.

The rest of the dialog is pretty much like the default test transform I described in the tools chapter (Chapter 7). The big difference between this and the FIR dialog is that `CIIRFilt` handles the multichannel case internally, so you have to pass it a channel number when you call the filter engine. I've added some of the same code that I described for the FIR transform earlier in this chapter to redraw the response of the filter so that you can see it as you change the parameters. I didn't handle quite every case, buy you can add the additional message handlers if you want. For now, if you make a change and the filter graph doesn't update, just click outside of any control anyplace in the dialog to get a fresh plot.

EQ Transforms

Although the filtering transforms described above are pretty handy for scientific applications, they're not what you'd want for audio unless you are simply stripping noise off the band extremes, for example. What most users want for audio is something that only boosts or cuts by some definite amount at some frequencies, not these simple-minded square box types of shapes, so I rolled another few transforms that you'll find far more useful, including the standard multiband EQ, a simple parametric EQ, and an arbitrary-shape filter designer, which is the most powerful of all. The multiband EQ is the simplest, so I'll cover it first.

Graphic EQ

The graphic EQ I built has the usual 10 bands and a nice trick that sets them up as a function of the actual Nyquist frequency. If your file is sampled at 44.1KHz, the bands work out to octaves, more or less, but if it has a different sample rate, the bands are divided ratiometrically using the same idea (constant percentage width) and come out to something other than octaves, since you always want to start with the lowest band edge at around 20Hz. I did the controls a pretty dumb way by using the resource editor to make a number of sliders that are hooked to class members. A smarter way might have been to dynamically create an array of sliders and derive a class from CEditSlider that would handle the control message reflections and update each control on the fly, as in MusiCad. However, that would have confused the issue here. This is still not supposed to be a book on Windows programming! This transform uses the new shape option, which is option 9 in CIIRFilt. It creates a smooth peak or dip around the center frequency as a function of an amplitude (i.e., the contents of pamp) in plus or minus decibels.

I didn't bother to add a filter section if the corresponding slider was set to 0dB. There are a couple of reasons for this. Obviously, doing nothing is always faster to compute, but the real reason is that these filters, just like the ones in any analog EQ, add some phase shift and ringing even when they are set "flat," so if you don't need it, don't do it. The result is an EQ that affects phase only if it also affects frequency response, which is more ideal than most analog implementations in many senses.

A nice extra is that I allow the user to set the Q. You should find this instructive in conjunction with the response plot and listening tests. Most people aren't aware that even 31-band EQs use quite low Q factors to avoid various ringing phenomena. Because of this, even if they have one-third-octave band centers, the bands themselves tend to be over an octave wide. If you want to hear the effect of a high-Q EQ, try tripling the Q precomputed in the edit box and see the effect on a sweep tone or hear it on music. It can be cool, but it's not usually what you want for music. In any case, the use of high Q values generally requires the ability to set the band center just where you want it or it's not very useful. This is what parametric EQs do, and I have an example of one later. Here, it's much more interesting to set the Q very low so that a single band can affect a broad range of frequencies with gentle slopes and little phase shift.

The OnOK() function used here is the same code as in the standard IIR filter transform. The only difference is the way the filter is designed for this use.

The most exceptional thing about the graphic EQ transform is that it does live control updates during preview. I thought it would be nice to show one possible way to do this on one transform. As described earlier, the normal preview simply moves the selection over a preview area, then applies

`OnOK()` while loop playing it. `OnOK()` performs an undo first, so when the user cancels a preview, the transform can restore the original file data.

There are two major problems with live updates. One is that you cannot simply reapply the transform over the data, since that data is no longer the original file data. The other problem is with controls like sliders and edit boxes. You may get a ton of messages before the user has really stopped changing something, and you don't want to get into too much of a flurry of work for each `OnVScroll()`, for instance. You might see 50 or 100 of these in less than half a second as the user moves a slider, but even if the user only moves a slider one tic or types one character in an edit box, you want to respond to it eventually. My "simple" solution was to add yet another boolean variable to the class to indicate whether a timer is running.

Whenever I get one of these control update messages, I check to see if the user is previewing and, if so, whether a timer is already running or not. If not, I start one; if so, I kill the running timer and restart it to get the effect of a retriggerable one-shot. I always wait one timer interval after the last control change to update the preview, then I command an undo and rerun in the timer message handler. By setting the timer timeout to one second or so, I'll eventually respond to a change, but I won't get too caught up in trying to track more quickly than I really can. The nice thing about this approach is that it can work with transforms that are much slower than real time without forcing the user to hear gapped sound. Instead, the worst thing that happens is that the user hears a sound I haven't gotten around to processing yet and gets a free a/b compare in the bargain. This works because WavEd is multithreaded, and the loop-play function just goes round and round regardless of what I'm doing to the underlying data it is working on. This might even be the one time it is useful not to synchronize access to data across threads!

Morphing Parametric EQ

OK, OK, I protested too much. This book is, at least in part, about Windows programming, since it is the most popular desktop platform these days and surely where a lot of DSP is done. Therefore, I went completely bananas and whipped up a full-bore UI for the Morphing EQ example.

The main difference between this transform and the other IIR-based transforms is in the UI. The underlying engine is about the same, the filter class is the same, and the boilerplate for hooking to `OnOK()`, `OnPreview()`, and the rest is all the same. Even the timer one-shot I use to deal with live control updates during preview is the same. Only the UI and the way I take data from it and jam it into filter settings is really very different.

I wanted a new control type, an edit box combined with a slider control such that anything you enter in the edit box affects the slider control and any movement of the slider control affects the data in the edit box immediately. It would be nice if it updated a number it holds a pointer to in real time and sent a Windows message back to the parent to indicate any control change. It would also be nice if it handled various types of scaling and offsetting, too, and you would want it to be reusable so you wouldn't have to write it over and over. I harked back to some MusiCad code that Alan Robinson wrote to see if I couldn't modify it to meet my needs. In the original work, he had tied together an edit control and a slider that were dynamically created for MusiCad's mixer view, so all I had to do was add a log scaling option and a new message to be sent to the parent. The message value is defined in `stdafx.h` and can be used by other controls, too. The original code used a nested class which, unfortunately, confuses the DevStudio wizards when it comes time to add message handlers and such to the embedded class; therefore, I de-nested the edit box class. It's not as pretty as it was, but from my point of view, it works, and that's what counts. In a way, the original nesting was appropriate because it created two Windows controls that acted like one smart control without exposing the details to the user or the programmer. The result was a class named `CeditSlider`.

I won't waste time explaining the Windows or MFC message-handling system, but I will explain how these derived controls work. They use message reflection, which MFC does automatically. If a control message is sent out to a window that doesn't handle it, the message "reflects" back to the control, but with a new message ID. Thus, a button can receive its own `OnClicked()` message, for instance, and if you derive from `CButton`, you can do something special about it right there. This is the basic trick for my new control classes: they handle their own messages, update a datum they are passed a pointer to at `Init()` time, and maybe send a special message home to mama whenever something changes. Just for fun, I defined a "standard" format for this message. `Wparam` has the control ID, and `Lparam` has the control's pointer to its data. I did a little trick to ensure that I started assigning control IDs on an even 256-bit boundary, so a little fast bit-fiddling can tell the message handler which control in which band was changed.

I needed another similar type of thing here, only this time it would be a check box that would send the update message and would be dynamically manageable. That was pretty simple. I just derived a new class from `Cbutton` and added and `Init()` to set things up, create the control, and hide the ugly details of remembering the button styles for a check box. Why did I need all this? I wanted a cool-looking tabbed dialog for initial and final settings, so it would be obvious how it works and so it would fit on the screen. Nothing in MFC really suits. I don't want a sheet-page "wizard" metaphor, because I can't easily put my other controls on that sheet, and I can't get the standard layout of OK, Cancel, and Preview no matter what I do. It also has some problems about what happens as you flip between the tabs, and it looks really difficult to change. Read the MFC source code! Therefore, I created my own dialog with a completely dumb tab control that I dynamically drew controls on as the selected tab changed. Here, the nightmare began.

The tab control is fairly well behaved; it just doesn't do very much. Every time you switch tabs, it redraws just the tab parts (actually, a full-width rectangle of just the height of the tab labels themselves, which erases anything placed in its path). It tells you when you change tabs; that's it. It doesn't erase the main client areas except once at startup, and it doesn't manage anything else going on in its client area. You have to do that. One other problem I run into pretty often using MFC is that in `OnInitDialog()`, even after calling the base class, nothing has been drawn yet. For that to finally happen, a few more messages have to fly around. There is no message you reliably get after everything has been drawn, either, so if you want to do something tricky, like draw controls over a tab control, you have no easy way of finding out where that is. The calls to get its location return a rectangle of all zeros at `OnInitDialog()` time since it hasn't been drawn yet. One dodge I've used to get around this is to start a timer in `OnInitDialog()`, handle the timer message someplace else, and finish the `Init()` after everything's drawn. I didn't do that here, and I'm sure it won't work right because of that at some screen resolutions or default font sizes. I'm leaving it as an exercise for the reader. Basically, you need the timer trick [except you already have an `OnTimer()`, so you'd have to enhance that a bit to check the ID], create your own fonts, and deal with conversions from dialog unit to pixel. Another possibility is to create the tab control dynamically, as well, instead of having it in the dialog template. This just moves the problem up one level. You now need to know how big and where the dialog main window is, which is not a simple task.

CParaBand

Every band in a parametric EQ has four controls: Amplitude, Q, Frequency, and Enable. It seemed wise to make another class that would represent one band and encapsulate all the different scaling and set-up issues for each control in that band. Thus was born `CParaBand`. It's a simple nonwindow class that has some `CEditSliders` and a special `CUpdButton` as members and sets up and creates one band's worth of controls, along with various static labels, in an `Init()`. It also contains the data members that the EditSliders and check box will dynamically update as any control changes occur. The only real problem is with Enable: The check box is smart enough to update in real time and inform the parent window, but it doesn't know how to tell `CParaBand` about the change, so the parent has to check for this case and call back when it occurs. This method is particularly ugly, but I couldn't think of a cleaner way of doing it without making `CParaBand` a window itself (perhaps deriving it from a static group box used for the border) or without giving `CUpdButton` a callback mechanism, which would make it harder to reuse.

Because the tab control is dumb, every time a user flips tabs, you have to erase everything in the client area and redraw the new contols you want. I did this by using `CWnd::ShowWindow()` on all controls when I get the message the tabs have been flipped. To make it simpler, all calls to controls in a band are contained in a single call into `CParaBand`. In other words, you don't actually create and destroy each control on demand; you create them all once then show or hide them as appropriate. A simpler way is to have just one set of controls, but change what they update and are updated by, but I hadn't thought of that yet. Thanks, again, Troy. By placing the morph check box right along the top of the tab control, every flip erases it, so there's a little logic in the tab–flip message handler to redraw this one. When any windows control is invisible, it conveniently ignores mouse and keyboard input, so it can remain in existance safely, saving creation and destruction overhead for tab flipping.

As in the other filter transforms, `CParaBAnd` has a `CDrawDat`. This one draws the response of three traces: one for the initial settings, one for the final settings, and one that shows the instantaneous interpolated settings during the `OnOK()` or `OnPreview()` morph. Without that last trace, it'd be pretty hard to predict what you'd end up with during and after morphing.

MorphDraw()

One UI issue here is that frequency and Q should interpolate exponentially to match the way human perception works. `MorphDraw()` converts down to a log domain, linearly interpolates, then goes back to exponential, accomplishing an exponential interpolation with a linear progress-meter input value. There might be a faster way to do this, but it works and is fast enough for this application. The decibels are already in log form, so they are interpolated linearly, and when they are exponentiated back into a gain value elsewhere, they are exponentially interpolated as well.

I think the hottest UI trick shown here is the color-changing edit boxes. I racked my brains, along with others, to try to solve a fundamental problem with edit boxes and live updates: You only get messages for each keystroke (or focus change, which you can't count on) so it becomes very difficult to deal with user input that will eventually be valid, but isn't while incomplete. You want to protect the underlying process from bogus inputs, but you sure don't want to pop a message box up while the user modifies an old value, for example. The user needs to know when a number is invalid and when a number has been accepted and used to update the underlying value and slider. As it turns out, EditSliders know what the legal ranges are, so they contain the necessary information already. The plan I came up with is simple to describe if not to implement. During typing, if the text doesn't represent a number that is in range or valid, the text and background colors are changed to garish ones, and the slider and underlying datum are not updated. The underlying transform process only sees the last valid input, and as a visual clue that something is wrong, the slider doesn't move. As soon as the edit entry becomes valid, the text and background switch back to normal colors and everything is updated. All this keeps you from having to waste precious screen space and user time on error messages, making MorphDraw() useful during real-time previews.

MorphDraw() is fun to play with. You can sweep a peak across a range by setting its frequency differently in the initial or final bands, or a peak can disappear at one end and appear at the other by using two bands, disabling one in initial settings and the other in final settings. Note that whether a band is enabled or not does not affect the fact that you use its slider values in the interpolation during the morph. You can change this if you like, but for now, it simply "interpolates" the Enable boolean and cuts the filter on or off, as appropriate, halfway through, if a band begins enabled and ends disabled or vice versa.

You can see some pretty interesting phase shift effects with this, too. One possibility is to set a number of bands at about the same frequency, with amplitude at zero and Q very high. This creates a lot of extra phase shift around the center frequency of each band without affecting anything else. If you play impulses through this, they'll be converted to boings as their frequency content is spread out over time. Other interesting things might happen if you do this and morph as well. Rapidly sweeping filters can even seem to modulate the frequency of the signal, since energy that was stored in the filter at one center frequency now comes out of it at another. In late breaking news: I should mention that whether you see the phase effects mentioned above with zero amplitude filters has to do with the sign of the filter's middle numerator S-Domain coefficient set up in `setfilt()` in `CIIR-Filt`. See the comments there on making either maximal or minimal phase filters by changing the sign of one number.

One difference between the Morphing EQ engine and that of the Graphic EQ is that a smaller block of data is worked on at a time. I chose $1/30$ of a second, but you can easily change this. It allows a smoother morph because the filter design is interpolated and set up again for each block, not just once at the beginning. This, in turn, computes more quickly than interpolating the filter design for each sample. I copied the transform engine code and the code for the control change during Preview from the Graphic EQ transform and made these two slight changes in a few seconds.

Chapter 11

Delay Effects

Delay-based efx are among those most commonly used these days, especially when computers are in the mix, because they are more easily produced digitally than with analog techniques and, perhaps because of this, aren't "worn out" yet. Delay-based efx include, but aren't limited to, flanging, chorusing, echo, stereo image manipulation, and reverb. This chapter provides some practical implementations for each of these.

Flanging

Flanging is most commonly accomplished by delaying a signal and adding it back into itself. The technique used in the first modern deliberate case of flanging by Jimi Hendrix gave it its name, although it's almost never done that way now. Although it is a prevalent "effect" in nature, it was difficult to accomplish with the analog equipment of the day. According to the popular story, it was accidentally rediscovered one day when, during the production of a tune, there was some difficulty getting two tape machines to "sync up." This method was often used to get more tracks than were available on a single machine. In this particular case, the same sync track was recorded on two different machines, which were played back at the same time and listened to to determine if the machines were synchronized. To get them in sync, the engineer would slow the machine that was ahead by placing a finger on the flange of the tape supply reel. During this attempt, when the machines were close to, but not quite, in sync, the flanging effect was noticed, and the rest is history.

Fortunately, it is far easier to produce the effect now by simply delaying the signal by a small, variable amount and adding the delayed copy back into the original signal. The delay is most often implemented as a circular buffer, as described in ("Circular Addressing" on page 205), and the delay tap position is generally a sinusoidal function of time over the range of approximately 0.1 to 5 milliseconds (ms). Longer delays usually affect the bass in undesirable ways. The depth of the effect is controlled by how closely the signal levels match. A 100 percent depth is achieved when they match exactly, giving the resulting comb filter notches an amplitude of precisely zero. The *width* of the effect is how much the sinusoid is allowed to move the delay tap around, the *speed* of the effect is the frequency of the sinusoid, and the *center* of the delay range controls the overall *quality* of the effect, with short delays sounding *high* or *close* and long ones sounding *low* or *far*. High width plus high speed can result in audible frequency shift effects as well. In stereo, you can produce two different flanged outputs from a mono source and send them into the stereo channels for an even more spacious or interesting effect. In this case, you use either the same or a different sinusoid to control the delay tap, or you can invert the sinusoid for one of them to get a very nice effect that also produces some interesting spatial cues to the ear. In so-called stereo output stomp boxes, two output channels are usually produced by adding the delayed signal to the input for one channel and subtracting it for the other. This is cheap to do, though not as general as what you could do on a real computer.

Although virtually all DSP special efx are programmed to work on blocks of samples for speed reasons, this tends to add an obfuscatory extra layer to the code, making it harder to understand. Because the real thing is provided elsewhere, this explanation and its C pseudocode assumes you work on the data stream a sample at a time via mythical `readsample()` and `writesample()`. I think you'll find that comparing this to the real shipping version is quite informative about the gyrations DSP programmers often go through to squeeze maximum performance out of a given platform. Listing 11.1 is the pseudocode for mono flanging.

Listing 11.1 A real-time flanger.

```
Flange(int nsamps, float samplerate, float maxdelay, float depth,
       float width, float rate)
{ // where
  // nsamps is number of samples to process
  // samplerate is the samplerate in Hz
  // maxdelay is is the max delay in seconds
  // depth is a number 0 - 1 (for 0 to 100%)
  // width is a number 0 - 1 (for 0 to 100%)
  // rate is the frequency of the delay modulation in Hz
#define pi 3.1415927 // doh, this isn't already in some compilers


  int i; // iterator, various uses
  int dindex = 0; // delay-input index for circ buffer
  int dmax = (samplerate * maxdelay); // size of delay buffer in samples
  int dmean = dmax / 2; // 'center' of the delay range
  int doutdex;     // index for a delayed sample
  float samptemp; // temp float for sample manipulation
  float theta = (2 * pi * rate / samplerate);
                 // phase (radians) to add to sweep sinusoid for each sample
  float swamp = dmax * width / 2;
                 // amplitude of the sweep sinusoid in "sample indexes"
                 // the divide by 2 is required since the sin() function
                 // has a range of +/- 1, or 2 total
  float *delay = new float[dmax];  // aquire a delay sample buffer
  for (i=0;i<dmax;i++) delay[i] = 0.0f; // make it silent initially


// setup done, now we can flange!
// this is what's in the stompboxes and digital delays


  for (i=0;i<nsamps;i++) //  big sample-by-sample loop
  {
   samptemp = readsample();  // get the next input sample
   delay[dindex] = samptemp; // put it into delay buffer too
   if (++dindex >= dmax) dindex = 0; // implement circular input index
```

Listing 11.1 A real-time flanger. (continued)

```
  doutdex = dindex - (swamp * sin(i* theta) + dmean);
                            // compute delay tap index
  if  (doutdex < 0) doutdex += dmax;
                            // delay circular buffer calx & sanity-check
  samptemp += depth * delay[doutdex]; // actual flanging right here
  writesample(samptemp);    // do output
 } // end sample loop
delete[] delay; // always clean up your mess!
} // end of flange function
```

This simple implementation is adequate for a real-time flanger [by changing the for loop to a while(pedalswitch) loop], although you would probably not call sin() sample by sample, but rather use one of the faster and more general table-driven waveform generation techniques as described in . When working on a general-purpose computer and using data stored on disk, the implementation is usually changed to work on blocks of samples at a time, rather than just one big block, which could far exceed available memory. These blocks are usually optimized for size to speed up disk reads and writes. Because the delay range of interest for flanging is short, a block of samples that is a convenient size for disk operations usually takes more total time than the delay needed. Most implementations use at least 16Kb of samples or so for a standard disk read or write. In this case, rather than creating a separate delay buffer, read buffers are often used, with the old buffer saved when the next is read to provide a contiguous delay buffer. This can be done in a circular fashion as well; the upshot is that the index calculations get somewhat more complex. The sweep waveform needs to be contiguous across disk blocks, so you have to use a sample counter other than i, and you should create an outer loop to deal with the discrete nature of disk reads and writes, as mentioned in ("Work in Reasonably Sized Blocks" on page 209). For a commented example of a highly optimized version, see the CD under the flanging directory. This version is implemented as described, alternating read buffers in a circular fashion and using table-driven waveform generation for the sweep. It can run considerably faster than real time on most Pentium platforms for 44.1KHz sample rate signals because so little computation is really involved.

As mentioned above, another output channel can be generated in several different ways. The cheap stomp boxes generate another output by using -= instead of += with the input sample (whether in analog or digital), which results in a different batch of notch frequencies with the same spacing. For a different effect, you can generate another `doutdex` by using the negative sine instead of the sine of the sweep sinusoid, for instance. This would make one output curve go "up" as the other goes "down." Another interesting variation is to use filtered noise instead of a pure sinusoid for sweeping the delay tap, which can be much more natural sounding if the parameters are chosen carefully.

Yet another variation for stereo input signals is to cross the channels — add or subtract the delayed signal of channel 0 to channel 1 and vice versa — and perhaps subtract each without crossing, too; in other words, generate a stereo output for each input channel and then mix the two stereo results into one stereo output. This has a lot of potential to make a sound source seem to move around the room, although there are better ways to achieve the perceived spatial results in a more deterministic fashion. By adjusting the numerous options, many wacky and weird effects can be generated with this simple structure; however, the pseudocode example given in Listing 11.1 produces a normal flanging effect similar to that Hendrix was fond of using.

Comb Filter Modifications

The delay and add strategy for flanging produces a comb filter. This filter has prominent notches at frequencies determined by the amount of delay currently in use. For instance, if you're producing output using a 0.5ms delay and adding the output of this to the input, the first notch occurs at 1KHz, where the delay equals a 180-degree (half-cycle) phase shift; the next is at 3KHz, at 360 + 180 degrees (three half-cycles); and so on at 2KHz (i.e., 1/0.5ms) spacing. This can also be seen as an extension of phasing, in which only one or two zeros usually are produced. In the phasing case, you have control over the location and number of zeros independently. However, you usually can't afford to produce each one individually, although with more powerful platforms becoming available, the idea should be kept in mind for use in the near future.

Another way to flange is simply to generate the comb filter zero by zero using cascaded second-order filter sections. This is a lot of work compared to the delay-and-add strategy, but it has considerably more potential to create a completely general flanging transform. If the zeros deviate from even spacing, it won't sound precisely like flanging, which may be just what's desired in a given case. As usual, it depends on what you're after.

As discussed in Chapter 2, this even spacing maps onto a simple shape on the cochlea. However for various reasons, you might not want to exactly reproduce the natural flanging sound found in nature; for example, it is boring, so you might want to modify the equally spaced zeros in the net response. The most common way to do this is use an all-pass filter to produce a frequency-dependent phase shift in addition to that produced by the delay. In this case, frequencies around the corner frequency of the all-pass filter will see an additional delay, so the notches around this frequency will be offset somewhat. This can be used to create a subtler effect than straight flanging alone. Another popular modification is simply to use more than one swept delay tap. This allows some of the zeros to be canceled by other outputs and makes the effect subtler and more complex. The extreme case of this turns into chorusing, but with a shorter useful delay range than is used for that case.

WavEd Flanger Transform

To illustrate some of the above possibilities, I wrote a flanging transform for WavEd that works very nicely. As usual, the boilerplate code is about the same as for all the other transforms. First, I created a couple of reusable classes for the basic functions that I'll use in the other delay-based efx. The first is a simple circular delay buffer that I placed in-line for speed, although it would be faster yet to replicate the code all over and not use the implied `this` pointer to get to class members. Like many of my classes, the actual initialization is done in `Init()`, rather than in the constructor, so it can be a static member of something else. You finally initialize it later when you know what you want. It is safe to call `Init()` repeatedly; it won't leak memory. It takes a length (in samples) and rounds it up to the next power of two. I did this to manage the circular wraparound with a simple `AND` instead of having to test the index and conditionally add or subtract the buffer length to get it back into range. It works with `float`s, but you can change it to use anything you like.

The other class generates time-varying control signals. I got fairly fancy here. Rather than just the simple sine wave with adjustable amplitude and frequency that most flangers use, I also made it capable of generating band-limited and scaled random variations. If you create an array of these, the outputs will be uncorrelated with one another, which is what you want. I used the run-time `rand()` function to create a seed for the `RandN()` provided in MathHelp. This means that each time you `Init()` one of these, it's going to start at a different place in the 2^{32}-length sequence the random generator makes. Each copy of the class has its own seed, so they are fully independent. I used state variable filters (another in-line reusable class) to shape the noise bandwidth because the range of interest is the very low frequencies, and state variables work better under these conditions. The quick study will notice that I didn't even bother to design "real" filters: I just set the damping to ≈2 and cascaded two identical sections when I discovered that one wasn't quite enough for the job. It works fine, and there's no rule that says you need a certain shape. In fact, other shapes can be very useful and interesting, by keeping most of the random nature, but emphasizing certain main frequencies from the noise.

One knotty problem was getting the output to scale correctly. Noise, being noise, is not very predictable. Also, you can't assume the filters do a perfect job of limiting the noise to a certain bandwidth because they have significant tails. Even though the noise has perfect equal energy per Hertz of bandwidth, filtered noise doesn't. I came up with a "brown" algorithm for scaling that seems to work pretty well, but I couldn't work out a scientific justification for it. Such is life, and sometimes you just have to be satisfied with what works. I made this class so that Init() takes a minimum and a maximum range, and it ensures that a number that is out of range cannot be returned from its Next() function. It will center its output on the minimum and maximum values you put into it, which means that it is safe to specify these values and then use the number returned by Next() directly, as for a circular buffer delay index, for example.

Now you have the necessary toys to program a nice flanger. I added yet another wrinkle to this implementation: that of delay line feedback, which can vastly increase the effective Q of the comb filter generated by the delay. Normally, delaying a signal and adding it back to itself creates a filter with infinitely deep notches, but the peaks are only 6dB high between them, since $1 + 1 = 2$, and 2 corresponds to 6dB. However, when you add feedback (i.e., the input to the delay consists of the original signal plus some portion of the current output), you can produce infinite gain or oscillation. I also let you invert either the delay output or the feedback, just for fun. Inverting the delay output shifts the comb filter locations by one-half of one of the teeth. This determines whether the response at DC would be nonexistent or +6dB. Inverting the feedback has a similar effect on it. Listing 11.2 shows that the code, minus the UI stuff, is short and sweet.

Listing 11.2 Flanging delay line with feedback.

```
// operate on the channel (c)
        for (i=0;i<amount;i++) // sample by sample, we go
        {
        tmp1 = Delays[c].GetSamp((int)CSigs[c].Next());
```

Listing 11.2 Flanging delay line with feedback.

```
if (m_Invert) tmp2 = -tmp1 * realdepth;
else tmp2 = tmp1 * realdepth;
Delays[c].PutSamp(pWorkBuf[i] + realfeedback * tmp2);
                      // update delay line
pWorkBuf[i] += tmp2; // do output
}
```

This loop is inside loop c, which loops over the channels. To keep them independent, I created an array of delays and control signal generators of MAXCHANNELS size so that I can do a completely independent flange for each source channel. There are other things you can do if you want to get tricky; for example, for a nifty stereo effect, put some of the flange output into a channel other than the one it originated from. Delays[] is the array of circular buffers that uses PutSamp(), which puts a new sample into the buffer and updates its index, and GetSamp(), which gets a delayed sample. The number put into the call is the number of samples of delay wanted. If you have an absurd or invalid request, the class protects you from yourself. The Csigs[] array contains the control signal generators. Earlier in the code, these are initialized to generate numbers in the legal range after you've told it how to generate them by your choices of sine, low-passed or band-passed noise, and corner frequency. The depth is realdepth, which the UI shows as 0 to 100 percent, converted to the range of 0 to 1.0, saving one multiply per sample. The same idea holds for realfeedback, which I allow to range between -1 and 1 so that it can be right-side up or inverted.

I added one other slick trick that is commented out for now. It turns out that you can reset the CEditSlider class ranges on the fly. Messages are generated whenever there's a control change, which allows you to implement the real-time Preview. I thought it would be extra-cool to let the width range be a function of the currently selected delay, so in the change message handler, I reset the width range based on the current delay value. It works fine, and even updates the slider position in real time to reflect the fact that even though the number didn't change, the slider now belongs in a new place because of the different fullscale range. The thing is, it isn't a necessary feature, but it's nice to be able to set a width far larger than the delay range sometimes because the code protects you from invalid numbers anyway. I'm sure there will be some place in either my code or yours where this will be very handy.

Chorus

Originally, chorusing was invented to make a single source sound like many — hence the name. Meanwhile, it has taken on a life of its own, probably because the early implementations didn't actually produce a chorus from a single voice very well, although they did sound more spacious in some sense. All this really means is that a bit more care needs to be taken with definitions.

Look at what separates a group of performers all playing the same thing from a single performer. No matter how hard they try, no two people can sing precisely the same — there will be at least slight differences in vibrato, timing, loudness, formant frequencies, and so on. The same is true for other instruments, but voice makes the best example here. Even the same person usually can't repeat a performance exactly. I have met a few very talented singers who could, down to the phase of pitch and formants, but this is very rare indeed and not worthy of much consideration for real-life situations, other than to note the astonishing fact that it *is* possible. Early chorus implementations attempted to produce the effect of many performers from a single source by adding in other copies of the original signal with various slight variations in pitch, amplitude, timing, and so forth. This was done in a way very similar to flanging discussed above, but using a longer delay with multiple taps swept at different speeds and amounts and having a time-varying set of signals to control each tap's mix-in amplitude. In fact, most stomp box choruses today sound more or less like multiple flanging, depending on the settings, since the result is basically still a number of somewhat more complex comb filters. The big difference here is what signals and how much you use to affect the various parameters. The cheap choruses use one or more sinusoids, whereas more elaborate choruses may use filtered noise to mimic more closely what really goes on in a chorus. A much more elaborate approach that sounds more like a really huge chorus uses frequency shifting in bands, variable amplitudes, and swept-tap delays to produce more variations. So far, no one has released a product that does this in real time, but on a computer you need not be limited to what can be done real time. I'm hoping some reader will cook one up using the Hilbert filter from the FIR transform and some balanced mixers from , along with some band splitting filters. I'd like to see it if you do.

You can design a good chorus based on some rules of thumb given in Chapter 2. Up to a delay of about 30ms, the ear doesn't detect a distinct echo, but rather perceives a separate, simultaneous, additional source. Very short delays produce flanging sounds, so most choruses use the range from about 10 to 30ms or so, perhaps adding longer delays as well once there are enough shorter ones to mask the echo effect. In effect, this is just like flanging, but with more computed taps, each using separate sweep sinusoids of slightly different frequencies and amplitudes.

You can also define a lower limit on the useful number of taps for a given effect based on other psychoacoustic data. For a small number of signals, say fewer than seven, the ear has no trouble separating and recognizing them as copies of the original, whereas for numbers around twice this, the ear has no hope of separating the signals into separate entities, providing a clue as to how many taps to use when simulating real choruses. After each tap has been computed, an additional improvement is to sweep the depth, or amplitude, of each tap separately. That's really all there is to it for the simpler chorus implementations. Slightly more complex ones may add filtering of taps individually to cause upsets in the even spacing of the incidental comb filters. Figure 11.1 is a block diagram of a simple mono chorus, with only three delays shown. A good chorus would have more; this is just to show the structure. The control signals could be either sine waves or band-limited noise. It is important that they all be different from one another for the effect to succeed.

The ramifications are somewhat different than for flanging. In this case, with longer delays and thus more sweep range available, the frequencies can change noticeably (as though the sample rate had changed, which in effect is what happens), if only temporarily, but with a mean rate still equal to the original sample rate (i.e., the sounds are dopplered somewhat). Although this may *scale* a signal upward and downward in frequency giving the vibrato effect, within each signal the *ratios* stay the same; that is, an octave is still an octave and harmonics stay the same in relation to one another. This tends to defeat the effect somewhat, since the ear easily detects this as a speeding up and slowing down, and it is one big reason the inexpensive stomp boxes don't truly sound like a chorus, although the effect can still be quite pleasing. To make it sound natural, the sweep frequencies should be in the range of natural vibratos — from about 0.3 to 8 or 10Hz — although frequencies outside this range are useful for unnatural or circus-like effects.

Figure 11.1   A simple mono chorus

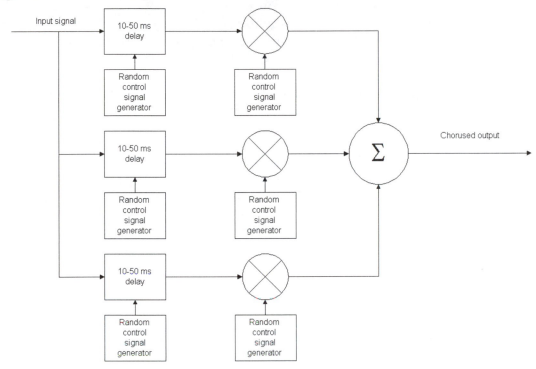

For a stereo chorus, instead of adding the tap outputs into a single channel, they are split across the output channels. Because the delays all vary in time, it produces the illusion that the fake sources are in motion in the perceived stereo output. I suppose that for more than two output channels, the concept should be extended. If the chorus is also stereo input, there are simply two (or *n*) sets of everything.

WavEd Simple Chorus Transform

The "simple" chorus transform in WavEd exemplifies the above theoretical description. I used almost the same boilerplate as always, but I finally got disgusted having to calculate rectangles for all those CEditSliders, so I added a function that allows you to create them by subclassing an existing static group box, using the dialog editor to lay them out. I didn't rewrite all the existing transforms, though. One of the sad realities of writing a book is that you actually have to finish it someday, so I don't have the luxury of going back every time I make something a little smarter to make all the other parts perfect and beautiful. I'm doing things in ways that don't affect the older code, while allowing improvements in the newer code. This is a worthwhile skill that allows you to see the evolution of software and see that there is more than one way to get it done. Some of you will probably notice that with this transform (and several others) I don't necessarily follow my own tips on how to make it run at top speed. It has several by-sample function calls and several places where I make copies of data unnecessarily. If I had written all the code in-line, the copies and function calls would be unnecessary, it would be hard for you to understand, and I would have a hard time figuring it out after leaving it alone for a while. Since the purpose of this part of the work is to teach you how to do various special effects, I feel at least partially justified in this. Anyone who knows how to write good code can apply the tips I give elsewhere and gain a significant speed advantage, perhaps at the cost of some generality or maintainability.

As Hal Chamberlin said in his book *Musical Applications of Microprocessors*, the numerous parameters available in this transform allow you to adjust it for many "wild and wacky" effects. Despite filling the screen with the dialog, I *still* couldn't make all the parameters available individually. A dynamic tab-count tab control and adjustable parameters by-tap would be nice, just for starters, but it would confuse some users. I compromised at what seemed like enough parameters to allow you to get a feel for the range of things, without going totally overboard. The first place to make improvements or achieve a wider range of efx would be in `CChorus.Init()`, where I made some assumptions to keep the complexity reasonable. After all, stomp boxes usually limit you to speed and depth controls. They make a lot of assumptions elsewhere for you that may or may not be right for what you want in a given case. This is better than a stomp box; it just isn't quite perfect yet.

My basic approach was to generate a CChorus class, which in turn contains a `CircBuf` and some `CPrCSig` classes, to generate the variable parameters for each tap, delay value, and amplitude. This class then has a `DoIt()` function, which works on a buffer's worth of data at a time. However, to do so, it makes calls back to the signal generation classes on a per-sample basis, as well as calls to the delay class, so its speed could be improved quite a bit. I've fixed it up with some code-only tunable parameters that do nice things like set the tap positions nonperiodically and, in the case of using sine waves for the control signals, dither their center frequencies a bit so that they don't all track precisely, which is just the sort of thing to avoid in a good chorus. You want to overload the ear's ability to separate sounds, and having identical control waveforms is bad news in that quest. I just picked some numbers for these operations; you can probably find better ones for your particular case. I more or less ran out of both screen space and patience on this one, and it sounds better to me than most other implementations as is.

At the last minute, I added a special case for stereo mode that exemplifies how *not* to go for speed. It splits the taps across the channels; that is, some of the chorus outputs from the left channel go into the right channel instead, and vice versa. This adds a nice stereo effect. I did it the most sluggish way possible using extra buffers and copies of data, so it's pretty slow, even on a Pentium II-400. You might figure out a faster way to do some of this even without in-line code. I just fudged it into existence for a research-grade test and example. I thought the sound was worth the wait, but if I were using this in a studio situation, I'd fix it up for better speed.

I have some suggestions for improvements. One would be changing the CChorus class to work on multiplexed inputs and output directly, especially for the stereo case. This would speed it up and avoid calls to the muxing/demuxing/mixing routines, which although fast, aren't theoretically necessary if you design things right. Another improvement would be to calculate tap modulation depth differently. What is really wanted is something that will allow you to set how much frequency modulation occurs regardless of the other settings. Right now the depth is computed as a fraction of tap spacing, so it changes with the number of taps, and the frequency modulation amount also changes with the rate of the control signal. A better UI design would normalize this out for the end user.

Complex Chorus

For a more advanced large chorus effect, you can split the input signal into various frequency bands and operate on each band separately before recombining them by filtering then adding the results together after chorus processing. The result is a more natural sound, since you can now do things that are slightly inharmonic. What happens to one band will not match what happens in another, so now harmonics do not stay in phase or frequency alignment. This more closely mimics what happens when singers with slightly different pitch and vocal tract sizes work together, at least outside of barbershop groups, which often strive to "phase lock" and "formant lock" to sound like one voice. The most advanced approach I've seen involves band splitting then frequency shifting each band separately via a balanced mixer, adding a simple, delay-based chorus to each band then recombining it all. This supposedly accurately simulates a huge chorus of thousands at the expense of being computationally expensive.

Figure 11.2 Complex chorus using frequency shifting

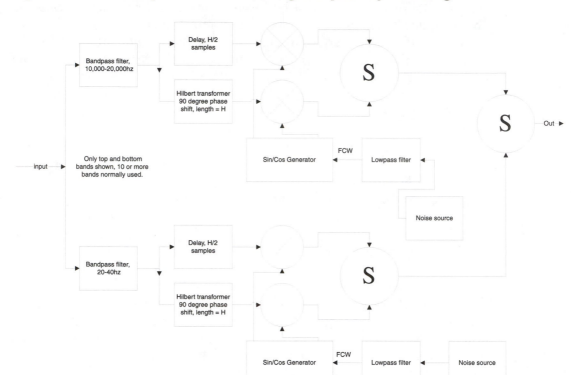

The idea illustrated in Figure 11.2 is that after you split the audio into bands, you *frequency shift* each band separately in a pseudorandom fashion with a balanced mixer. By making each input relatively narrow-band, you make it easier to create the 90-degree delay it needs. You can shift the input band either positively or negatively by allowing the analytic probe waveform to be either positive or negative in frequency. For a natural quality on voice waveforms, you should probably set up the pseudorandom shifts so that they shift about plus and minus one-half step (~6 percent), and the center frequency of the variation should be around the normal human vibrato rates. After frequency shifting each band, a lot of practitioners also like to use the normal variable delay for each output separately, then finally combine them for the result. This is a lot of work, but it can really produce an impressive effect.

It occurs to me that once the audio is split into bands, you would seem to have enough frequency modulation using the simple chorus classes already present and could skip the Hilbert transformers (the 90-degree phase shift) and balanced mixers. I hope someone will try both ways and let the world know. For me, the simple chorus does all I want it to do, and it seems to do a good enough job of simulating a big chorus. Other effects classes in this book include code for splitting signals into bands, so the reader can experiment with these approaches easily.

Echo

The simple delay has long been a popular effect, perhaps made most famous by Les Paul and Elvis Presley. Originally this was accomplished by using a tape loop with variable speed and record–playback head positioning, but today this is done this pretty much the same way as flanging — delay and add signal back in with a sample delay buffer and circular pointers. The big difference is that some of the delay tap is also added back into the input, multiplied by a feedback factor that usually is less than 1.0. Most of the advances you might make involve a better type of UI to set the delay time and feedback parameters more easily or automation that discovers the pulse rate of the underlying music and adapts either to it or to some simple (sub)multiple of it. Sometimes reverb effects are attempted by using delay–feedback modules together, but they often sound pretty unrealistic because the reflections for reasonably simple chains don't "paint" a logical room shape on the cochlea.

WavEd Delay-Echo

I created a sample delay-echo transform for WavEd. Most of the comments that apply to the other simple transforms are also true here. Along the way, I noticed that at certain exact control heights, EditSliders have a problem with the range from 0 to 100. It turns out that floating-point numbers just aren't good enough, and sometimes I'd see something like 2.98 in the Edit window when pulling a slider down all the way. I was confused by this until I scrolled the edit box and saw 2.9837293e-7, which is much closer to the zero I expected. I changed EditSliders to use `doubles` for the scaling numbers and internal temporary variables, and it *seems* to be fixed, but there's probably that special case just waiting to blow it up again. One neat UI trick I did here was to have the tap delay sliders reposition to their new fraction of the total when I set the total maximum delay. Also, setting the total delay high then low after setting a slider high could put the slider out of range, so I added some clipping logic to EditSlider's `SetScales()` while I was at it.

The delay-echo part of the transform is pretty simple. The only odd thing is that I give you three independent taps, within which you have separate control over how much makes it into the output and how much is fed back into the delay. You can set in feedback without also directly hearing a tap; in other words, watch it, this can be confusing. Also, if feedbacks sum to more than 100 percent, you make an oscillator, which can be unpleasant. One nice thing is that a feedback of 100 percent comes awful darn close to being exact: the amplitude of the repeats stayed the same for as long as I was willing to test it.

Stereo Image Manipulation

Most popular music recordings use the simple technique of panning to place a mono source on a stereo soundscape. Although this is better than nothing, there are now better ways to be exploited. Most mixing boards do subtractive panning, in which the signal in the direction away from that in which the pan pot is turned is simply reduced in volume. The best boards have a "constant power" approach, which also increases the power of the channel toward the pan pot just enough to keep the total power constant. Although the ears' mechanism tries to force a mono source to be perceived equidistant between the speakers, it can be pulled off-center slightly toward the louder side. This effect fails between the extremes of center and either side and is not very convincing. The perceived image tends to "jump" at some point all the way over to one side, not smoothly pan across, as the pan pot is turned, and different listeners perceive the effect differently, which is usually not desired. Research at Bell Labs discovered that the effect can be markedly improved by delaying the channel whose power is being reduced. This works more closely with the way the directional hearing mechanism of the ear works. Delays up to about 10ms are commonly used, which is more or less intended to cover a range of distances between speakers. This works pretty well if you're content with the old stereo imaging vision, which still makes you strain a bit to perceive the soundscape, but it also led to a more in-your-face "virtual 3-D" approach that can provide a realistic sensation of sounds moving around in relation to the listener or can simulate listener motion through a soundscape. This technique more properly falls under the topic of reverb, due to other similarities, so I'll talk about it in that section.

Reverb

I can define two useful classes of reverb: those that make an attempt to simulate a real room and those that don't. Either can sound good, and it's up to the producer to decide which to use, or whether to use both, in a given case. Reverbs that successfully simulate real rooms can be quite convincing and can open up whole new possibilities in virtual stereo imaging. A subset of this, variously known as q-sound or virtual 3-D sound, is produced by using only the first few reflections created in a "logical" room. The listener perceives the imaging effect without noticing too much time-smear, as would be produced by a full reverb simulation.

WavEd's 3DR Transform

I put together an example of a virtual soundscape generator for WavEd called 3DR. It's listed as Dyn under the delay transform menu. It could almost be called a combination effect, since not only does it generate a 3-D soundscape, it allows you to move objects around all the while.

Not only that, but the Eigenmode back-end filtering simulates walls and air absorption. Although this could be used as a stand-alone reverb, that was not the intention here. To really be good at that it would need to do more processing — such as simulating the high and low frequency losses as sounds bounce from walls or travel through the air — in the front end as well as the back end. The point is not really to simulate reverb. What I'm going for here is to show a transform that cues the ear to source positions of sound. The ability to move them around dynamically makes a better, if slower, demonstration. It runs slowly because a new impulse response for every combination of input and output channel must be computed on a by-sample basis, which involves a lot of CPU work. I also introduce a practical application of a sparse convolution. Basically, all reverb transforms involve convolving a room response onto a source signal — usually a separate one for each combination of input and output channel (e.g., right source to right ear, right source to left ear, left source to either ear, and so on). The total number of different responses is the square of the channel count. These room impulse responses are "sparse" (i.e., reflections only happen at discrete times or distances), so most of any one impulse response is identically zero. This implies that normal convolution techniques would perform the work very inefficiently, since most of one of the inputs produces no output. Therefore, when generating the impulse responses in the first place, you must format them differently than you normally would to use a more efficient technique than straight convolution.

Two lists are generated instead of the usual single list. One specifies amplitudes and another specifies how much delay is associated with a particular tap. The lists are generated by code that does ray tracing via mirror symmetry — a technique that can handle only rectangular spaces — and are short because all the zero entries are omitted. The original code is from an old Dr. Dobbs article, from which I removed the usual errors (see the comments in CImpResp) and to which I added code to make it more practical to use. The code generates a number of images of the original room in a 3-D array around the original image and computes distances to the virtual listener based on these duplicates. A little thought will indicate that this must be done with an odd order or else there would be no center (original) room that contains the direct source to listener paths.

Order 1 is just the original room, and the only distances calculated are from the sources directly to the microphones. At order 3, the original room is wrapped on all sides with 26 virtual rooms for a total of 27 rooms (3^3), and distances that specify a single reflection from a surface are also computed. This works because of a geometric identity. At order 5, all the paths that require two reflections to reach the ear are computed (5^3 total), and so on. As you can see, this problem blows up quickly computationally. An even order implies reflections only from one side of the axes — for example, left, top, back — instead of from both sides of each axis. In this case, the code blows up for even orders, so I limited the acceptable user input to odd numbers.

As luck would have it, even lowly order 1 isn't bad for cueing the ear to the direction of a source or sources because the correct amplitude and timing for the direct path are generated. This is far better than the simple constant power pan found on most mixing boards. At order 3, information about the locations of virtual walls is included without sounding too much like a reverb. Order 5 does a pretty decent job of reverb simulation. At order 9, the simulation is nearly perfect but very slow to compute. Remember that with most rooms, walls absorb some sound, as does air, and you don't need all that high an order to represent the sounds you can actually hear, which fade quickly with distance and reflection count.

Once you've computed the necessary impulse responses, you have to convolve them on the input signals to generate the output. The class CRSource helps with this. This class owns some CImpResp classes and takes care of a buffer of past samples. One CRSource is used for each input signal and contains enough CImpResp classes to generate each output channel, at which time, it's a fairly simple matter to loop through the input channels and output channels and use the impulse responses therein to generate the output signals. The sparse nature of these lists of non-zero delay taps means that not too many multiplies are needed, because not many delay taps are non-zero. It works out to 27 delay taps for third order, 125 for fifth order, and so on. Actually, those are worst-case values. It turns out in a lot of cases that there is some further symmetry in the room or source–destination placements and that some of the listed taps can be internally combined because they have the same delay value. This is done in the `CImpResp` class and often reduces the computational load. Some simple code sorts the list and then looks for adjacent entries that can be combined. The sorting operation also means that you will go through the delay buffer in order, which is generally faster at convolve time. There is a big trade-off here, however. When calling the compute function once per sample, as you must to simulate motion, it takes longer to regenerate the lists than it does to do the convolutions, so I added some conditional code that only calls the sorting/duplication removal code when the lists are big.

I used some MCGuins to allow graphical user input specifying source and destination locations versus time. It turns out that for anything like fast motion, you have to interpolate between adjacent pixels so as not to jump around too quickly in position. In fact, that is not the only thing that should be interpolated here. Right now, this quantizes the tap positions at the sample interval, which is not precise enough to accurately simulate fast motion. A quick look at this might help you understand.

Suppose that you're working with a 44.1KHz sample rate and you want to simulate motion at 100 feet per second, or about one-tenth the speed of sound, which should produce a nice Doppler effect. If you run the numbers, you find that a direct path length might change 0.02 inches during a single sample time. Sound, however, would have traveled about 0.3 inches during a sample interval. Because the delay taps are quantized to sample intervals, you can't accurately simulate this small motion. You'd have to have almost a 15 times higher sample rate for this simulation to be reasonably accurate; 30 times would be even better. What actually happens is the source location appears to jump 0.3 inches at a time in a jerky manner about every 15 samples, instead of moving smoothly a little bit on each sample. This leads to the familiar artifacts of undersampling. You're just not getting enough information about the signal, and little phase glitches are inserted as the virtual source position jumps around in .3" increments, even on a pure sine wave source signal. You could cure this by interpolating between samples or up-sampling first, but the transform runs slowly enough as it is now, so although you can move the source fast enough to get a Doppler effect, you also pick up aliasing-type artifacts when you do so for every input frequency component above about 1,425Hz — the original Nyquist limit divided by approximately 15. It obviously gets worse as you try to move source or destination locations even faster, and it gets better when you go more slowly. One reason it gets better when you go slowly is that, although you're still limited to minimum jumps of 0.3 inches or so, which for reference is about a half-cycle at 22.05KHz, they happen less frequently and insert less total artifact power. You get some artifacts even for lower frequencies because of the instantaneous jump in tap locations, but these aren't as bad.

Notice that unlike many other reverb implementations, there is no way to set a "dry" gain or to set the amount of reverb other than by manipulating the room reflectivity, size, tap cutoff, and so forth. This is a pretty true simulation of what really happens when you move toward and away from a source, and lo and behold, when you get far away, the source gets softer. You should also note that this inserts a path delay even for the straight path. You could probably modify the code to minimize this, but you can't eliminate it entirely because there is only one shortest path length, so even the other direct paths must be longer. This has ramifications when used on one of several files that are to be mixed later, so beware. In that case, you may want to process each file before mixing or just process the resulting output file.

Recently, I added code for what I call an Eigenmode back end. I use a version of this on my fancy commercial reverb. It recreates the major room modes, floor to ceiling and wall to wall, with some fairly simple circular buffers of past data. Its input is the ray-traced front-end output. I did a little trickery on the back end to sort of simulate the selective absorption of sounds as they bounce off walls, and I coupled the major modes together a little bit to do a first-order approximation to diagonal paths. There's a delay line for each wall (after all, sound can move both from left to right and right to left at the same time, separately), which I simulate with a wall simulation filter at each bounce.

Basically, all this does is repeat (more or less correctly) whatever the front end passes in. If the front end is suitably complex (try order 3 or 5), the ear is fooled into thinking you have a logical room and accepts the not-quite-right longer time response without complaint. This is far faster to compute than simply ray tracing to a much higher order to get long reverb times. I mentioned this way of fooling the ear in Chapter 2. By the way, because it does a fair job of simulating the major room modes, you'll find, as architects have, that it can be darn hard to get good sound in a rectangular room. If you're not careful, many of the modes for left–right, front–back, and ceiling–floor can pile up together and make some very nasty resonances. Try using prime numbers for the room dimensions to prevent this.

Improvements

I spent a lot of time on this code to make it fast enough to be worth using. I even added an adjustment for update rate so you can recalculate the impulse response less often for those times when things aren't moving too fast. As a result, I didn't get around to adding other possible niceties. It doesn't currently notice control changes during previews, for instance. Other possible improvements might be to add separate filtering for the original path and delayed paths and to get rid of the preset selection shortening, as the amplitude envelope has. This would prevent "compressing" the motions into a shorter time during preview, so you could set the update rate more accurately for your intended application during a preview. You might also hook up some UI controls that decide whether to sort the impulse response and combine duplicate taps. I used to keep this code in all the time, and it sped things up when I didn't recompute too often because it moved through memory in order. However, when updating every sample or so, it slows down dramatically, so I commented out the calls.

The following exercises are offered for the diligent reader.

1. Add filtering to simulate wall losses and add the head-related transfer function to the front end. To do this, you'll have to add an array to the TD and TG arrays created by the CImpResp class so that you'll know which way a particular sound is supposed to be coming from. The information is there (in the i, j, and k indices that are used during the impulse response calculations), but it is not kept right now. This will dramatically increase the realism in all modes.

2. Add the correct filter types to the back end. Most objects that absorb do so mostly around 1KHz. All I do now is some cheesy filtering to simulate bass going out through wooden walls and treble being absorbed by the air. There's also a filter that lowpasses sounds going the wrong way before entering the ear, but it's not fancy enough to really work as well as a true head-related transfer function. You might want to substitute a dip at 1KHz or so of whatever amplitude for most of the loss implied by the gain coefficient.

3. Add motion updating to the back end class. This would mainly require adding a new init-type function that only recomputes tap locations, but

leaves the delay buffers alone, and calling it from inside the `reinterp()` function. Additionally, you'd have to move the call to the backend `DoIt()` function inside the by-sample loop instead of letting it work on amount-sized buffers as it does now.

4. Force some variables to be prime numbers (regardless of user input) so that it's easier to make a reverb sound good, even if, theoretically, less accurate. This tends to help spread out the room modes so they won't all add up at just a couple of frequencies. You might start with forcing each front-end tap delay to the nearest prime number of samples. Do the same sort of thing in the back end or for exact room sizes. You'll be amazed how much this can improve things.

The WavEd 3DR transform is a lot of fun to play around with, and it has practical uses even with its limitations. Try fooling around with the order set to 1 at first. This only handles the direct paths, but once you have it working the way you like, it will be easier to wait for order 3 or 9 transforms to finish computing!

Chapter 12

Amplitude Effects

Some interesting and worthwhile effects can by created by simply varying gain over a signal, especially if some thought is given to a smart strategy for level or gain setting. Common amplitude effects include compression, noise gating, limiting, enveloping, and normalization. Often these are combined with one another and with other processing. This chapter will just deal with amplitude effects used alone.

Simple Compression

I implemented a simple compression scheme for WavEd. The boilerplate code is just about the same as all the other transforms, as usual; the main difference is the DSP engine and the UI. This compressor is a feed-forward design because it's easy and accurate to do in digital-land. In analog-land, just about the only workable design is feedback. In that implementation, you measure the level after the variable gain element and use it to control the gain. Feedback implementations therefore always have some slop since they cannot have infinite control loop gain without creating oscillations; that is, in a feedback implementation, the output level always goes up some when the input level goes up. In feed-forward compressors, you can make the output level any desired function of the input level.

This simple implementation uses a first-order filter after an absolute value operation to compute the current level. The fancy part, if you will, is the two sets of coefficients: one for attack or for when the current sample is greater than the current level estimate and one for decay. This allows users to set different attack and decay rates, which is nice because you usually want a relatively fast attack to prevent clipping and a relatively slow decay to make the effect less obviously audible and to allow some signal amplitude variation. If this is not done, the result sounds unnatural, because sounds that normally have a loud initial peak and then ring down no longer decay as they should. Once you have a current level estimate, you generate a gain value that, multiplied by the current level, achieves your target level. If the signal level is lower than the noise gate level, you reduce this gain to provide noise gating. This is where things get interesting.

It turns out that having what amounts to infinite loop gain (as if a feedback compressor) means that every time you are in attack mode with a zero attack time, you do what amounts to clipping. Every time your level estimate increases, a gain is calculated such that it exactly hits the target level. If your level estimate is allowed to have instant attack, the net effect is that of clipping at the target level, since each new sample and level estimate create a gain such that the result will be exactly the target. This adds distortion and sounds bad. One approach is to prevent the gain from moving quite all the way to the target level in a single sample, and that's what I did here. The new gain estimate and the old gain are simply averaged to move halfway to the target each sample so that you do not clip at the target level during rising signals that exceed the current level estimate. This amounts to an additional first-order low-pass filter with coefficients equal to 0.5. I really should have calculated these based on the sample rate instead. Diligent readers take note! Because of a compiler bug in DevStudio v6, when I added an additional level estimator, so the noise gate could have its own settings, this simple code quit working. I'm trying to help them track the bug down, but with no luck so far. A variation of this bug has existed in several revisions of the compiler, and it appears that some combinations of FPU operations and if() statements eventually results in the FPU stack being pushed too many times. This doesn't happen if you code in assembly and can count to eight. Often, rephrasing your code can cure it, but I couldn't find a phrasing that would do so this time, so I took the averaging code out. This may be the same bug that was reported in the *WDJ* December 1999 issue (Volume 10 Number 12), in "Bug++ of the Month."

When your level estimate indicates that you should be doing noise gating instead of compression, you need a way to smoothly change the gain from whatever it is down toward zero, and when you come out of noise gating, you have the same need going in the other direction. Remember that multiplying one signal by another signal produces sum and difference frequencies, and if you let gain change quickly, or there's excessive ripple on the level estimate used to generate gain, you cause distortion products in the output.

For most cases, the ideal solution is exponential — moving a constant number of decibels per second, for example. However, you really want to avoid clicking, and you want to get all the way down to zero and back up as fast as you can without creating nasty artifacts. At a constant number of decibels per second, you'll never hit zero gain — it's like the frog that jumps halfway each hop (Zeno's paradox). This is all worsened somewhat because at the gate threshold, the gain is at its maximum allowable value, when changing it potentially produces the greatest artifact power. I overcame most of these problems by creating another variable that you could think of as the noise gate gain modifier, which I allow to move linearly from 0.0 to 1.0 or 1.0 to 0.0 in the number of milliseconds the user supplies. This solves the problem of moving too fast and greatly reduces the audible artifacts. It also solves the problem of getting to zero. Now, there's just one more wrinkle. You generally want a noise gate to cut on, or restore audio, much faster than it cuts off, since you don't ever want to bite off the front of your signals. I accomplished this by multiplying the noise gate increment by two while it's coming on, so it happens twice as fast. Perhaps I should have added this factor to the dialog settings? Another possibility here would be to delay the data output until you know a signal's future, then cut the noise gate back on a little before the signal actually arrives. This approach would have no artifacts other than the extra delay, which doesn't matter for this particular application because you're working with files, but it could matter if you were working on a real-time data stream, where even a short delay might be objectionable.

I recently added a nice wrinkle to this transform by adding a little more logic and a separate, faster level estimator for noise gating. The plan was for the noise gate estimator to have instant attack and just the right decay time so it wouldn't chop off the front ends of sounds and it wouldn't be fooled by low bass signals alone. The extra logic also holds the normal level estimator at its last value, instead of letting it decay to zero or near zero while in the gate-closed condition. That would be wrong, so I keep the previous level as the best guess of what the signal will be the next time it appears. A restart after gating becomes somewhat smoother. The other advantage here is that the gate decay can be a lot faster than the normal decay, which means that a noise floor won't have time to be boosted at the end of a signal before the gate finally cuts it off.

I call this a simple compressor despite the apparent complexity because, compared to some implementations, it is very, very simple. I know of a hearing aid company that spent almost 10 years trying to improve a fancy analog feedback compressor, noise gate, and limiter. They wound up with separate fast and slow level estimators that would contribute only under certain conditions and a variable control loop gain before they finally got the precise behavior they desired. In this case, if the fast level estimator grew significantly above the slow estimator, it was used instead. The slow estimator prevented the gain from rising very quickly to high levels and boosting background noise when no signal was present, and it tracked the average signal better, while the fast estimator still acted as a limiter if a sudden loud sound came along. Because the quick, loud sound doesn't affect the slow level estimator very much, it also doesn't cause the gain to be reduced for much longer than the loud sound exists, so recovery from loud, short noises was improved. The sky really is the limit with even a simple concept like this, and a lot of nonintuitive (at first) things happen with particular signals. The upshot is that if you want a compressor that is ideal in some way, you need to think about signal statistics, artifact generation, and so forth. "Ideal" is tough to define without some interaction and testing.

An important variant of this technique is a "combo" processing effect called compression in bands. It uses filters to separate the input signal into chunks of bandwidth, compresses each individually, optionally refilters, then combines the results so that energy in one band doesn't affect energy in another as it does when full bandwidth compression is used. Compression and noise gating can be done more aggressively without producing as many audible artifacts. For instance, if the filters are "perfect" and less than one octave wide, all the first-order harmonic and intermodulation distortion that might be produced by aggressive compression simply fall out of band and are removed in the postfilter, although you still have to worry about clicks, which have energy at all frequencies and make most filters ring annoyingly. This technique adds a whole slew of possible new variables to the game. You can use it to encode signals at lower bit rates because each band can be encoded separately and at a lower sample rate. Because you can apply the noise gate more aggressively, a band's output often is precisely zero, which can make bit rate reduction coding even easier. Because the output dynamic range is now a lot less, you can use fewer bits in adaptive delta pulse code modulation (adpcm) mode with less distortion. In this case, the postfilter takes place after reconversion, so it filters out the adpcm distortion somewhat as well.

Normalization

When processing audio, it is often desirable simply to adjust a signal to a standard level, called normalization. The fun comes in when you start to define what you mean by "level." Do you want peak level, root mean square (rms) level, what? If you don't want clipping, perhaps the best choice is peak level. Sometimes you want a certain average power and don't necessarily care where the peaks will fall. Sometimes you might be willing to accept a certain amount of clipping, since below a certain percentage, it is not very audible. I wrote a normalization transform that lets you do all these things.

The big difference from compression is that the gain generated does not vary in time, and because I'm working in digital-land and can know the future, so to speak, by reading a disk, I can do a better job of gathering signal statistics before modifying the signal itself. The transform builds a histogram of how many samples fall at each possible level (to 16 bits resolution) in addition to the usual rms and peak sample calculations. This allows for the case of accepting a certain percentage of clipping, which can be a far easier way to get what you want than by diddling with the rms level. A possible improvement is to "soft" clip the peaks that do clip. For instance, if clipping is detected, go back in the stream to the previous zero crossing, compute a gain for this half-cycle that would exactly clip just one sample, and use that gain only for the one zero crossing. The sudden gain change will produce some artifacts, but doing it at a zero crossing reduces these way below what you'd get if the signal was allowed to clip at maximum power.

This transform uses the timer trick I mentioned earlier () to initiate collection of statistics. It is old code that I ported in from MusiCad, which is why the boilerplate code is different, but it shows that there is more than one way to hook up a new transform, and may help you better understand the other transforms. I'm not saying this way is better, just different.

Once the statistics have been collected, they are displayed in the dialog. The user can choose a method, as well as the desired output level for that method, and either Preview or transform the selection. Because the percent clipping option is new to the business, I'll explain how it works; the other methods are trivial. When going through the selection gathering statistics, I take the absolute value of each sample and create a histogram. I used 32,

768 bins, initially set to zero, and for each sample I simply increment the corresponding bin, giving a probability density function, more or less, for this selection. When the user enters a desired percentage of clipping, I walk the histogram from the top down, adding the values of all the bins above until that percentage of samples is above the current point. The gain that would make that point just clip is calculated, which becomes the magic number. I know how many samples constitutes the desired percentage because I also know how many samples were in the selection.

I recently added some code to this transform that uses the DrawDat control. It displays the histograms that I calculated for the percent clipping version, which makes it a bit more of a learning tool and built-in test equipment. You might want to play around and look at histograms of random and gaussian noise, sine waves, and music to get a feel for what you might learn from a histogram of sample values. If you use a short selection, you'll learn about low-confidence sampling in statistics, since there won't be enough samples for the histogram to become smooth or be consistent.

At this point the transform doesn't do soft clipping. It's conceptually pretty easy, but consider the case where clipping occurs on the first sample of a buffer. You've already written the previous buffer, so it's tough to get back to the previous zero crossing and redo the gain from there, so it's actually a pretty complex process. You would have to delay writing each buffer until the next was processed and ensure that one buffer was always long enough to completely contain at least a half-cycle at the lowest frequency present in the signal — either that or use more than two buffers. Again, I leave this as an exercise for you to complete.

Limiting

Several very different methods actually go by this name. Clipping is obviously limiting, albeit rather harsh limiting. Other schemes (and analog tape decks) simply reduce the gain sample by sample above some input level to prevent clipping, similar to taking a log of the sample. Tape media limits gracefully as more and more of the magnetic domains saturate. This produces artifacts, too, but not as badly as flat-out clipping. Another limiting technique is the soft clipping described above. More commonly, a limiter is just a special compressor that only takes effect above some signal level (a level ceiling), reducing the gain above that level and leaving it alone if the signal is below that level. The amount of limiting is then adjusted by controlling the input signal level. You could convert the compressor transform into a limiter easily using the same level estimation process by setting the gain a bit differently. I didn't write a limiting transform, even though it would only take about one minute of work because I *hate* limiters. They are overused in radio stations and album production, and I just don't like the loss of dynamics or the artifacts. Compressors have their uses for speech signals that have to punch through noise, guitar special efx, and so on, but I can't think of very many good uses for a limiter. Radio stations and album producers use limiters to get a higher perceived average level on a signal that has to go down a channel that has a power or maximum amplitude limit. For instance, TV commercials are usually very compressed and limited so that they sound louder in an attempt to get your attention. Limiting in bands sounds a little better than full-band limiting, though.

I think most audiophiles believe the world would be a better place if there were no limiters in it, but at the risk of my audiophile credentials, I'll show you how to make one. All you have to do is take the compressor transform and add an `if` statement to prevent the gain from continuing to increase when the input level drops below some value.

Amplitude Envelope

A very useful tool for computers in music production allows you to modify the amplitude of a section of data versus time. This corresponds to the common level slider control on a mixing board, but on a computer, you can do this much more accurately and with more repeatable results once you figure out how to get the user input. As usual, the DSP aspect requires far less work and thought than developing a decent user interface. The main DSP concern is not having gains change too quickly versus time so that artifacts are produced, which could also happen if you just don't have good enough resolution in your gain-setting widget. A little analysis indicates that you don't need to get all that fancy. If you draw an envelope on the screen at a decent width and height and use a linear scale (so that it includes zero in the range, always), you'll have no significant artifacts. My friend and sometimes employee Troy Berg helped out on this one. He developed a set of extensions for the `CDrawDat` class that allows graphical user input, which makes generating a transform for amplitude versus time adjustment more or less trivial. These are `CGuin` and `CMCGuin`: one is for single-channel drawing and the other contains multiple Guins for fancier applications. Even though you can draw a number of traces in each Guin, the UI gets cluttered fast when you do, so the signals are now separable by channel for this demonstration.

This one time I had to modify the boilerplate code. Normally, the selection lengths for previews are limited (temporarily shortened from what the user had selected on the file) so previews are more snappy. In this case, however, the user enters time-based information, specifying a gain value versus time, while seeing the envelope of the existing signal versus time on the same plot. The transform then interpolates the gain curve over the selection length. If this were allowed to change, there would no longer be any obvious relation between what you put on the screen and what you'd actually get, so I simply removed the selection-length-changing and -restoration code from this transform. No matter how hard you try, nothing is ever perfectly reusable!

For amplitude enveloping, you always want to include zero in your gain range so you can completely silence a selection or a portion of one. This may be one case where operating linearly makes more sense, since zero gain can't be expressed as a dB value. The Guin returns a number between 0 and 1.0, which is multiplied by the value you type into an edit box, where it's safe to use the decibel convention. However, what actually happens to the data is based linearly on the curve shape. The initial Guin user input curve is drawn at half-scale, and the gain edit box defaults to 6dB to give a gain of 1.0; – 6dB x 0.5 = 1.0. Remember that in linear-land, if you set the basic gain to 60dB, the default line will give 54dB (half as much) of gain.

It would be nice to draw the existing amplitude envelope of the data you're going to modify at the same horizontal scaling the user input curve uses, so the user can easily see what they are working with. This makes it very easy to get just what you want, compared to mentally trying to scale a dialog curve to what you see in the normal view, as some wave editors force you to attempt. Some of the underlying document code, the same code that is used to collect data for the normal view, is encapsulated in UpdateAmp(), so you can call it from various places. Because it provides more speed and less flicker, XORPUT is used for drawing, so you don't want a lot of vertical lines or sharp corners. The Windows' line-drawing algorithm often draws parts of lines over the same pixels, which leads to artifacts in which you see a single dot at a sharp peak with the lines leading up to it missing. However, using two horizontal pixels per datum would cut the basic resolution in half, so before drawing, a first-order low-pass filter is applied to the envelope data to make it look nicer.

CGuin is derived from CDrawDat (for a history of CGuin and CMCGuin, see , "DrawDat to Guin to MCGuin," on page 224), so you can use it in the same way: draw a button in the dialog editor of the desired size and position, add a member variable of type CButton, then change it to a CGuin or CMCGuin. I then take this out of the AFXDATA section to protect it from class-wizard. Now make a call in OnInitDialog() to initialize it with InitResp(), since by now you presumably have the information to do this correctly, which you didn't have at creation time. For this case, some of the parameters don't mean anything — you're not going to draw frequency scales or use the data members that assign frequencies to pixels, for example, so just shove some bogus number into the sample rate to satisfy the semantics. An additional parameter in the InitResp() function of CMCGuin specifies the number of Guins to draw. Set this to the number of channels in the file so that you have a separate Guin for each channel. Although you can draw up to four of each type of trace on any one Guin, this leads to a cluttered screen display that's hard to interpret. Next, call UpdateAmp() to draw the envelope. Now you can extract the data and use it as follows.

```
pWorkBuf[i] *= m_Draw.m_Guins[c].m_UserTraces[1][(int)findex[c]]
* gain;
// gain is from edit box

// pWorkBuf[i] is the current file sample being processed
```

m_Draw is the CMCGuin static member class. You need to select which Guin then which trace in that Guin you want the data from. Because trace 1 is used as the default green user-drawn curve, you need to specify that as part of the doubly subscripted array of traces and data. Finally, you get to ask for the actual point within the array that you want. The variable findex, which is incremented by a previously calculated step size for each sample, is used to interpolate between the arbitrary number that is the size of this array (it depends on your screen resolution and how you drew the original button) and the arbitrary selection size. This finds the closest point in the graph corresponding to where the transform is currently processing. The usual problems and artifacts involved with using a floating-point number as a counter apply here, so to prevent problems like fence-posting, you need to limit this index so that it doesn't go outside the array, which can happen if there is no exact floating-point representation of your step size or as the result of cumulative error. The error, although minor, may put you outside of the valid graph array bounds.

CGuins and CMCGuins notify the parent window via the same mechanism worked out for CEditSliders. A control change message is defined in stdafx.h and fired off to the parent when the user modifies the curve. As is the usual convention, the wparam of the message contains the ID of the control (in this case, that of the underlying button you drew) so that the parent can tell where the message came from. A diligent reader could modify this scheme so that the modified channel comes back in the lparam of the same message, although this transform doesn't need that capability, since its only purpose is to trigger the Preview update mechanism.

As usual, the code is still evolving. In general, it's smart to plan for changes as needed. Don't make it so it never breaks; make it so you can fix it easily!

Chapter 13

Combination Effects

Combination effects use more than one technique in tandem to produce the desired result. Often, delay-based techniques and filters are used together, or one or both are combined with modulation. You can get as complex as you want or set the limit at the number of computer cycles you're willing to burn. Most of the best effects use more than one technique. Sometimes the techniques are used in tandom, such as the case of a distortion transform followed by filtering. Other times, one DSP transform is used to control another, for instance when an envelope derived from the signal is used to set the frequency of a filter that is applied to the signal later, as in a "touch wah".

Boxcar Filtering

The first combo transform I'll discuss is really just a fancy filter. It qualifies as a combination transform because of the way it's used and the way it's organized, with an averaged time-limited signal length used to modify an entire signal. Boxcar filtering is a time-proven technique for averaging a periodic signal over successive periods to separate it from the noise or for separating a predictable noise from an unpredictable signal so it can be subtracted later. It's pretty simple. You have to know the periodicity of the desired signal before you begin so you can allocate a boxcar buffer of that many samples. This version only works if the period works out to an integer number of samples, but a variation of the table-driven waveform generator used for indexing could handle noninteger multiples, or resampling could be used to satisfy this requirement. As you go through the data, you simply add each sample to the boxcar you've allocated, while indexing modulo its size. In other words, as each new sample comes in, you add it to the next slot in the boxcar buffer, wrapping around to the beginning again when you reach the end, over and over. At the end of the process, the data is averaged modulo the boxcar period, and anything that works out to be periodic at the boxcar length or a submultiple of it tends to grow, whereas anything not related to this period tends to cancel or average itself out, since it will not be in phase with itself on each wraparound of the boxcar length. You can then divide your entire boxcar by the number of times the index wrapped around to get an accurate sample of what a signal periodic at the boxcar length would look like without the noise, on average. This post-scaling isn't necessary unless you need the exact average signal amplitude. Another way of saying this is that it amounts to dividing each sample bin in the boxcar by the number of data samples that were added to it during the accumulation, just as you would do any time you are generating a mean value. Here, we are just generating a lot of them. A fancier technique keeps a running average over all the data by treating each bin in the boxcar as though it is the accumulator in a first-order filter of the type described in . This allows the boxcar to track slowly changing conditions without being thrown hither and thither by noise. Fancier filters could, of course, be used — at the cost of increased complexity and more memory — but the simple first order type usually suffices. The filter design trade-off is that the slower the boxcar tracks, the slower it responds to changes, although it becomes less sensitive to noise. If you let it track fast, it acquires the periodic signals quickly but tends to be polluted with signal you don't want. The CBoxXFM transform example I provide here also allows for a couple of neat options. One option lets you use the boxcar output to subtract from the input signal synchronously in "running" mode. When set for 735 samples at 44.1KHz, this becomes an almost ideal hum and hum harmonic remover, if the hum level was constant. It just happens that 44.1KHz is an exact multiple of 60Hz. The other option, available when you're doing an average over the entire selection, autonormalizes the resulting boxcar data for you to full scale, which is handy for seeing tiny signals.

When averaging the entire selection, the single boxcar is simply substituted for the signal repeatedly over the selection. To use this mode as a hum remover, you have to do a little more work. For instance, I made a duplicate copy of some data in a new file, ran the boxcar on that to get a hum sample, inverted this using the amplitude envelope transform at its default gain of 1.0, copied it to the clipboard, then pasted it back to the original file in mix mode, which accomplished the subtraction. I perhaps could have added this to the original boxcar transform, but I decided not to. This mode is not normally quite as useful as the running average mode, which will track despite small differences in sample clock to signal ratios. I didn't put the "full Monty" on this filter's UI. Obviously, it'd be nice to have a DrawDat control and perhaps the option to fill it up with the boxcar, leaving the data unchanged. Things like this should be fairly easy for the dilligent reader to add for themselves.

Distortion

A good distortion effect requires a combination of techniques. The popular guitar amplifiers and stomp boxes do a good bit of pre- and postfiltering and have a nonlinear element to produce a pleasing and useful effect. Any nonlinearity produces harmonics as well as sum and difference frequencies between the original components. Some of these sound nice and some don't. The sums and differences can be very annoying, especially when chords or multiple notes are played, because they don't tend to land on the 12th root of two notes notes (the standard scale tuning used throughout the western world), much less musically desirable notes. Most often, both the input bandwidth and the output bandwidth are limited and shaped on both sides of whatever introduces the nonlinearity. Because a lot of the most annoying difference frequencies land in the low frequencies and corresponding sum frequencies (the scratchy-sounding stuff) land at high frequencies, the approach is relatively workable. The very best fuzz boxes work on each guitar string separately to avoid the more nasty intermodulation problems, but you don't often have that luxury on the Wintel platform with only two input channels. If you want the more musically pleasing even harmonics, rather than mostly odd order and intermodulation distortion, you simply make the gain different for positive and negative half-cycles of the input.

In the distortion transform I wrote for WavEd, the boilerplate code is about the same as always; no new tricks were needed. Well, that's not quite true. I wanted my new "tilt" filter shape to be available, since it closely matches the filtering used in the popular Marshall (and many Fender) series of amplifiers, so I added it formally to the CIIRFilt class. This transform uses fourth-order input and output high-pass and low-pass filters, as well as a tilt before and after the nonlinear part. The tilt is set up to have a center frequency that is the geometric mean of the low-pass and high-pass frequencies for either case, so it appears to act on the whole bandwidth.

For the nonlinear part, I allow the user to draw separate user graphs for each polarity of the input, using the `Cguin` tool. This way, you can have what amounts to separate processing for each half-cycle and have effectively different gain for each polarity, simply by sloping the two lines differently. This is part of how the famous Marshall sound is made. I once fixed one of these for a friend and noticed that the phase splitter that drives the output tubes was deliberately unbalanced about 20 percent; there were different resistor values for each half. Being an audiophile at the time, I *just knew* this was wrong so I "fixed it"; I assumed that this was the result of a previous incompetent repair. Well, that didn't last too long. He brought it right back, saying I'd messed up his sound; he was right. An imbalance for positive and negative half-cycles is what produces the most useful types of distortion (even-order harmonics), and I'd removed this! We live and sometimes learn. In addition, significant filtering is involved in both the inputs and outputs of an instrument amplifier. Some of this is obviously deliberate, and some is subtle. For instance, in a tube amp, the output transformer cannot pass very low frequencies and loads the output tubes heavily if low frequencies are trying to be output. This is the case when you have a DC offset, which is a natural result of having different gain for half-cycles, depending on polarity. The longer one tube tries to be "on" the more the output transformer will try to force it back to the zero state. Speakers of course, don't pass DC either.

To simulate an overdriven, unbalanced push–pull output stage well, you have to postfilter. Virtually all guitar amplifiers have a large amount of high-frequency tilt in the input stages. Sometimes this is greater than 18dB per octave and always boosts the high frequencies, while cutting the low ones. Fenders use a bit more tilt than Marshalls do, in general, since this makes more effective use of amplifier power and sounds better for the usual Fender use in country music, which uses less distortion (or used to in the old days). At any rate, a boosted high-frequency content makes the results of any following distortion more pleasing, since the low-frequency difference products are minimized, and the very high frequency output products fall out of the band the speaker reproduces. Oh, and by the way, replacing the power supply filter capacitors in a guitar amp with larger values is similarly verboten. In the case of Fenders, these are *deliberately* too small. They design their output stages to draw very little quiescent current, so the power supply caps charge up to very high voltages between notes, allowing very high (but unsustainable) peak power to be available. As the caps bleed down under load, the amp limits at a lesser power, which means that even when you overdrive a Fender, the attacks are not so clipped and some dynamic range is preserved. They figured out a good time constant for typical guitar players, and this is part of the sound by design.

To simulate this in digital-land, you have to prefilter and tilt to simulate the usual amplifier preprocessing, use a nonlinear element to simulate tube overload, then postfilter to simulate the effective filtering caused when the output tubes try to draw grid current through their input coupling capacitors. This causes the output transformer and speaker to limit the output frequency response. I didn't quite do the output transformer right, but it is darn close. I can make it sound like either my Fender Champ (single-ended 6v6) or Marshall (push–pull 6550s) when either is overdriven, just by making sure the settings are right.

You'll notice when you run this transform that there is a high boost on input and a low boost on output by default and there's band limiting on both. What's really important is that with the defaults, the different polarities of input have different gain, which reduces after reaching the clipping point. This is not exactly how the tube output stages work, but it is close. You can get base clipping by modifying where the curves start their initial rise from and you can get differing gains by adjusting how quickly each rises. The case of no distortion is a simple diagonal line for each polarity from the lower left to the upper right. You can simulate overdrive by making the curves reach the clipping point early in the horizontal span, which is what the defaults do. If you want a high end that sounds nastier, you might want to insert some up and down motion in each curve before it reaches the clipping point. This will add copious extra high frequencies, which will mostly be nonharmonic. I get more satisfaction messing around with the slope after the clipping point because it more accurately simulates what happens when a real output tube runs out of current carrying capability while the output transformer is demanding more just to stay at the same level. This is why the default curves slope downward after reaching a maximum.

The output filters will have some ringing and overshoot, and the tilt may have some flat-out gain at some frequencies, which will probably not sound good, so I added an edit box that allows you to avoid clipping after postprocessing by changing the gain before converting back to the wave file (probably 16-bit) format. I didn't add an input gain edit box because (1) I was running out of screen space and didn't want to redraw the whole thing, (2) there are about two or three other ways to get this in WavEd anyway, (3) you can draw your curves offset to the left and get the same thing, and (4) it would be hard to figure out what horizontal position corresponds to what input level, with the first being the most important reason. You could add an input gain edit box if you want to simulate extreme overdrives, but you'll wind up boosting all the nasty artifacts from your sound card and other recording lashup if you do. I suggest that it'd be a good idea to pass your signal through a good digital noise gate before using this transform, anyway, for best results; a digital noise gate can make a signal exactly, not just close to, zero, and will prevent boosting hum and noise to ridiculous levels in the absence of signal, as the analog gear often does. Many analog noise gates work fine, but when followed by a high-gain fuzz box, they seem ineffectual. One reason is that analog noise gates still have a little output noise, and the cable between the two can still inject some hum, which the fuzz box will then gain up to full scale.

Another nice addition would be the ability to programmatically add nonlinear curves (I suggest log-shaped curves). You can add user points to a CGuin programmatically, so all you'd have to come up with is desirable shapes. As Hal Chamberlin points out, the ability to transform a signal through a table has many important uses. You can do frequency doubling or create very different sounds as the amplitude envelope of the input progresses. There is a whole lot more to this than immediately meets the eye. If I'd made it fully general, I'd have put the zero point of the curves in the middle, vertically, so a positive input could create either positive or negative output. You can modify it to do this easily in OnInit(), but I thought it would be confusing in this discussion of its basic functionality.

I did use linear interpolation between the CGuin pixels, since a stair-step function implied by discrete pixels would add a lot of cruft that you might not want. This slows things down a bit, but it's one of the faster transforms, so it seemed reasonable at the time. Now a straight line is truly straight, if that's what you want.

If straight lines aren't what you want, you might try the curve smoothing I added as an option. Once I saw how cool this was in the arbitrary waveform generator, I couldn't resist returning and adding it here, too. It treats the contents of the CGuin as a signal and applies a low-pass filter to it. The filtering used here is somewhat different from that described in the next section, since the problem is different. You don't care about the mean curve level at all here, but you really care that the curve begins just where the user drew it: a little off is no good, since that would add base clipping or distortion at very low levels, for instance, when it wasn't intended. This hopefully will help create a trend in which DSP is used to aid user interface code.

Imagine how hard it would be to draw various smooth curves, even with an interpolating type of draw control, such as `filterdraw.ocx` used in the *S*-domain example. You can't get a straight line there to save your life; it's very nonideal for certain shapes. Other interpolation schemes don't pass the curve through the user points, which I don't like, myself. At any rate, this allows you to draw curves more easily that have slight nonlinearity similar to tube-type amplifiers, so it's definitely helpful. It would be nice, I suppose, to add a UI variable for the amount of smoothing, as I did in the arbitrary waveform generator. A DSP coder's work is never done.

Arbitrary Waveform Generation

Although not truly a special effect on its own, it's nice to be able to generate an arbitrary waveform on occasion, especially to test other transforms. Sometimes the right test signal can save you time by showing you in some obvious way what the transform is doing wrong. With this in mind, I added a generator for user-drawn waves to the Generate menu in WavEd. It has a `CGuin` to draw curves and the necessary table-driven waveform generation logic described in Chapter 6, to replicate the waveform at the desired rate and output amplitude. Because the `CGuin` only has straight-line interpolation, which means your curve will always have sharp corners, I added two options that allow you to make relatively purer waves. Also, it's pretty hard to draw perfectly balanced waves that don't wind up with some net DC content, which is not usually what you want, so I added one option to postfilter the generated wave to remove both DC and unwanted harmonics and another option to smooth the user-drawn curve before use. They're both effective but work in different types of domains. Directly filtering the output lets you think in terms of output frequency components. Smoothing the curve is similar but doesn't remove components in a frequency-based way, exactly, and this filter isn't as sharp as the fourth-order postfilter, but it sure looks cool on the screen. Remember that you can still create components above Nyquist that will alias back down into the passband. A little playing around with this transform should show you how hard it is to predict a sound's quality from looking at the waveform.

If you want to synthesize music the hardest way, you could use this transform to generate a waveform, use an amplitude envelope to put an attack–sustain–decay on it, then filter it, perhaps with the morphing parametric EQ, so the high frequencies fall of quicker than the low to simulate a natural source. This transform is hooked up to the real-time Preview, or it wouldn't be nearly as useful because it's not obvious to most people how a change on the screen affects the perceived sound quality, so it's nice to hear the changes as you make them. I suppose this could be improved further by adding an instance of the fft class and drawing the resulting frequency components on another trace of the CGuin.

The user curve of the CGuin probably isn't precisely a power of two in length, so a DWORD counts and wraps around (as described in Chapter 6, "Tables in Sound Synthesis"), then I scale the resulting bits to whatever the horizontal size of the CGuin is when it's drawn. Now full scale in the DWORD corresponds to full scale in the CGuin user curve. I precompute a scale factor up-front to avoid having to use a divide for each sample. I left off using the very top bit of the DWORD because for a lot of cases, compilers don't do the right thing for DWORDs, and instead treat them as signed numbers. To do what I wanted, I had to cast it to float anyway, after removing the top bit. I didn't even try it with all 32 bits because I've been bitten by that one before. I went straight for what I know is the reliable answer, so there's *only* one part in 2^{31} frequency resolution available. Instead of using just the top few bits of the DWORD, I used all of them to interpolate the user's CGuin curve points. If I didn't interpolate, there would be little stair steps, which would add high-frequency artifacts. None of this is required to generate sine or cosine waves out of a 1K sample table, for example. These are little tricks and wrinkles specific to doing this particular task well. You might accomplish the same thing with a faster run time if you initially interpolated into a table a power of two in size; here, it did not matter.

Band Splitting

In some cases, such as clipping, you can really benefit by first splitting the input signal into bands. If you split the signal into bands of less than an octave, clip it, then filter again and recombine it, the output will be compressed but with very little distortion. This is because most of the distortion produced falls out of each band and is filtered out. For instance, the lowest harmonic of the lowest frequency will fall just above the top of a suboctave band. You can work out the same sorts of relationships for sum and difference frequencies for intermodulation distortion.

This has interesting implications for data rate compression, as well. If you can get away with clipping in bands (and you can) and it still sounds OK, then you may only need 1 or 2 bits per band to encode a signal. Because each band is less than full bandwidth, the sample rate can be reduced for each band before encoding. This is part of how MP3 works. Next, something like adpcm is used to encode each band, and to get the most efficient use of limited bits, they are dynamically assigned to the bands depending on what's going on in each. For instance, a band that is silent doesn't need very many bits to encode this information.

The more effective single-ended noise reduction schemes are more or less just noise gating in the bands. This helps because in any one band, the narrower it becomes the greater the chance that there is only noise content; thus, there's more chance that you can force the amplitude in that band to zero without affecting the desired signal. With a full-band noise gate, every signal component combined has to drop below the gate level for noise gating to occur. If you are working in bands, the high-frequency bands can gate out the high-frequency noise in a signal during the time that it has only low-frequency content, for instance.

The example splits the input into the number of bands you specify then applies the compression transform to each band separately. I added a new boolean to the compressor class so it could be used as just a noise gate, so you can experiment with that too.

This transform works about the same way as the morphing parametric EQ (Chapter 10, page 182). I created a class that handles drawing a set of controls for a compressor. It contains the control variables and a compressor for each possible audio channel and creates as many of these as bands specified above. It decides which ones to show depending on what the tab control is doing. Next, I added various message handlers that update everything as you work, so the basic code is about the same as in the other transforms. One difference here is that you have as many work buffers as you have bands. I could have gotten by with two if I'd modified the filter engine to work out of place.

Suggested Improvements

The UI for the noise elimination demo can be pretty daunting to set up right. It would be nice to go through the selection once at startup, gather statistics on each band, and present them the user (or automatically set the controls). You'd want minimum, maximum, and average energy in each band so that you could set the target levels intelligently and not mess up the overall balance of the music, set the gate level just above the noise floor in each band, and so on.

It should be easy to add postfiltering by duplicating the code that does the input filtering. Just create another array of filters, set them up, then apply them after whatever in-band transform you use. This way, you can set the compressor to zero attack, which generates clipping, then filter out the distortion products before reconstructing the signal, for instance.

There's a known bug in the transform that I discovered after things had gone past my patience limit. I didn't manage the band centers exactly right, and I'd already spent a lot of time adjusting the filter parameters for just the right overlap. See the comments in `DoCFs()` in `CBandSplitXFM` for more on this.

Chapter 14

Audio Data Compression

Many times, the goal of digital audio processing is data compression so the audio can be stored or sent over a channel in fewer bits. Most of the development work so far has been with speech in mind, but a subset of the techniques also work for music. Recently, serious work has gone into compressing music data as well as speech, resulting most notably in the new MP3 codec. In all cases, some intrinsic redundancy in the signal or data is sought in order to save bits by leaving it out of the transmission. The method of compression can take many different forms. Most people are shocked to discover that normal Huffman-type coding, such as that used in popular compressors like PKZip, not only don't compress sound data very well but often make the data files bigger than before. This is also true of methods like run-length encoding, because the redundancy in the signal doesn't show up in the byte-by-byte time domain, which is what these coder-decoders (codecs) work with. In 16-bit audio data, all the possible byte values occur about equally and apparently at random, so these approaches wouldn't buy you anything. One way of phrasing the problem is that you want to get the signal into a domain where these types of codecs *do* work; however, not much has been done with this idea so far. Most techniques use *a priori* knowledge of the types of redundancies present in a specific type of signal, remove them, and encode the remaining information directly in a compact form.

For instance, a single-pitch period of speech, perhaps a few hundred samples worth, can be adequately coded and reproduced by characterizing the effective filter formed by the vocal tract, sending the filter parameters (approximately 10 numbers), and exciting a duplicate filter at the receiving end with an impulse, achieving a many-to-one compression ratio. This technique is called linear prediction, and is the basis of most low bitrate speech vocoders in military and commercial use.

Other techniques include reducing the number of bits per sample in various lossy ways, including reducing the sample rate to its absolute minimum, and taking advantage of some sample-to-sample correlation that is present in most signals. The fanciest lossy compression schemes also take into account psychoacoustic phenomena. Wouldn't it be nice if a lossy compressor had its losses mainly in areas the ear finds difficult to detect? For the most part, MP3 hides whatever loss or distortion it generates by using the ear's inability to detect certain sounds in the presence of other sounds. This is our old friend — the ear's critical band phenomena — coming back to visit, in a good mood.

I won't cover all these techniques completely because they have been developed over many years and chronicled in IEEE articles in *Journal of the* (choose a DSP/Computer/Math specialization) *Society*. Each method would require a book of its own to fully describe the approach and the empirical tweaking used in its development. The important thing is to get an overview of the methods that have been successful in the past, perhaps as a guide to developing your own techniques. Unfortunately, much recent development in this area has been either top-secret military or licensed proprietary work.

Logarithmic Quantization

About the simplest and most widely used method to code audio data is logarithmic quantization. In normal quantization, each a/d or d/a step has equal weight. In logarithmic quantization this is not the case. As the absolute value of the signal increases, the step size also increases. With "pure" log coding, the lsb step size is always a constant fraction of the total sample amplitude. This is hard to do, so piecewise linear approximation is implemented instead. U-law is used in the U.S., and A-law is used in most of the rest of the world. These methods are so similar it's a shame that the world doesn't get together and use one or the other. The net result of this approach is a noise floor that tracks the signal level. When the signal is soft, the quantization noise is softer because of the small step size; when the noise level increases, the signal increases, so the noise is masked by the signal, mostly. The technique allows eight bits about the same dynamic range representation as 12 or 13 bits and is widely used in digital telephony. Compared to the other schemes described below, it doesn't give very much compression, and the signal to noise ratio is limited to about 36dB. However, it is widely used because it's been around for a while and needs no computation to implement, so it's cheap; it's usually built right into the a/d and d/a converters used for telephony. I provide a (rather cheesy) code example for U-law encoding and decoding on the CD in `Source/compress/ulaw.cpp`. I've used this before to deliberately degrade a good signal to telephone quality for testing. It's not an ideal implementation from a speed point of view, but it's accurate to the standard. A fast implementation would simply use two tables to convert from linear to log and back again.

adpcm and Variations

The name alone, adaptive delta pulse code modulation (adpcm), tells how this compression technique works. Most of the variations on it concern what is adapted, whereas the simplest case exemplifies the principles on which all the variations work. In the simple case, an attempt is made to predict the next sample of the signal, and the difference between this prediction and the actual next sample is sent across the channel. To work, the predictors at both ends must see *only* the channel data — they cannot stay in sync with one another otherwise. The simplest useful predictor is simply the previous sample value. You can see how this might be so with audio and speech, because most of the energy lives at the lower frequencies, which implies that sample values usually slowly increase and decrease to follow these low-frequency sinusoidal components. In other words, on a sample-to-sample basis, most audio and speech signals are highly correlated. The net effect of a good predictor is to reduce the dynamic range required to send difference signal across the channel, and many special-purpose predictors do a better job than this with *a priori* knowledge of certain signal characteristics. This dynamic range reduction in the predictor error can buy a reduction in the number of bits required for each sample.

So far, you have a predictor and a way to send the difference between it and the input signal. It turns out that sounds don't often change overall level as quickly as the samples change: The envelope moves more slowly than the instantaneous sample-by-sample changes. To take advantage of this, adpcm uses a small number of bits per sample whose weight or step size adapts with the signal level. This way, when the signal is loud the bits have more weight, and when the signal is soft they have less weight. The use of a small number of bits always creates a noise floor, but with this technique, at least the noise floor tracks the signal level instead of being constant. When the noise is loud, so is the signal. This has more or less the same effect as compressing and expanding a signal around a noisy channel; in fact, it's precisely equivalent. The noisy channel in this case is created by using too few bits per sample to do a great job of reproducing a large dynamic range. Both sides adapt the step size or bit weights upward whenever the bits sent across the channel are big, and they clip and reduce it whenever the bits represent small values. With a table that describes how to do this, you can even simulate a fast attack/slow decay type of compression, which is how it's usually done. It makes you wonder why this equivalence between normal compression and expansion with pre- and postequalization and simple adpcm hasn't been mentioned in the literature. Of course, the real power of this family of techniques comes in when the predictor gets a little smarter than the implied integrator/differentiator.

Figure 14.1   IMA ADPCM decoder

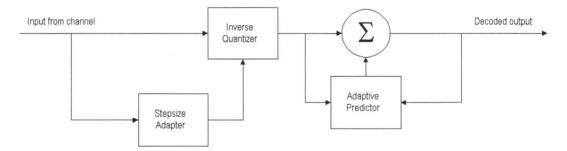

In all adpcm schemes, a copy of the receiver is embedded in the transmitter. Figure 14.1 shows blocks labeled inverse quantizer, step size adapter, and adaptive predictor (actually a misnomer since, for this version, the predictor doesn't adapt. It always uses the previous output as a prediction for the next.). Suppose you are producing a 16-bit output signal from a 4-bit input signal step by step with this decoder. Initially, the step size is set to its lowest possible value, which would be one, for a full-scale range of plus or minus seven steps with 4 bits. Sign magnitude is used in the standard, so you have plus or minus zero codes as well, which is a waste, as I'll point out later. The step size adapter contains a variable that defines the current step size table index, which is updated with each new input sample. If the sample is large, the step size is increased; if it is small, the step size is decreased, usually via a pair of tables. One table contains a list of values that seven steps (or full-scale input) represents, and the other controls how this table is indexed. The first table therefore contains all the allowable step sizes, and the second table contains a list of numbers that determines how much to increment or decrement the first table's index based on the current sample. In the IMA reference algorithm, the first (step size) table therefore contains 7, 8, 9, 10, 11, 12 … 130, 143, 157 … 27086, 29794, 32767 in a total of 89 approximately log-spaced entries. The second table, whose index is the absolute value of the current channel data, contains -1, -1, -1, 2, 4, 6, 8. These entries decide how much to adjust the index into the first table with each new datum. For example, if the sample is big, the step size increases by up to eight entries per sample in the first table, and if it is small, the step size decreases by one entry per sample in the first table. All that's really been created here is essentially a smart d/a converter that has variable gain. Its gain at any instant depends on the history of samples that have been fed to it. To handle various channel errors, a simple if() statement won't let the table index go past the end or before the beginning. The predictor in this implementation is simply an integrator. This is a fancy way of saying that each new input sample from the inverse quantizer is added to the current total and the result used as the output. Again, a little bit of logic here clips the integrator to full scale, or ±32,767 for 16-bit output.

The integrator (and matching differentiator in the transmitter) are the simplest "predictors" that work well, and it's easy to see why. Most speech and music signals have most of their energy in the low frequency range, so "flattening" it before transmission makes best use of the channel. In addition, it's good to have an integrator low-pass filter the channel noise on output because high-frequency noise is usually the most objectionable.

Figure 14.2   IMA ADPCM encoder

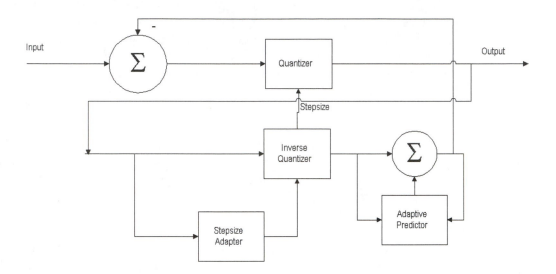

You can see that most of Figure 14.2 is identical to Figure 14.1. Only a couple of blocks have been added to the top to produce an encoder. The embedded receiver runs off the encoder output and thus duplicates at the sending end all the information available at the receiving end. All the encoder does is subtract the receiver's previous output from the current input and quantize the difference using the same step size both receivers are using at the time, ensuring synchronization between the ends of the channel. Because the receiver integrates and the transmitter encodes the *difference* between this and the actual input, the result at the transmit end is differentiation. The astute reader will also notice that this scheme tends to correct for error buildup, since the difference between the actual receiver output and the actual new input is sent each time, thereby continuously correcting the quantization errors.

This implementation has about two to three percent rms distortion when used on speech and somewhat less when used on music. With only a 4-bit sample, this is pretty good. By increasing the size of the step size table and using a brute-force search technique for the ideal small table values, I was able to reduce this distortion to about half with a 512-entry table for step size and the same eight-entry table (with different values) for the step size adapter, depending on the sample rate. The main problem with adpcm implementations is slope overload. If a sudden, large signal comes along, the implementation limits the slew rate until the step size can adapt to track it. This is another big reason that both ends must contain the same receiver; otherwise, errors will accumulate under these conditions. Music, especially the popular sort, has a more even level than speech, which is why adpcm works slightly better for it. Because it never dies away completely between pitch pulses, music tends to keep the step size high enough that slew-rate limiting is less of a problem. Noise also is masked better by the human perceptual system because of the increased auditory complexity compared to speech alone.

I've described the simplest possible variation that actually works — the IMA reference implementation. There are many others; for example, you can have any number of input/output bits, have any sample rate, use any predictor, and use any method to update the step size. The essential point here is to use *channel data only* or the two ends cannot stay synchronized with one another using this method.

One interesting possible wrinkle on this scheme that would eliminate slew-rate overloading would involve using two's complement numbers for the coding. You would then have one extra code, minus full-scale, that you could use for other signaling. I would use this to indicate slew-rate overload at the transmitter, at which point I would simply send the original data bits as the next chunk, providing a variable data rate, but not too many more bits, since overloading is relatively rare. Both transmitter and receiver could stay in perfect synchronization with this plan, since you also could use the new value to imply a step size for both. In the IMA scheme, even codes of plus or minus zero don't really represent zero at any time — they represent ±one half of a least significant bit at the current step size; that is, rounding rather than truncation is assumed. This creates a no-signal condition called bobble, since the transmitter sends out a continuous stream of (plus or minus) zeros, which show up in the output as noise at one half lsb amplitude. With the two's complement scheme, a zero code really would mean zero, so the idle noise would go away.

Because many other coding schemes produce data with sample-to-sample correlation, adpcm is often used with these to reduce the signal bitrate required.

The C example adpcm.c in the Source\compress directory on the CD is an example of the IMA standard adpcm encoder and decoder. This example was pulled from interrupt-driven running code in TMS320C31 DSP board my company makes. Simply pull out the pieces and reuse them. I added a trick that gets rid of some rounding errors and improves on the IMA implementation a little. See the code for comments.

CVSD

Continuously variable slope delta (CVSD) modulation can be considered a variation on the simplest type of adpcm. Instead of coding signals into chunks of bits, say four per sample, like IMA adpcm, CVSD usually runs at a higher sample rate but produces only 1 bit per sample. If the bit is a 1, it means "go up," and if the bit is a 0, it means "go down." The same integration scheme is used, and the a/d converter can now be a simple comparator. In fact, this one has been around since the days when it was the cheapest way to digitally encode sound data. Basically, leaky integrators are used, and at each sample time, only the sign bit of the prediction error is sent. The adaptation for this scheme takes the previous few bits into account and sets the integration rate based on whether they are all the same or not. It's not used much anymore and is more or less equivalent to adpcm. The main attraction is that before digital signal processing came into wide use, it was an inexpensive way to get into the digital domain. With a little math, you could convert a CVSD bit stream back into "real" samples and proceed without the use of what was a very expensive a/d converter at the time.

Sub-Band Coding

An important class of coders, including the popular MP3 audio codec, breaks the input signal into bands before coding it for transmission across the channel. Because each band has less bandwidth than the total, it can be sent with a lower sample rate than the original total, although obviously the total sample rate still comes out the same. However, as I mentioned elsewhere, when working in bands, you can filter out certain types of distortion and noise products at reconstruction time, hiding the fact that fewer bits were used to encode each sample. The MP3 codec takes this even farther by using a very clever (and fancy) scheme. It takes into account the critical bands of the ear for masking and dynamically allocates bits among the bands at coding time, using the most bits on the perceptually most important information. By noise gating aggressively in bands, using adpcm to encode each, and arranging the data so that something like PKZip can effectively compress all the resulting runs of zero data, you can effectively reduce the bit rate with low perceived losses.

Although I didn't provide a code example for this, you're most of the way there with the transform in WavEd for compression in bands. You can use its filters to band-split, then adpcm-encode the results in each band. I would build a frame, since for every sample required in the lowest band many are required in the highest band. The alert reader will realize that you can simply decimate from the original sample rate according to the band width. It doesn't matter if this results in aliasing or inverting the spectrum as long as the net bandwidth fits in under Nyquist. Up-sampling at the other end will turn the results right-side up again.

Linear Prediction

Linear prediction is a family of techniques that attempts to remove redundancy in signals by working in the same domain as the original sound production model. It is the basis for virtually all very low bit rate speech coding, and a lot of development, both theoretical and practical, has gone into it. The fundamental idea (when applied to speech) is that voiced speech can be modeled quite accurately by exciting a filter with impulses at the pitch rate. All you have to send across the channel for voiced speech to have a speech coder are slowly changing filter parameters and the rate at which to excite the filter with impulses. The filter parameters can't change very fast because they are physically related (literally) to the way you hold your mouth, which can only move so fast — quite slowly compared to an 8KHz sample rate, for example. It turns out that you can also model unvoiced speech (i.e., the consonants) in about the same way, except you need a simpler filter and the excitation used is white noise, instead.

Of course, what I've just said in one paragraph can hardly cover all the details of decades of intense research and empirical tweaking, and there are more variations on this basic technique than you can shake a stick at. One problem with the basic technique is that a pitch pulse isn't a perfect impulse, and making the assumption that it is reduces the naturalness of the resulting speech. One workaround is to send information about the actual pitch pulse across the channel as well. Also, at least until quite recently, pitch tracking of speech was very difficult and prone to error so that averaging and median-smoothing techniques were required to generate the pitch estimate. It turns out that a slight timing jitter in pitch pulses is one of the things that makes speech sound natural, so the smoothing needed for defense against large errors also took away some naturalness, lending a robotic quality to the result. I recently patented a speech pitch tracker that does not have these problems, making accurate pitch tracking a reality, but so far no commercial vocoders are based on it. Linear prediction is the practical king of speech bit rate reduction so far and has found wide use in applications where low bit rate really matters, such as military secure speech and in internet telephones. The only thing that might get you a lower bit rate with decent quality would be a formant vocoder, but this adds the error-prone formant tracking operation to the error-prone pitch-tracking operation, and none of these, so far, are practical outside of laboratories. I am hoping to produce a practical one soon. It turns out that having a good pitch tracker makes formant tracking much easier, bringing it into the realm of the possible with a little help from a neural net. With formant tracking, we enter a domain where formant frequency interpolation is possible and reasonable, so only the starting and ending formant frequencies for a syllable need be sent across the channel along with pitch information, greatly reducing the bitrate. Further, it turns out that for any specific speaker, a long term plot of their first 3 formant frequencies in 3-space resolves into 2 planes; most of the available formant space is never used by a given person. Sending the locations of these planes across the channel first would mean that now only the plane (one bit) and the location within it (a few more bits) needs to be sent to convey all the formant information, rather than complete filter specifications. I expect this to reduce the bitrate needed for speech well below 600 bits/second, but still to give "toll quality" results.

The best free demonstration code I've found for linear predictive coding is for the TMS320C3X family, available on the Texas Instruments Web page. It implements the TI version of CELP-10, which uses a standardized pitch pulse, instead of an impulse, and tables of possible filter coefficients that can be quantized into fewer bits by sending the table index for each filter coefficient instead of the coefficient itself. It uses the Gold–Rabiner pitch tracker and lattice filters along with Durbin's algorithm to encode. Unstable filters are prevented by having only stable combinations of coefficients in the tables and picking the nearest one to the computed optimal value. The sound quality is pretty good, although the averaging in the pitch tracker is evident. This code would probably have no trouble keeping up with in real time on any current Wintel platform.

I didn't include the TMS320C3X code here because it is copyrighted, but a little searching on the Web for speech compression or encoding will turn up this and several other examples in a short time.

Linear prediction doesn't work with music unless you have access to each instrument's sound separately. There's no simple model for a whole ensemble of instruments playing at once. You could consider MIDI to be the ultimate linear predictor for music: All that is encoded is a sound model and what notes to play when. The basic idea is the same.

Baseband Voice Reduction Folding

I just couldn't resist putting in a few words about something my Dad and I did a long time ago, since it's kind of cute and might inspire some new thoughts. At that time, we were interested in a relatively simple problem: how to get two phone conversations plus one teletype channel into the same phone line, well before the days of DSP. The teletype was, of course, 110 baud and didn't eat a whole lot of bandwidth, so the real problem was getting the two voices to fit, cheaply, with analog gear only (it's all we had then). We deliberately "aliased," if I may use that term, in analog-land by multiplying the speech signal by a 1.2KHz sine wave, folding the higher frequencies down over the lower ones, filtering out everything above the 1.2 kHz folding point, and sending the result across the channel. At the receiving end, we multiplied the coded signal again by this same frequency and folded it back up again (along with spurious energy from the original lower frequencies). The result was very understandable, if a little distorted because of the artifacts of having an extra upside-down copy of the low frequencies translated up to the higher band, but it worked because of the normal redundancy in speech signals. Because the coded signal now only took up about 1KHz of bandwidth, there was then room for two of them plus a teletype channel in the normal 3.5KHz telephone channel bandwidth, using the regular frequency division multiplexing scheme in wide use at the time at AT&T.

We tried to sell this to the then monopoly for any royalty at all, but they wouldn't go for it. They found a way around our patent claims and did something very similar, so we never got a dime. I guess we should have taken that small flat fee they offered; it would have been better than nothing. I suppose this sort of thing is why patents are falling into disfavor. Whoever has the biggest lawyer wins anyway, but I don't like the trend because the only way to protect your inventions is to keep them secret. This suppresses information and research results, which holds back new development.

Chapter 15

Noise Reduction

The big problem with good noise reduction is figuring out which part of a signal you want to remove (noise) and which part you want to keep, in some automated way. It can be quite difficult for even the human ear to tell whether a short burst of nearly white noise is part of a cymbal crash, a snare drum hit, or just background noise, and humans have that magical stuff called understanding and context. You can tell whether it's a cymbal crash or noise by when it occurs in a rhythm and by attack and decay rates, even when the sound in question seems to be buried below other sounds occupying the same spectral area, as shown on a computer screen. In general, computer software or analog hardware cannot. This has resulted in a plethora of different approaches, some of which work better than others on a particular signal or a particular type of noise, some of which hardly ever work but are popular nonetheless because they are either famous or rarely harm the signal, and some that just don't seem worth the effort to implement at all. One big difference between all the approaches is whether the original, uncontaminated signal is accessible. At first you might think that if you have this, all your problems are over, but they are not. I may be at the transmitting end of the loop and know my signal is going to be degraded before it gets to the other end, such as recording analog records or tape, sending it over RF (radio), or anything else like that. In that case, I may want to modify the signal ahead of time in various ways to make it stand out better against channel degradations. A two-ended noise reduction processor, one for each end of the channel, is appropriate for this class of cases. Just as often, you'll have somehow acquired a signal in degraded form that you want to fix. Obviously, this is usually the tougher of the two possibilities. Sometimes the problems can be made easier if you have various information about the signal, its noise characteristics, or both. Usually, this information must be supplied by human judgement. The state of the programming art just isn't up to doing all that a human being can where perception is concerned. Yet.

Double-Ended Noise Reduction

There are a number of popular double-ended noise reduction techniques, as well as a few obscure ones, as well as some other possibilities that as far as I know, have never been widely exploited. The single most popular double-ended noise reduction scheme is pre-equalization. It takes advantage of the fact that nearly all sound type signals have a 6 dB per octave slope in their average power spectrum, with most of the energy occurring at the lower frequencies. In contrast, most annoying noise either has most of its energy at the higher frequencies, or is constant energy versus frequency (white). By flattening or whitening the original signal spectrum prior to sending it over the channel, one makes better use of the channel's bandwidth. On the receive end, the filter needed to make the net channel response flat again is a high frequency cut, which will also cut high frequency noise components from the channel. This scheme is used in everything from tape recording to record mastering to FM radio transmission.

The next most popular double-ended noise reduction scheme is level compression at the transmitting end, followed by level expansion at the receiving end to restore the original signal dynamics. This approach works because most channel degradations, including noise, are additive to the signal rather than multiplicative. The idea is to boost the desired signal as much as possible all the time at the transmit end so that it's much louder than the noise while traversing the channel. At the receiving end, when the original dynamics are restored, the noise level ends up being turned down whenever the signal needs to be reduced to its original level. This is a nice result, since noise is usually most obnoxious when the signal is soft, and therefore, the signal to noise ratio is the worst, assuming a relatively constant channel noise level. Sometimes it is a good assumption, sometimes not. For instance, when working with analog tape, you can pretty much assume the noise is constant and at a relatively low level compared to the maximum signal. The assumption falls apart for various types of AM radio transmissions and other situations like pops on vinyl records, where you must make other assumptions. For analog tapes, the most popular, well-known noise reduction techniques are probably the Dolby system and the dBx system. In the Dolby system, the higher frequency parts of the signal are compressed somewhat to make sure they fall at a uniformly high level. This makes the best use of the tape's signal carrying capacity. On playback, an expander works on the same frequencies to put the correct dynamics back. Whenever this expander turns the high-frequency level down, the tape noise in that band is turned down with it, and it's the high-frequency hiss that most people find the most annoying. The dBx system is a full-bandwidth system that does nearly the same thing, but compresses over the entire spectrum, also adding a little additional high-frequency boost at transmit time and removing it at receive time. Tape decks already equalize because most acoustic signals have most of the energy at low frequencies, so "normalizing" the average slope to equalize the energy versus frequency curve makes the best average use of the tape's signal-handling capabilities. One possible problem with these schemes is that they reduce the level variation in the transmitted signal and use the tiny level variation left to know how to re-expand the dynamic range. This is good when the channel has constant gain, but in the case of a tape dropout, it will be made even worse at the receiving end when the expander operates, so this methodology only works really well when the channel has fairly constant gain. A similar flaw is shown when the channel has loud impulsive noise. Since this raises the power coming out of the channel, the expander will assume incorrectly that it should boost the level even further, making the noise worse.

I can't give you code examples of either of these methods, since that would violate someone else's patent, but I'm sure you can work up your own if you really want one. There are even plenty of analog components and chips available for which the license fees already have been paid. Other possibilities might include sending an out-of-band signal along with the desired signal in order to detect things like tape dropouts, but this hasn't been done much in the past because with tape, all the bandwidth it has is usually none too much, and there's none to spare for this sort of thing. In the case of "official" dBx, dropouts limit the usable compression/expansion ratio to about 2:1, which is still pretty darn good when expressed in decibels. I built an analog version of this once for my recording studio using the approved chips and was able to use more like 3:1 with very good results on a clunky old 1/2" 1950s Ampex tape machine. One nice thing about that old tape deck is that it put only four tracks onto 1/2" tape at 15ips, so the dropouts were minimal to say the least. It allowed me to continue using this fine old piece of gear for many years after it was officially obsolete.

In one sense, you could consider PCM (Pulse Code Modulation, e.g., converting the signal to a digital format), or even just FM, as a double-ended noise reduction technique. In both, you trade off increased channel bandwidth for lower noise if the signal is above some detection threshold. I'm sure you know what happens with FM when the signal drops below this threshold and your receiver doesn't have muting — full-scale noise. Pretty much the same happens with digital bits. In both cases, you can get essentially perfect reproduction as long as the signal is above some detection threshold. The downside is that digital bits take a lot more channel bandwidth to transmit. For instance, to transmit a 4KHz bandwidth, PCM might take 64Kbits per second, which is a lot more than can easily go down a phone line with current modem technology. Even at that, you wind up with only a 36dB signal to noise ratio with the common 8-bit companding codecs. To do this just as well as a decent local phone line, you'd need to up the sample rate even further and add more bits per sample. The big advantage of the digital approach is that no matter how long the phone line or the number of repeaters, it never gets any worse than that. The same is mostly true of FM. Both of these types are, however, susceptible to multipath or channel echoes corruption, and in this case, it's the original signal's own energy doing the damage.

For the FM case, multipath corruption causes distortion; for the PCM case, it causes bit errors, which usually sounds even worse. When either of these techniques sees a signal that is too degraded, performance is worse than if the signal had been sent linearly in the first place. In fact, this phenomenon is common among all "dumb" double-ended techniques. At least in the case of PCM, you could add error correction bits and hope it is enough. However, in nearly all cases, it really doesn't help all that much; the error bits also are wiped out because of the burst nature of most errors and the bit error rate rises very, very quickly near the detection threshold. The effective schemes add a lot of different delays to the signal that spreads the error correction bits and signal bits around in time so that one noise burst doesn't corrupt the entire signal and prevent reconstruction.

To summarize, most double-ended noise reduction schemes consist of attempting to fully utilize the channel bandwidth, even if the original signal does not, then undoing whatever was done at the receive end in hopes that this will also reduce the channel artifacts. Amplitude compression/expansion, spectral equalization, and PCM all fall into this model. You might not think of it, but even computer disks are a type of analog media; the signal thereon is analog until it comes through the read channel and a comparator, so even a computer disk uses this type of scheme. If you knew how close to the noise your disk data is before being cleaned up into a digital signal, it would probably scare you. However, a near miss is a hit, so the scheme works well.

Single-Ended Noise Reduction

More interesting to most people is single-ended noise reduction, and here I can at least give you code for some of the examples without going to jail. WavEd has an effective noise-removing transform for compression/noise gating in bands that can be set for noise gate only. In the single-ended case, you only have access to the degraded signal, so you have to figure out a way to tease the signal and noise apart or, indeed, to tell which is which. A popular scheme that works fairly well for music is called dynamic noise reduction (DNR). It's based on the theory that if the music contains a lot of high-frequency energy, it will mask the noise perceptually anyway, and nothing needs to be done. However, if the high-frequency content is very low, it's probably all noise and should be removed. This is implemented with a tunable low-pass filter that is tuned according to the high-frequency signal energy. If the energy is high, the filter corner is set at the top of the band and it does nothing. However, if the high-frequency energy drops, so does the filter corner frequency, and high-frequency energy is successively removed from the signal starting at lower and lower points.

I can't give you a code sample for DNR because of patents, but the tools are there on disk if you want to create your own. Of course, the devil is in the precise details of how sharp to make the filter, how it tracks detected high-frequency energy, and what you could think of as the gain constant for the filter tuning.

Adaptive filters have various uses in noise removal. Sometimes the noise isn't noise at all in the sense that it is completely unpredictable. For instance, sometimes the degradation is AC line hum and harmonics, in which case, a notch filter will remove it along with the signal content at those narrow frequency bands. An alternative is to train an adaptive filter on the AC line hum to predict it, and then subtract it from the original signal. This would handle cases where the hum was not at a constant frequency, for instance, such as the case of a tape deck with wow or flutter. In other cases, it's the signal that is predictable, not the noise, so you just do the opposite thing. Sample code for this already exists in WavEd in the adaptive FIR filter.

When used with malice aforethought, that is, an intelligent plan, it can be very useful in certain cases.

For truly predictable noise like hum and its harmonics, you can create a replica of the hum very accurately and then simply subtract it from the combined signal. This has the advantage of not affecting the original signal components that happen to lie in the same frequency areas. WavEd has a boxcar transform that does this.

The simplest noise reduction scheme is the noise gate, such as that described in the compressor example (Chapter 12, "Simple Compression"). Used without the compressor, it reduces the level to zero whenever the signal drops so low that you can assume it's safe to completely ignore it because everything left must be noise. An even better way to do this is to split the degraded signal into a number of bands. This works well with the human perceptual system (see the discussion of critical bands in) because signal in a noisy band will largely mask the noise, but a noisy band with no signal is annoying even when there is plenty of signal present in other bands. In fact, some of the popular noise reduction schemes, such as that in CoolEdit, work this way, although they are a good bit fancier. In the CoolEdit (David Johnston) scheme, the noise reduction transform is first trained on a sample of "just noise," and statistics are gathered on the noise floor in each band. During a run, the energy in each band is compared against this information. If the energy in a band is near the noise floor, that band's energy is forced to zero, so the noise in that band is completely eliminated. The trick of gathering noise statistics by band is, as far as I know, unique to this implementation. It is a very good idea, and full credit should go to its originator, David Johnston.

Johnston's implementation uses FFTs to split the signal into bands, which has some good and bad points. A good point is that this largely mirrors the ear's critical bandwidth situation. A bad point is that artifacts that are created when suddenly setting a band to zero before inverse transforming. As it is, I've gotten quite a bit of good use out of it in the recording studio. You might improve the scheme, and maybe even make it faster, if you use plain old filters and match the bands more closely to the critical bands of the ear. For low frequencies, I'd use IIR filters and accept some phase shift artifacts for the sake of speed, and for the high frequencies, I'd use FIR filters, because they're short and will not create artifacts other than an easily canceled delay. You could also use a noise gate much like the one in the compressor example, which would not create any tweedle tone artifacts whatsoever in the output. Of course this sort of scheme still leaves all the noise in bands that also have signal energy. There's just not a lot you can do about that unless either the noise or the signal is very predictable. If that's the case, the problem reduces to predicting one or the other and simply subtracting.

Types of noise that are not white and constant amplitude include record scratches, which are impulses that are often louder than the signal. Other places impulse noise is common is in radio channels. Many times, you can remove the worst of these by simply detecting the impulse as being too high above a running rms signal level estimate and blanking that interval or storing some signal from just before it and replacing the noise impulse with a little previously stored signal. Many of the vinyl record pop removers work in just this way.

As far as I know, however, none of the schemes do something that seems pretty obvious to me. For either the record or the radio case, equalization is involved. This tends to smear out channel noise impulses in time so that they don't stand out above the signal as much as they might. I'd suggest to any fanatic that the proper thing to do is remove this equalization first (or get to a domain where the noise is most impulse-like and least smeared in time), do your removal there, then restore the equalization to the way it was. You'd catch more impulses and each would be shorter, so you'd remove less of the signal in the process. If enough readers contact me and want this, I'll write it, but it seems the days of vinyl records or AM radio are past for most of us. In fact, for the collectors out there, yet another technique might improve old records: Discover the mappings between the perfect original sound and the sound you get with a stylus, which could include various resonances in the tonearm of your player and the original cutter, for example. You might be able to get a handle on these by using a part of the signal on the original that you know is an impulse (say, a drummer striking sticks together or a short record scratch for the vinyl parts of the degradation), finding the transform that reverts it to the original impulse, then applying it to the entire signal. This is generically called deconvolution. Sometimes you don't even need a reference because you can figure it out by ear or by characteristics that are predictable, but you're on much more solid ground with some sort of calibration signal. In theory, this technique could correct just about everything wrong with a recording except distortion, which can't be fixed with purely linear techniques.

WavEd User's Guide

I didn't include Help files with WavEd, so I should take you on a quick walkthrough and explain a few things for you. WavEd is very similar to MusiCad, so you can also check the Help files on its Web page (www.swva.net/mtnaudio).

WavEd was designed to reuse the skills of anyone used to working with Windows applications. It has just about the same features as the rest of the desktop publishing applications, but it works with signal data instead of words and pictures, so you have a lot of ways to play the data as well as look at it. The biggest problem most people have in using this program is in making the assumption that it must somehow be intrinsically different from other editors you've used, but it's not. Most of the key shortcuts and so forth work just the way they would with Word, DevStudio, and so on, with some notable exceptions. For instance, pasting takes on a whole new dimension. If the file being pasted to has a different number of channels, sample rates, or whatever than the data on the clipboard, you have the option of a Paste-Insert (like text editors) or Paste-Replace (overwrite or mix in). If several wave files are to be synchronized or remain in sync when you mix them together later, you want to avoid various length-changing operations by default, although you might want to snip or add a bit here or there to make a sloppy sync better, once you become an experienced user. The same kinds of problems can apply to cutting and undoing Edit operations. WavEd often has to pop up a dialog and to ask what you really want to do, unlike a simple word processor, which can make more assumptions about what is logical to do when you hit a key shortcut, so the shortcuts still work, but the program may have to ask for more information to complete the job.

WavEd was stripped out of a full-blown IDE project-oriented environment originally designed for recording studio work and to automate operations on groups of synchronized wave files at a time. Now, however, it's a straight MFC, MDI, document/view application with just a a few advanced features left in that allow various doc/view pairs to be aware of one another for synchronized scrolling and such. A lot of the nonalphanumeric keys have functions, including the numeric keypad, cursor control keys, Insert, PgUp, and so on, that are more or less intuitive. The Scroll Lock key controls whether the current view is slaved to and drives the other locked views when panning and zooming. Because the basic view is a splitter window, you can split up a view that wouldn't fit on-screen when zoomed way in and scroll and zoom the panes separately to get a selection exactly right, for instance. It also allows multiple views to be opened on any document, which in this case means a .wav PCM file format.

Table A.1 describes the nontransform shortcuts and operations that are not indicated in the menus. The transforms or signal generators also have the normal Windows shortcuts, which you can find by looking at the characters that are underlined on the menus.

As for mouse operations, the usual click-drag makes or modifies a selection. In the full-blown version, a double right-click spawns the editor of your choice on the selection in the clip file, but you'd have to add this back to the code if you want that functionality. WavEd was meant to be simple, for developing and testing plug-ins, rather than a fully commercial product.

WavEd provides the ability to open or record any PCM-standard wave file for efx development. You can put just about any data into a wave file. You should not limit your thinking to sound data; it's just a convenient format to store almost any type of data because WavEd parses it for you. At the moment, the maximum channel count is limited to four, which is twice the number supported by most other wave editors. A maximum channel count allows pre-allocation of memory for the viewpoint cache and increased speed with big files. However, if you're not too worried about memory footprint, you can redefine MAXCHANNELS in the code and recompile to get any number up to the maximum a wave format file can support, or 65,535, which should be enough.

As of now, however, WavEd does not support playing files with more than four channels, of which WavEd plays the first two. You'd have to add code to Cmmplaythread() to get more channels of output. Normally, this would not be meaningful because multiple sound cards would be required and they would not stay in sync with one another. This is one of the many mmsystem weaknesses MusiCad addresses by using its own hardware and drivers. However, you could cut from a multichannel file and paste back to a stereo file using the channel map provided, in order to choose which channels to play. That allows you to use multichannel files in WavEd then excerpt channel(s) from there into files that the other tools can read.

In essence, WavEd gives you are some dumb but potent tools. Just as a hammer can be used for something other than driving nails, you'll find that you can do just about anything you want with these simple tools by combining them in various ways, once you get used to the approach. All you need is a plan, and there's little you can't do.

Table A.1 Shortcut keys.

Key Shortcut[a]	Action
Ctrl-C	Edit-Copy
Ctrl-X	Edit-Cut
Ctrl-V	Edit-Paste
Ctrl-Z	Edit-Undo
Ctrl-A	Edit-Select All (and zoom to scale)
Ctrl-N	File-New
Ctrl-O	File-Open
Ctrl-S	File-Save
Ctrl-M	Open-File MRU list
Ctrl-L	Show/hide level meters/scope mode
← →	Pan the file. The number keypad is faster
↑ ↓	Change vertical scale. The numeric keypad is faster
PgUp, PgDn	Scroll one view's worth at current zoom scale
Home, End	Go to that end of the file
Insert	Edit-Paste at cursor
Delete	Silence selection
Scroll Lock	Lock/Unlock this view from global pan/scroll
Spacebar	Play this window or stop recording

Table A.1 Shortcut keys. (continued)

Key Shortcut[a]	Action
Esc	Safe Panic button, stop play, stop record, redraw, etc.
Alt-V	Toggle view mode: time or frequency
Alt-C	Cascade all windows
Alt-T	Tile all windows
l (L)	Loop-play selection (can use F2/F3 while doing this)
s (S)	Play selection
F1	(Will be Help someday)
F2	Set selection start to current mouse cursor
F3	Set selection end to current mouse cursor
F5	Zoom in maximum, one sample point per pixel
F6	Zoom in (×2)
F7	Zoom out (÷2)
F8	Zoom to maximum file length
F9	Zoom to $4/5$ of window and center selection
right-click	Context menu for this view

[a] None of the shortcuts are case sensitive; capital letters are easier to read in tables.

To use a transform or generate test signals, you need to open a file and highlight a selection. The transform will operate only over a selection on the currently active view. Pasting and the like use the selection start for their start but ignore the rest of the selection, since the data to be pasted determines its own length. When you want to record, please remember to check the record start and stop settings on the dialog bar before beginning. Usually when things don't do what you expect, this will be the cause. That's why it was moved out of the Settings menu so it would be on-screen all the time. Unfortunately, this new bar also requires the IE4 common controls, so WavEd needs to be on a fairly up to date machine.

WavEd Tutorial: Getting Started

Waved is an unusual beast that needs no installation and doesn't need tons of megabytes on your hard drive to operate. Your files may be big, but WavEd is tiny by today's bloated application standards. It's statically linked to all the libraries it needs other than the common controls DLL so that it will not have versioning problems with changes in the future. It still doesn't take all that much disk space, despite this. Just copy it, then run it. I made a shortcut to it and put it on my desktop.

If you have any experience with other Windows editing program, you're going to find this pretty easy to get going with. Only the type of file content and the display options are different: You can play as well as look at content. Just like all other editors, if you want to reformat or modify content, you have to select the content you want to change first. All the normal key shortcuts work for this, as well as the obvious click-drag for making selections. I suggest you create a new file for starters. This works the same way and uses the same key shortcuts and toolbar icon as most other editors. The one exception is that the program needs to know the parameters of the file you're creating, such as channel count, sample rate, and length. The latter is not terribly important now, since you can easily change the length later, but it seemed reasonable to let you select this at the beginning as well. You'll be prompted for the information (the defaults are fine for this tutorial), and for a filename and location in the usual Windows way. A view window opens on your new blank file in the usual way as well. I usually create a directory on my biggest, fastest disk called something like Waves so that my files are easy to find later, but for now you can use any directory you choose to store it in. Just remember where you put it so you can move or delete it later — wave files can get very big!

Now that you have a file to play with, it's time to play! Most of the things you'd want to do will require that you have a non-zero-length selection on the current file content. Two easy ways to accomplish this are to use the Ctrl-A shortcut (select all) or to use the normal click-drag of the mouse to select some portion of the file. For now, use Ctrl-A, which should select the entire default 10-second length of the file and automatically zoom the screen view to make it all visible. You won't see anything much in this window yet, because new files are created blank. Selections change the background color to white over their length so you can see where they are.

Once you have some or all of the new file selected, make a sweep tone so you have something to play with. Go to the Generate menu and select Sweep either by using the mouse to pull down the menu or Alt-G-S. You'll see a dialog that has parameters you can set for the sweep. For now, the defaults should be fine. Just click the OK button and your sweep will be generated and should show up on the screen in the view window. You may see some slight display artifacts, which go away when you zoom in. If you're pretty sure your sound card is set up and working, try hitting the Spacebar. You should see a play cursor move across the view and hear the sweep tone. If this doesn't work, look at the Settings menu to troubleshoot, or try adjusting the sound card mixer settings with its utility. If you hit the Spacebar again during a play, the play should stop and the play cursor disappear. If you have the normal low quality soundcard, you will probably hear birdies or extraneous tones during the higher frequency portion of the sweep. This is not a flaw in WavEd, and is pretty common among soundcards, even though they claim 16-bit quality.

The escape key is a safe panic button anytime you want to stop something. If play works, as seen by the cursor moving across the screen, or if you need continuous play while adjusting your sound card, make a selection on the file and use the L key to invoke a loop-play of the selection. This will play over and over until you stop it with the Spacebar, the escape key, or any other play function key.

If you'd rather record something than just generate test signals, you first have to make the correct sound card settings so that input is from the correct source and at the correct volume. This is made a bit easier by having a little real-time oscilloscope available in the program. To invoke this, use Ctrl-L (for Levels), and a display should pop up showing you what the sound card is seeing. The ubiquitous "any" key will cause this to disappear again when you are actually ready to do something. Before recording, check the Rec Start/End setting in the toolbar to see that it is set the way you want. Record can be limited to your selection, the whole file, and various other options. If you have the Manual Stop option selected, Record will continue until you stop it with that ubiquitous "any" key, even if the recording goes off screen. It will continue to record and use up disk space until you stop it or your machine crashes.

It's always good to know how to stop something before you start it. Now that you know how, you can record by invoking the Record command in the Edit menu. I didn't make this more accessible with a toolbar button or single key shortcut so that you would not accidentally wipe out existing content, as I sometimes have. As you record, the waveform should appear on the screen. This is also a handy way to set levels, since there's no penalty other than your time to rerecord if they aren't the way you want them. Disks are nice this way: Nothing wears out any faster because you are recording.

Now that you have content, either the sweep tone or something you've recorded, experiment with different ways of looking at it. If you want to see a sonograph, or spectral plot, try the Alt-V shortcut to flip the display mode back and forth. You can set the spectral view parameters to your liking in the Settings menu, but again, the defaults should be fine for now. If your recording is at a very low level, the spectral view may be mostly black. Try using the + and - keys on the number pad to adjust the brightness of the spectral display.

Now, get yourself back into waveform view mode using Alt-V if necessary (none of these commands are case sensitive; I'm using caps to make the letters easier to distinguish in print). You can now look at other things that are adjustable in the view. You can scroll back and forth with the scroll bar or use any of the arrow keys on your keyboard. You'll find that the cursor keys scroll slowly, whereas the number keypad arrows move faster per keystroke. If you have Play While Scroll enabled in the Settings menu, you should hear the sound as you scroll by it. The Home, End, page up, and page down keys also do the obvious things here. If you want to see the waveform amplified, use the up/down arrow keys. The gain goes up and down with the arrow up and down keys, which go faster if you use those on the number pad. (This doesn't affect the file content, only your view of it. In general, it is true of this program that file content won't be changed unless you explicitly ask it to be changed or you are warned and have a chance to Cancel first.) In spectral view mode, these same keys "amplify" the spectrum so that you can see a magnified view of the bottom part. As you move your mouse cursor around, you'll see some helpful information in the Status Bar Update. The settings in the Settings menu control the format of this information.

Often you need to see more or less detail on your waveform. Horizontal scaling is changed via the zoom buttons on the toolbar or their key shortcuts. The leftmost zoom button (Z) zooms in to one sample per pixel (which is as far in as this program goes), the button marked O- zooms in by a factor of two, the O+ button zooms out by a factor of two, and the M button zooms so that the entire file is visible in the window. The button I use the most is the funny-looking button to the right of these, which zooms to the current selection length, with a little data showing on either side. It centers whatever selection you have at $^4/_5$ of the window size, whether it was visible before or not. Zooming works the same way in either view mode. Play around a bit and get comfortable with it. You'll notice that when resizing a window, the zoom factor doesn't change in terms of samples per pixel, which is one of those design decisions I had to make; however, after resizing a window, you can always change the zoom factor as desired. When multiple windows are open, you can tile or cascade by using Alt-T or Alt-C. You'll notice that these nonstandard shortcuts usually use the first initial of the command name so they're easy to remember.

This might be a good time to look over the normal editing functions, like Cut, Copy, and Paste, which have standard toolbar icons and are also connected to the same shortcut keys as they are in any other editor. But here, there's a big difference. In a word processor, cutting always moves the text past the selection up in the file to cover the removed content, and pasting always inserts the new content and moves anything after it down the file, out of the way. This program can do that too, but often you'll be working with groups of files that need to stay in time synchronization, so WavEd has to know what you want to do when you cut and paste. There are options for the usual Insert and Remove, but also for cutting to silence without moving the rest of the file content and for pasting on top of existing file content rather than moving it out of the way. You can also have a Paste mix in with whatever's there or Paste with an automatic repeat if you want. This is handy for making a click track by pasting a single bar or group of bars into a file over and over, end to end. In conjunction with the Loop-play option, which allows you to fine-tune the selection while listening to it loop, it can be a real time-saver for musical uses. The insert and delete keys also invoke Paste and Cut.

If you don't like what you've done with these editing functions, the program has an Undo function that is like most others, and as usual, it's hooked up to the Ctrl-Z shortcut key. The program warns you if you're about to do something that can't be undone, so don't worry about it. The Preview function in some transforms also uses Undo to restore your file after the preview, so it may also generate the warning message. Unlike many other editors, however, it has a limit on the number of Undos possible, since you easily could fill up your disk space during an extended edit session. If you want to change the default limit, as usual, go to the Settings menu. If you are sure you want to keep your changes, you can purge all Undos by using the Purge command in the Edit menu. This was one of those no-win design decisions. I got a lot of complaints when I didn't have the limit, and I still get some about having it. I wanted to make this a safe place to play around, so I went with the limit and the warning.

Now that you know WavEd works conceptually like every other Windows editor you've ever used, it's time to move on to matters specific to waveforms. First, I'll show you how to filter your sweep tone or recording. Select some or all of your file, then select Transform-Filters-Graphic EQ from the menu bar. You'll see the familiar graphic EQ front panel. I chose this for the tutorial because most people already know how a graphic EQ is supposed to work, and this is one of the simplest transforms. A resulting frequency response curve that updates in real time shows up below the controls. Unlike most graphic EQs, this one lets you control the sharpness of the bands via the Band Q edit control. A reasonable default is calculated for you. Try changing this number (the range is 0.1 to 100) after adjusting some sliders to see the effect. Better yet, this transform, like most, has a Preview function that loop-plays your selection and lets you hear what the transform will do nondestructively. It isn't instantaneous, but eventually the program catches up with any changes you make. If you watch the progress meter in the application's toolbar, you'll see how it attempts to keep up and when it is finished. Drag the dialog off to the side so you can see your waveform, and you'll see that as you preview, the waveform display also updates in real time, which can help you avoid things like clipping.

Most of the other transforms work more or less the same way but do different things. The only ones that don't have real-time updates in preview are transforms that would update endlessly because they are slow, like the three reverb. It will be obvious if a transform does this or not by watching for activity on the progress meter or by paying attention to the sound as you change controls. To see what else is available, walk through the Transform and Generate menus with a selection in an open file, invoking each in turn. You can always cancel or Undo the results of any transform.

Once you've used WavEd for a while, other features start to come in handy. One I use most often is Open All MRU. This reopens the entire most recent file list from the file menu. There's a practical limit to how many files you can have open at once, but I've never actually run into it. The amount of memory you have would determine this, and at some point it would suddenly become very sloooow due to virtual memory swapping in windoze. With Open All MRU, all files open minimized as icons at the bottom of the screen (which saves disk read and drawing time at this point). At this point, you can play with some of the special multifile WavEd features. Use the Restore button on some of the icons to get normally sized windows. Now try the Alt-T and Alt-C shortcuts, which tile or cascade the open windows. You'll notice that if you scroll one window, all the others scroll to stay time-aligned. You can alter this behavior for any particular window by using the scroll lock key — this is what the locked or unlocked indication in the title bar of each window is all about. You can even start a selection on one window and end up in another! Any view window may also be split into two panes. Left panes of different views scroll along with each other if locked, as do right panes. In other words, when locked, left panes are only locked to other left panes, and right panes are only locked to other right panes. When unlocked, you can zoom or scroll a pane without affecting the others of its handedness. Different panes may also be in different view modes, so you can see a waveform view in both modes at once. One thing I find window splitting especially useful for is making long selections with very accurate endpoints. You can zoom left and right panes to one point per pixel, then scroll one or the other a long way off in the file. You can then set selection endpoints to single-sample accuracy at a zoom scale that wouldn't normally allow the entire selection to be on-screen at once. You can do this with a click-drag, since you can start a selection in any window and finish it in any other, but I usually use the F2 and F3 keys to set the start and end to the current mouse pointer location because you can set one end of a selection without affecting the other end this way. Often the most useful way to do this is to use the F2 and F3 keys while loop-playing a selection so that you can hear what you're doing the next time it wraps around.

WavEd can loop-record over a selection, and in this mode, each new loop writes over the one before it. This may initially seem useless, but consider putting the view into spectral mode first, and you now have a real-time sonograph! You can also let it record continuously until some event you're waiting for happens, then stop it before it is wiped out, potentially using a lot less disk space to capture something rare.

You could use the Paste-Mix function to give you the same effect as a mixing board. WavEd doesn't have the full-boat project metaphor with a defined output file (which someone could easily add back), but you can still do this sort of thing manually. You can also move channels around between files of different channel count and so forth with these simple hammer and tongs type of functions. In general you can do just about anything in WavEd. You might need a plan, or even a temporary file to hold intermediate results, but you can probably get what you need done.

I designed WavEd mostly as a learning tool, rather than as a practical production wave editor, but you could certainly use it for this. Readers who are also good programmers could use an ActiveX control called `pre-sets.ocx` that I provided on the disk and use DevStudio to add presets to most of the transforms for a lot more usability.

Appendix B

WavEd Programmer's Guide

For those of you planning to program with the WavEd platform, I'm providing an overview of how it all works, why it is the way it is, and a little history. This is the only part of this book that discusses the 90% of WavEd that isn't DSP transforms, and hopefully it will help with any programming efforts you make. Especially when debugging, it's nice to know how the rest of any system is supposed to work, and extremely desirable to understand something about how the author or designer thinks, as well as what they think. Although the code itself is obviously definitive, it can be hard to see the overall picture from code and comments alone. A little flavor of the person who wrote it is often very handy when trying to understand things such as "why'd they do it that way instead of my favorite way?" Once you've got some flavor of my design attitude, it should help you understand other things that I don't have space to describe explicitly here.

As I design, write code, debug, or do much of anything related to computers, I have a mental model of what's going on inside the box. In essence, I become the computer internally, and execute the code mentally. It has become so natural that I'd forgotten that not everyone does this himself or herself. I want to strongly encourage the practice! It is key to being both fast and good when it comes to computers, and those things in turn are very conducive to nice paychecks. Since computers are nicely deterministic, any conflict between my internal model and what actually happens is without doubt the fault of my model, which also encourages a healthy humility. After awhile, one realizes that it's best to fix the problem rather than the blame, which makes it less painful to adjust one's internal models as needed. That said, let's move along and get to an overview of the entire WavEd program and the design decisions I made that make it what it is. I'll throw in a little bit of history and color along the way, just for fun. While reading this, the best thing you could have along is a computer with DevStudio in it and the WavEd project loaded. Using the file and class navigation windows, you'll be able to hop straight to any code I'm discussing and see the real thing with very little effort. Without this sort of navigation aid, it would be hard to do projects this complex at all; it's very helpful, even necessary, to keep everything straight even for the original developers.

Some History

For many years, even before such things as PC's, my engineer friends and I had discussed how nice it would be to bring computers into the music recording business. It just seemed to be a natural application to us, and the advantages a computer could bring to the table were pretty compelling. At the time, only analog tape was in use, and this media is noisy, unreliable, not random access, and wears out noticeably after only a few plays. In doing takes in my analog studio, I noticed that I'd better save any content that had lots of high frequencies for last, because during the process of getting the other takes, this sort of content would lose quality quickly just by passing over the playback heads. At the time, computers simply cost too much for anyone but the government or a big corporation to consider owning one, so the idea had to languish for a couple of decades. At that time, a good pair of 16 bit A/D converters and D/A converters (not even normally commercially available) cost more than a PC does now.

Things change.

I developed MusiCad, with a little help from "my kids", Alan Robinson and Troy Berg, to overcome the very severe limitations of mmsystem.dll and soundcards on PC's, with the goal of providing "Desktop Publishing for Music". This was triggered by the advent of the first CD writers, which at the time cost only about $3000, too much, but in range for a studio product. We went with our own audio hardware design and opsys drivers, since `mmsystem` doesn't provide any consistent support for synchronous full-duplex recording, (e.g. playing one thing while recording another) which is the most fundamental feature needed in any recording studio environment. Yes, I know that nowadays some applications do support this capability using mmsystem. However, the authors of those apps inform me that they have to go way down on the metal and essentially write new code to support each and every different soundcard, after parsing its driver name. It's almost at the point where it would be better for them to simply write a new driver for each card and talk directly to that rather than `mmsystem`'s "friendly" interface. And they still can't guarantee that both record and play will start up in perfect sync every time, or stay that way. In addition, it is still pretty hard to find a soundcard at any price that actually has anything close to true 16-bit quality, despite the market-speak. I have measured quite a few, and most either have severe aliasing problems (just play a sweep tone through yours and listen to all the spurious frequencies — birdies — that come out), or noise problems (most are closer to 50 dB Signal to Noise Ratio than 93 dB, when honestly measured). Further, most have a user interface plan that's more suited to playing music from an internal CD than any serious recording situation; they tend to have poor programmatic control for their internal mixer application for instance, and no two mixer interfaces are just alike. Since full-duplex recording depends on mixer settings that are usually a lot different than the normal use of a soundcard, this lack of ability to programmatically set things right on demand is a fatal problem. This situation is beginning to improve as the rest of the industry catches up to where we were in 1990 or thereabouts, so don't bother flaming me with how great your new xyzzy brand soundcard is. You couldn't get that one a decade ago, and many current manufacturers' claims of 16-bit (or better) quality are still laughable, and perhaps should be legally actionable, on the test bench and to the ears.

We have Microsoft and Creative labs to thank for these limitations as well as other "features" such as a different number format for 8 bits than for 16 and so on. It looks like Microsoft just went with what Creative Labs was doing for Dos when it came time to "design" mmsystem, rather than thinking things out.

A system interface should be designed to abstract the functionality of whatever hardware is being driven, not limit what new features can be added, or at least not leave out ones that were definitely desirable even way back then. I'm actually being pretty kind here. My real opinion on this situation is not exactly printable. Of course, I have the benefit of hindsight, and they did not.

So, moving from MusiCad where we had our own hardware and drivers, to WavEd where we're dependent on mmsystem.dll was a bit of a challenge, but easily fit into the original design. All we had to do was write an `mmsystem` version of `CPlayThread`, and then decide which version to instantiate at startup. But much of the cool functionality of MusiCad just had to go out the window, or this would have become a book about how to handle full duplex or the mixer interface for each soundcard on the market, separately. We lost nice things like punch-record while loop play, full duplex sync recording, and since this is not a book about professional recording studio techniques, we left out MusiCad's Mix Project doc/view and some other stuff related to real time mixing of audio files.

WavEd Overview

WavEd is a more or less standard MFC Doc/View MDI application. This means that any decent book about Windows programming with MFC should be applicable to the basics, and I mention my favorite one below. The main reason for this is that I had a Doc/View pair laying around from the MusiCad project that was relatively easy to port to this project. For accuracy reasons, I'm using both MusiCad and WavEd here while describing things. However, one is a virtual carbon copy of the other except as regards some features I don't talk about here, so everything I say is relevant to WavEd, child of MusiCad, now all grown up.

When we wrote MusiCad, we had structured it such that there were several threads so we could keep both the visual and audio UI's responsive while DSP or mixing work was being done. After all, it can be time-consuming to process several gigabytes of data for a single command, no matter how slickly I optimized the basic design and assembly language helper functions. In WavEd, we still have two threads running at all times. One is the visual UI thread, the one we are born with, which also does all of the DSP and editing work, and the other is the play UI thread, which handles all audio I/O between files and soundcards. The play UI thread is basically a slave to the other thread, doing whatever it requests. These two threads communicate primarily via some programmer-defined Windows messages, which are defined in `stdafx.h`. For instance, if some view code wants to get something played in response to user keystrokes, it can make a windows message with a pointer to a `playinfo struct` in it and post it safely to `CmmPlayThread`, which will do the actual work and then `delete[]` the `playinfo` structure. Yes, that's right, we do this by `new[]`ing memory in one thread and `delete[]`ing it in another, passing the structure pointer across threads in a Windows message. And, it actually works! We even keep the pointers to `playinfo`s in a queue in the play thread and don't `delete[]` them until we actually process them, since the other code may generate more requests than we can handle instantly. You just have to be careful that anything that does queue removal also does the `delete[]`. We have not had a problem with messages getting there, or in wrong order, or any memory leaks due to this. Amazing what one can get away with if one is suitably careful.

There are obviously times when `CmmPlayThread` has to access document data, and since it has a pointer to the document(s) it is working with from the current `playinfo struct`, it can simply call the document functions it needs. This causes the spectre of cross thread synchronization problems to raise its ugly head, and there's a couple of non-obvious ones lurking in runtime library functions that are supposed to be thread-safe, but actually aren't. To handle the basic thread synchronization issues, every instance of the document code has a single critical section that is entered before every file access operation. This prevents obvious threading problems such as one thread seeking a file to a new position, and then another tries to read the file assuming the old position. However, it turns out that you can't count on a file's local ram write-data buffer being flushed properly to disk before an access is made by another thread, even if all the actual file accesses take place inside critical sections — you must do this flush manually. Unfortunately, the call to the `flush()` function is time consuming, so we don't want to call it more often than necessary. We accomplished this via some global `BOOL`s that flag whether the various threads are working with document data. If more than one of them happens to be true at a time, the file access code does an explicit flush of any file ram buffers after every write operation, while still locked in the critical section. And, that's it! This is all it takes to keep everything happy with multiple threads, but it requires a little co-operation from the programmer to make it work. This is why we set `gbEditworking` in every transform before doing any file operations, and clear it after we are done. The scheme of having some of the same code used by two different threads does often confuse DevStudio's internal debugger, but I'm used to that, and frankly, the scheme works so well I don't expect to be using the debugger on it again. In WavEd, the edits happen on the visual UI thread; in MusiCad, they happened in a separate edit thread. It makes no difference to this scheme how you do it if you keep the appropriate `BOOL`s updated properly as you work with file accesses.

There are quite a few global variables used in WavEd, which track things like settings, or the current user selection information. Other globals are used to flag functions in some class other than the caller to operate in some special way "just this time", for instance gbAutoOpen is used to modify the behavior of the file-open function in various ways. We needed this because many of these sorts of things in MFC are windows command message driven, so we couldn't just add a calling parameter. Others are used to track what threads are working on files right now, the current working directory, and all the other obvious things for which one might use globals. For convenience, I put these into a file called globals.cpp, and declare them external in a file named globals.h, which is included in stdafx.h like just about everything else is. Now, every non-trivial Windows application has to have some global variables, if for no other reason than to allow for a global settings dialog to function. Purists may call them something else, but if anything can get to it, I call it global, and forget about being politically correct to whatever fad is current; I've outlived too many fads to care. These global files are a good place to look around and get a feeling for what sorts of information the program keeps available to all of its parts. By using DevStudio's Find in Files function, you can search on their names and quickly find out how, where, and when they are used. I also made a decision early on to put every class header and virtually all #defines into stdafx.h, rather than using the #pragma once directive or the newer GUID-based include scheme. This makes all the various dependencies a lot more clear, at least to me, at the cost of triggering a full rebuild whenever anything significant changes. In earlier versions of DevStudio, this was actually nice, since partial rebuilds didn't always work correctly for some unknown reason. So, just about every other piece of code simply includes stdafx.h, which in turn includes all the class definitions and other header files. One ramification of this is that I had to be careful not to use any duplicate names. In my mind, that's a good thing, not a bad one. And each new transform you write won't force you to figure out what you need to include this time; everything is in stdafx.h, so you just include that. Hopefully, you will notice that I used all uppercase letters for the objects of #defines, to make them stand out better when reading the code. Other naming conventions include a passing nod at Hungarian notation, with a g being used to indicate a global, p used to denote a pointer, s for structure or string (it's usually not too confusing because we still have a normal name after the prefix) and so forth. We didn't bother to flag integers with an i, or floats with an f, since we have decent memories and it's usually obvious which we have anyway. I most often didn't use the m_ notation for class members. Why? It's hard to type! Virtually everything that isn't a global is a class member, so the notation is a waste of my time, assuming I can remember the names of the few local variables I always declare at the tops of functions; I generally eschew the new freedom to declare variables in the middle of code. In general, first letters of words used in naming are capitalized, as is the normal practice, and a C fronts every class name, also the usual convention.

It may not be obvious, but I tried fairly hard to come up with good, unique, descriptive names for everything important.

The Document Class

In the MFC document/view paradigm (20 cents!), there are two main classes that handle opening and viewing file data. One, usually responsible for handling the actual file and perhaps edit and undo functions, is the "doc", and the other, which drives the screen and handles the user, is called the "view". There are also an application class that mainly handles pushing Windows messages around, and several other minor players such as the frame window classes, and document template classes that specify which doc mates with which view. If you didn't already know this by heart, stop right here, go get a good MFC book, and read that first. None of this will make much sense to you otherwise. Nearly all of this program is coded using the normal MFC plan, and if you don't know what that is, you'll have a much tougher time understanding some things.

The `CWavEdDoc` class is multiply derived, and contains most of the WavEd-specific code, such as edit and undo functions. Next up in the chain is `CWVDoc`, which has a lot of wave-file-specific things in it, and above that is `CDocument`, the normal MFC document base class, and all its antecedents. There was one major thing we had to change about the way MFC does things here. MFC really, really wants you to read or write an entire file during your `serialize()` function, and it "helps" with this by making an undocumented temporary "mirror" file. The archive it hands you really points to this mirror file, instead of the file the user thought they opened. MFC simply does a brute force file copy to the mirror file to implement this, more or less. During a program run, it is this mirror file that gets worked on. When you save your changes, the original file is removed and the mirror file renamed to have the original file's name, again, more or less. The mirror file is actually slightly smarter and only copies what's needed to and fro. Unfortunately, what's needed to view a whole file is the whole file. Well, this is a smart idea for word processing or other small documents, but perhaps not for huge wave files — or programs that might open large groups of such files. Normally, the Windows assumption is that a user should be able to revert to the original by simply not saving changes at the end of an editing session, or be safe even if the power fails. This is normally accomplished by using this mirror file to work on, and only renaming it back to the original filename if a save is explicitly done. Well, we had to cheat this mechanism, and only use it for the wave file's tiny header information, which we keep up to date with every edit function. This saves both cycles and disk fragmentation. You should see what a mess most other wave editor applications make doing it "the Windows way" in just one editing session, not to mention how long they take to open or save a big file. You can see this by running a disk defragmenter both before and after a session. We started with the venerable ChkBook example, which has some errors; it doesn't test the archive direction flag correctly. But it did show how to open and then close the mirror archive file right off, and then re-open the file via a normal CFile for partying directly on the file data in-place, a huge speed and fragmentation advantage for this type of program. Since the program does support Undo's, we felt the advantage of far better speed (Remember 66 mHz 486's with 20 mS 200 Mb disks?) was worth the slight reduction in safety. The other important issue is that when we wrote this, most users didn't have enough disk space for duplicate copies of all the wave files in a project. Even though most people have plenty of disk space nowadays, it's still shameful to waste and fragment it in my opinion.

The WavEd document is designed to work with simple PCM "riff" wave files, period. It does not know how to handle codings other than type 1, or straight PCM. We just had to limit this so as not to get tied up in all this rare proprietary format stuff with no end in sight, preferring to support only the format that is required for burning CD's anyway. Having said that, we do allow some non-standard things, for instance, numbers of channels other than one or two, data sizes other than 8-bit or 16-bit, and the like. We do this through the normal wave header variables, which after all use a whole SHORT to specify the channel count, and others for various other things; why not use all the bits we're given to make this more generally useful? As such, the wave file format can be used for just about any numeric data stream a whole lot more efficiently than most other formats, especially ASCII text, for instance. This plan has come in handy when using these DSP transforms to look at non-audio data, or to produce some lengthy numerical result from a transform that isn't really audio related. Make it into wave files and you can use a lot of existing tools to look at and analyze your data — especially if it's big. Every non-trivial waveform-handling program has a major emphasis on speed, for obvious reasons.

Riff header parsing is done in CWVDoc, and we don't do it the Microsoft approved way. At the time that just looked like too much work to do, and we didn't see the advantage, if any. Perhaps someday we'll change that (or some nice reader will and send a copy). Riff files are segmented into "chunks" which have various headers or tags followed by the actual chunk data, and a program is supposed to ignore any chunks that it doesn't understand and continue parsing for chunks it does. This allows riff files to contain all sorts of things, if one parses them according to the plan, but it rules out working in-place on files that contain mixed-data types. So, instead of doing it the recursive Microsoft way, we just read in the maximum legal wave file header length and parse that according to some pretty strict rules about what it's allowed to contain, and where. At a time when most attempts to record wave files to CD's had all sorts of problems and nasty front and tail-end clicks when the wave files had been processed by most popular editing apps, files from MusiCad always worked perfectly, by sticking with the lowest common denominator format. No matter what it looked like when you opened it, if WavEd opens it, it's going to be converted into the simplest format on the first edit change. You win some, you lose some. Heck, I think we mostly won on this one.

CWvDoc handles file header parsing, and does all the reading and writing of wave files for the rest of the program. There are some pretty-nifty read-write routines here that will allow you to work in the number format of your choice (mine is usually float) regardless of the underlying file data type. The ones to use are GetPackedBuffer() and WritePackedBuffer(), which auto-convert number formats to whatever you want, and take care of figuring out where sample #n actually is in a file. As the name suggests, these routines work with packed data, which is multiplexed by channels, the same as the underlying wave file format. These are pretty efficient in the main, but improvements are possible for a number of cases. The big one is what we do right now if you call us with say, an 8 bit buffer that you want to write on say, a 32 bit file. Obviously to do the conversion, we need more memory than the buffer you passed us. So, for now, the program uses new[] and delete[] to get some memory for the conversion process — not that cool, since these functions will be getting hit a lot if you're doing anything important. The only alternative to this would be to allocate some buffer upfront in case we ever need it, and keep it for the duration of the program...perhaps just one for the entire program, or one per document. The problem is we don't know what buffer size your code might want to call a read or write with, and we didn't want to put in any artificial limits on things like this. You "pays your money and takes your chances" when you do software design! It is currently more optimized for working in number sizes that are larger than the underlying file, and that is what one usually wants to do to avoid arithmetic noise anyway. One thing I should warn you about over and over is that most of the buffer conversions are done in-place. After you pass a buffer to a file write function, the data in it may no longer be usable by your transform, having been converted to the file format rather than whatever it originally was. Well, at least I warned you once! When doing file reads, the functions do this conversion to whatever type you wanted before you see it, and the correct buffer size for what you asked for is assumed.

Most transforms will want to separate the channels from one another for doing actual work. Here, a few optimized assembly language routines are handy for multiplexing and demultiplexing. They are listed in convert.h, and the ones that do muxing and demuxing have md in their names. These can be used to pluck a single channel's data out of a multiplexed stream, and also to put a single channel's data back into one, by having parameters that specify how much to increment a source or destination pointer after each move. You could even move channel 3 of a four-channel file directly to channel 0 of a two-channel file if you set the parameters correctly. Be warned that these are industrial strength chain-saw-like tools that will do what you asked, very quickly. So, if you put in negative pointer increments, or ask them to move 4 gigs of data, they will try to do precisely what you asked, possibly with some unintended consequences.

Since we needed exclusive read-write access to any file we're going to open, CWvDoc also handles some munging around if you happen to try to open a read-only file, allowing you to try to mark it read-write or to make a copy of it to elsewhere first and then mark it read-write. CWvDoc also immediately updates the views whenever a file write occurs. A little thought will tell you how tricky this might be — we might have done the write from CmmPlayThread's thread, and this call will propagate in the "wrong" thread all the way to the view update code — not good.

Windows can be very unfriendly if the "wrong" thread uses its GDI objects. So, if we want real-time view updates, the view must check for this case and handle it. WavEd is just about the only good realtime updating program there is for wave files.

Another nice thing CWvDoc does is to allow reading past the actual end of a file — in which case you get zeros. This allows the view to set selections past the ends and do things to even zero length files, such as show a default view of one. It will also automatically append silence to a file if needed to satisfy a write request that begins past the current end of file. You may notice the delay if you scroll out to the ten-minute point of a new file, set a selection, and start recording in it. In this case the program has to hustle and write all those "virtual" zeros out first, making them into real zeros! If you're doing wacky things like this, expect your first take not to work terribly well.

Moving down a layer in the inheritance chain, CWavEdDoc does some WavEd-specific things, such as providing undo and the basic edit functions which both create the original need for Undos and are re-used to actually accomplish them; a slick little trick that avoids some code duplication. If you check out the code for AddUndo() and Undo(), you will find that various types of undo can be created to handle the needs of different types of transforms. For instance, most transforms don't change file length, but some may, and a different type of undo is needed to handle that possibility. In general, these undo functions use the global selection information in conjunction with a type parameter to decide what to save, and some information stored in a temporary file for each undo to know how to restore. CWavEdDoc keeps the list of undo temporary filenames, which is lost if the program doesn't shut down normally, leaving some GarbageName.tmp files laying around in the WavEd directory. They are safe to delete if this happens to you, they only hold undo information that can't be recovered under the current scheme. Some of this code is very ugly, from supporting both the Windows clipboard and our clip file. One thing that might be improved here is that we currently do all clipboard and undo operations in 16 bit, which might not be that great if you were working on floating point files, for instance. At the time it was written, 16 bit was the best there was, and is still the most often used. A diligent reader might want to change this code to make these operations reflect the underlying file type. It shouldn't be all that hard to do and would probably result in less total code.

DocWalker

CDocWalker is a cool little class we brought over from MusiCad. The basic idea came from Mike Blasczcak's *Revolutionary Programming with MFC 4.0* book, which has become Mike Blasczcak's *Professional MFC with Visual C++ 6*, from Wrox Press. This is my personal all-time favorite Windows programming book. What CDocWalker does is plow through the document template(s) of an application, and make lists of pointers to all of the documents the application has open. This was absolutely necessary for MusiCad; there we had a project metaphor much like DevStudio, and often needed to do things to groups of files at once, usually all files of some particular type. With only slight modifications, WavEd's CDocWalker can keep track of more than one document type at a time, and provide access to these global sorts of operations with very little pain and suffering. The way we actually did this isn't quite as nifty as it might have been, requiring a different call into each document for each "universal" function rather than sending some operation code to a single call, but it works, and works well, just the way it is. If you find yourself wanting to do something that knows about all templates, docs, or views, this is the place to go fool around; a doc can enumerate its views for you once you have a pointer to it.

This class is what gets you the information you need to begin the process.

We have one as a member of the app class, called `AllDocs`, and the document code causes updates of the lists as they are opened or closed by telling it to `Walk()`. The other reason we sometimes have to tell this to update is a variable it keeps for the longest file's length, used for the zoom-full command. So, any length-changing transforms you do should call `All-Docs.Walk()` afterwards, so this variable can be kept correct. Hint: this type of thing is really, really useful in any MDI app where you want to do things to groups or classes of documents, or have a project metaphor with multiple document types as well. Thanks, Mike!

The View

Boy, is there ever a *lot* of code here. Some of it is pretty, some not. What we were trying to accomplish was a flexible view of any kind of data stream, and we almost pulled it off, I think.

It also wanted to be fast as blazes even on the 486's we started this project on, and that added a goodly amount of complexity to things. Firstly, we wanted to have a splitter window — users needed to be able to make sample-accurate selections over ranges that would not allow the entire selection to show on screen at one point per pixel, for instance. MFC did make some provisions for this, but not enough, since we wanted to keep track of the panes in a deterministic way. When many windows were split for instance, we wanted all the left panes to scroll together, and all the right panes to do likewise, but not to have a right pane scroll affect any left pane. Well, it turns out that when a user un-splits a window, neither you nor the user has much control over which pane is left (in *either* sense) with the garden-variety MFC stuff. The solution was our own splitter frame class that ensures that a "right" pane will never be orphaned on the screen, and we use that instead of the stock MFC widget for our frame window class. `CWavEdView`'s 2400+ lines of code handle most of the messages that fly around the application due to the user, provides most of the user interface, uses most of the app-global variables, and in general makes WavEd what it is, for better or worse. Looking at all this code, you'll be able to tell that encapsulation in its various forms is not necessarily one of my religions. Personally, I like Mike's idea that a document/view pair should be synergistic with one another. A Windows app simply must have some global variables, whatever you call them, otherwise you can't do settings from some dialog and then have them affect everything, something every windows program of any size has by design.

In this program, we decided to call things what they really are and put all our globals in their own file (`Globals.cpp/.h`) for easier maintenance. In a passing nod at Hungarian notation, we prefix all these with a `g` to indicate their global-ness. In general, they're safe to read at any time, and might be safe or even necessary to write if you know what you're doing. These include the global selection information, the default wave header to use for new files, and other such things. One, `gbAutoOpen`, is used to flag a document/view you're about to open or create to "just do the best you can with this path and give me a pointer to the result or `NULL` — and put no error messages up on the screen no matter what." This is handy to open a group of files quickly if you have a list of paths, without causing the user to have to wait or answer a bunch of questions. It's used in things like the `Open All MRU` command or when creating a new blank document/view in a transform to write on. But with all things like this, it's a good idea to leave them the way you found them when you're done; they are shared with your roommates and you want to keep the kitchen clean, right? Just about everything uses the `gsZoomInfo struct` to find or change the current selection, so you'd never want to leave this different from what the user drew for any length of time, for example.

Well, I suppose I should really start with what happens in `OnDraw()` — the meat of any view.

Actually, I have to start a bit before that. We wanted a black background for this view, and so had to override the `OnEraseBackground()` function. We found some "interesting" problems with this when using any other color, especially when the application moved to various different machines. It seems even a "solid" brush winds up being dithered on some machines, and this gives a very raggedy look as the view is scrolled — the dithering doesn't line up as it should on the new part revealed by the scroll and you can see a ragged line. We found this artifact happening even if we used a `RGB(0,0,0)` black brush, a color that should exist dither-free on any system! Hence, our dodge, used in `OnEraseBackground()` is to use `::PatBLT()` with a special "BLACKNESS" flag instead of a brush. Well, we wanted black anyway, but if you want to try something different here, expect to have a struggle. Who knows, maybe they've fixed whatever that was in one of the revs of MFC that's happened since this was written?

Moving to `OnDraw()`, we find that we first have to figure out what to draw, in what mode, and where. This is made more complex by the fact that an MFC scroll view still doesn't support a viewport larger than an unsigned short integer in any dimension, much too small for a wave file. In our case, we might have a 2 gig file and be zoomed in to one sample per pixel, or be zoomed so far out that the same huge file is one pixel or less in width — a 32 bit range rather than 16. We therefore had to write a little tricky code to map a much larger range onto the smaller one. And here, above all places, we were after speed, speed, and more speed — and we got it, too, having the fastest wave view on the market without using some crufty disk file to cache zoomed viewpoints. It isn't perfect, and methinks we might have used xorput about one time too many, but it gets the job done.

First, `CWavEdView::OnDraw()` checks if this is simply a selection update, which we can do quickly and then return. We use some variables that are updated in `OnScroll()` and `OnSize()` to help us quickly find where we really are, do the selection update, and get out. If we are not here to do a selection update, we instead fall thru and see what view mode we are in. We decided to let this be an int, so we could have many modes, and to keep this view by view, so we could potentially have more than one view mode on the same data easily. If we're in straight waveform mode (don't ask why it's mode 2 rather than 0!), we then look at our zoom scaling to decide whether to just read the file, or do something fancier. If we're at under 32 samples per pixel, we just do a straight file read and draw, which is the fastest way. If we're zoomed out farther, we switch to a different way of doing things. First of all, we keep a cache of view data to handle cases where the user is just scrolling back and forth and not changing anything. This is far faster than reading the data off disk and computing `zoompoints` (which are actually short vertical line segments representing maximum and minimum sample values over the interval) when we have a cache hit. This cache size is something that is affected by the global #define of MAXCHANNELS in `stdafx.h`, as well as some other #defines there. See the first few lines of the constructor for more on this. This is basically the issue that sets a limit on how many channels per file can be supported, traded off with how many files can be open at once, before the memory footprint gets too big for your machine to easily handle.

When zooming out so that many samples occupy each pixel, what we really need for drawing data is a good estimate (at least!) of the maximum and minimum data values that occurred during that pixel's worth of samples. This turns out to be sort of a knotty problem that most waveform viewers fail to solve, at least somewhat. It just takes too long to read each and every sample to get that max/min data if you are zoomed way out — you might wind up reading the entire file just to show a screen of data — or one vertical line's worth. So, the brute force approach is basically out for a practical viewer. What we did instead to approximate the same results was to read a few buffers worth of the file, but less perhaps than the entire pixel's worth, and then sample *randomly* within that data to get our min and max values. Sampling at just every nth sample causes some pretty nasty artifacts — there may be a huge energy at some frequency, but you only happen to catch it at zero crossings, or just at positive peaks for instance. Either way, you get garbage when what you want is a reasonable representation of min/max over the data range in that timeslot. This is just another way of demonstrating the Nyquist phenomena. The method we use here isn't perfect, maybe not even the best I could do, but I like it better than most of the other methods I've seen. One popular wave editor simply reads the entire file an extra time when you open it, and makes a shorter file that has this sort of information in it for use during runtime. It unfortunately doesn't seem to know how to auto-delete the resulting file, however, and so it trashes up your disk. Also, that shorter runtime-created file only has really good info for one zoom-resolution and so it doesn't work perfectly in all cases.

Since this data is expensive to get, both in disk time and computer cycles, we decided to have a viewpoint cache, even though having this cache brings all the problems associated with keeping it up to date with file writes possibly happening in another thread. This cache is very simple, just a contiguous range of viewpoints kept as a circular buffer so that as the user scrolls off one end or the other, the update is simple to do — and you can't have a miss in the middle. If it weren't simple, it probably also wouldn't be fast enough. We don't, but could, for instance save about half the disk work on a miss by compressing/expanding the existing cache data and only filling in the missing points at the new resolution. Lots of times we'd just as soon take the miss as a way to really refresh cache. There's still this one little bug...drawing behind dialogs. Kudos to anybody who finds and fixes it. I think it's actually a windows bug, or some wrong style bits someplace.

The other currently enabled view mode is the spectral view. For speed this uses the C-Lab FFT and its special ability to write directly to bitmap strips. These can then be `blt`'d to the screen.

It's the only code example of using our FFT, currently, and does not really do it justice. This is not cached, as a bitmap image represents just too much data to be reasonable. Instead, we made the FFT very fast, and the drawing fast, and right now, it's actually the drawing that takes most of the time for most FFT sizes and zoom scales. Even re-reading the disk seems fast in this mode, because the data is usually in system cache, and we only need, at most, one FFT-size worth of data at every pixel position even when zoomed way out, rather than an estimate of all the data that pixel represents. You'll probably get a lot of other use from the FFT. It's optimized with some assembly language code that we provide also in .obj form, so you can still use this even without owning Masm. There's probably still room for some improvement in this FFT's speed, but on the whole it seems faster for the sizes it was designed for than any other Wintel implementation I can get my hands on.

Other things in the view handle keystrokes when the view has focus, the usual right click invoked context menu, setting the selection and updating status bar text as the mouse moves. Look at OnKeyDown() and OnMouse-Move() to find this stuff. You'll note that we violate another major assumption that most windows programs make — we actually use the keys on the keyboard for what they say they do, for one thing, like the page and scroll lock keys. The other odd thing we do is to use single keys for shortcuts — even alphabetic keys. We did this because you don't do much typing outside of a dialog (actually, none) in this program, and why leave 101 buttons unused and let you use just the two on the mouse? Troy Berg hates mice, and types fast, and so he's led us into putting a key shortcut in here for just about everything. Most of the normal ones that are used in many apps also do just what you'd expect in this one, and in general that is accomplished via the accelerator table/menu/command message facility. It's only those special single keystroke handlers that live here in the view.

The view window often simply ignores your first mouse click, especially if it was inactive when that first click occurred. I did this to prevent your (maybe painstakingly set) selection from being changed too easily as you flip from window to window. Don't you hate it when you move from window to window in DevStudio, and the only place you can click on the underneath window to bring it forward also scrolls it off the left margin, or pulls up some dumb menu? I think that is extremely poor, even sorry, UI practice. I can't think of any printable words that actually express what I think of it on most days. At any rate, we support window tiling and cascading from keys like alt-t and alt-c as well as menu commands, and this makes it very easy to get to any window you have open. By the way, we've gotten so used to having these handy shortcuts that we've also added them to DevStudio itself and many of our other tools as a time saving good idea.

The main reason I expect anyone will be getting into this view code, other than to study it for any "steal me" slick tricks, is going to be to add transforms to WavEd. This is pretty simple to do, and is described elsewhere. Basically, you just add a menu entry in the WAVEDTYPE menu, then call classwizard up with the view code, and add `OnCommand()` and `OnUpdate()` handlers for the menu command. You'll see a bunch of examples of this at the very bottom of the view code. We pass our `this` pointer into the transform dialog, from which it can get any other pointer in the program easily. One might consider that in fact all of WavEd is simply there to support these transforms, and you wouldn't be far off — by design.

The Transforms

We've been handling transforms blow by blow in another place. I just wanted to point out that there is indeed more than one good way to do these, and some that I didn't describe in the text do things a bit differently. Sometimes this was just because we wrote them before we got smart about the best way to do certain things, and sometimes a particular transform just has a different set of needs. For instance, the normalization transform needs to do a lengthy collection of statistics before you can use it. But if we didn't draw our dialog right away, you'd think something was broken. So it sets up a short Windows timer in `OnInitDialog()` which, when it goes off, triggers the statistic collection process *after* everything has been drawn, which seems to happen after `OnInitDialog()` returns and some messages fly around. It turns out that this is the only way to reliably do this sort of thing, and even this isn't perfect, since there may not be a timer available.

In general, all of WavEd was written especially to support doing odd transforms. In transform land, you can get pointers to any or all documents and views, the app and any globals, or just about anything, and party on it. You can create your own new document/view pairs from there as well, a handy technique, sometimes. You could, if you wanted to, start another thread from a transform and have it finish up work after the dialog goes away, or do just about any windows programming you'd like to do. This is specifically why there are not a lot of private members in the rest of WavEd — I knew I couldn't anticipate everything that might be wanted in the future, and I was right — I get suprised at what new thing I want every time I tackle something really different. That happens pretty often, by choice. I like trying new things, and it's also what I get paid to do, in general.

The Utilities

There are a few of these around, really reusable "swipe me" code. Some are just little dialogs that let users specify channel mapping and such - you'll want to see how we did some of this tricky stuff like creating the right number of pages or dynamically titling them. Some are major general pieces like the FFT class or the filter classes. Some are just handy little control handling classes like `CEditslider`. We named everything with nice long filenames, so it ought to be obvious what's what. I just wanted to give my encouragement to anyone who wants to reuse any of this code. If you make some really nifty improvement to any of it, heck, why not send me the new, improved version in an email while you're at it? The address is clab@swva.net, just in case you can't find it elsewhere in the book. This is how a lot of this code was built up in the first place, and it'd be nice to keep the improvements coming as we all learn and grow. I'll put all this stuff up on a web site so readers can download the latest. I suspect that the main utility (pun intended) here is going to be learning some of the tricks we used to get things to happen the way we wanted, despite Windows and MFC. There are simply too many of them to write about them all here, and not all of them are necessarily the best way of getting "whatever it was" done.

However, I find that looking through WavEd or MusiCad is quite a bit more productive for me than looking through the average windows programming book, since I know that a certain problem was solved, and approximately where to look for the answer, something I don't get with most books. And here, you can set breakpoints and such and really see what's going on, kind of the ultimate example code. Well, you can't do this in the assembly helper modules, but you can safely assume they work as stated in the header file descriptions. They are also in use in a couple of major commercial applications, and I'm sure the guys who paid me $1k each for these would've reported any bugs in them by now.

The App

`CWavEdApp` doesn't do terribly much as of right now. The main idea is to set various things up in `InitInstance()`, recover settings and other state information from the registry, and to clean up in `ExitInstance()` when the user quits. At the time of this writing, a really annoying bug in DevStudio 6 svcpak2 prevents `ExitInstance()` from being hit when in debug mode, and it breathlessly announces all these resource leaks — which are of course, its own fault for not calling my cleanup code. I find this annoying since your settings don't get saved while debugging, and I'd rather have that most times. I'm sure you'll notice this if it is happening to you, but heads-up! You can decide to save most of the global variables if you like, and you might even implement undo across sessions by saving and restoring the temporary filename lists. In the sometimes-Byzantine message routing done by MFC, the app class is the last resort for processing a message. This makes the app class a good place to put handlers for messages that aren't specific to a particular doc-view pair, such as for invoking help, or other global sorts of functionality that would be inappropriate to handle elsewhere. One important bit of code in the app class provides the global progress meter. This is special, because this implementation also pumps messages, effectively creating a pseudo thread in which messages can be passed. This is the key to how transforms can be canceled while in progress. Every time `OnOK()` calls the progress meter to update it, messages such as the one generated by the user hitting the cancel button are passed. The implication is that a dialog's `OnCancel()` handler can be entered before a call from the `OnOK()` function to the progress meter returns, and all transform code is designed to account for this. What is done in `OnCancel()` is to check to see if `OnOK()` is currently running, and if so, as flagged by `gbEditWorking`, the `bQuit` flag is set. `OnOK()` will check this flag every time around the loop, and clean up and quit itself if it finds it set. If the `gbEditWorking` flag isn't set when `OnCancel()` is entered, it simply destroys the dialog in the normal way.

Postscript

I hope this will help you as you tinker around inside WavEd. It can be quite a nice toy to play with. I personally have used it to teach courses on DSP, and I have several versions of it around that are customized for the needs of various consulting customers. In my opinion, it's a great platform for fooling around with new DSP techniques and getting them debugged. We use it *extensively* in this mode, even when the resulting code is to be changed into assembly for some embedded project. That is a comparatively simple job once you know what to convert! There's just no substitute for jumping in and getting your hands dirty when you're trying to learn or invent something new. This program and the pieces of it make life a whole lot easier than starting from scratch once you get over a bit of a learning curve with it. Most all of it just plain works fine, and can be trusted to help you find out what's wrong with your new code; built-in test equipment, if you will. As I said above, since writing this forced us to figure out how to do some pretty nifty windows UI tricks, I often use this code as a repository for how-to information for other projects. I was kidding with my editor, Berney Williams, that I should now write "yet another" Windows programming book based on this very same code, but minimize the DSP parts just as I attempted to minimize the windows parts in this one. No book I have (and I have special buildings full of them) details actually solving all the problems encountered when writing a complex application. Most just have a trick here and there, with no coherent plan, no "why would I want to do it this way versus that", and only trivial examples out of context that don't completely solve your real-world problems. Perhaps if he or I get pinged enough on this, that book will be written.

Go forth and program, learn, and enjoy. I hope you'll someday take the pebble from my hand, grasshopper!

Appendix C

Transform/Generate Effects API*

Making Your Own Effects Modules

CoolEdit can support any number of transform and generate effects. New effects can be written by following this API. An effects module is nothing more than a DLL with the following exported functions that CoolEdit calls to do the reading and writing of audio data.

When CoolEdit starts, it checks the program directory for files ending in .XFM. If it finds any, it checks to see if the file contains the function QueryXFM() and, if so, it calls that function. If the return value is XFM_VALIDLIBRARY, then it is assumed to be a valid effects module DLL.

The only functions that are required are QueryXfm(), XfmInit(), XfmDestroy(), XfmSetup(), XfmDo(), and some sort of DIALOGMsgProc() for the Settings dialog. A custom structure should be defined to hold all effect-specific data. This structure must also be communicated to CoolEdit so that various variables in the structure can be filled in automatically if you are using the Presets functions. The COOLINFO structure contains many functions that can be called at any time. These functions can be

---

accessed easily by using the macros defined in xfmsdll.c. Functions include reading and writing audio data, progress meter control, cutting and inserting blank data, preset handling, graph handling, FFT functions, and functions for calling other XFM modules.

An internal data structure is used, and defined by the author of the DLL, for storing any relevant information pertaining to the effect. The structure's format is communicated to CoolEdit through the szStructDef element of XFMQUERY during QueryXfm(). Never use any global variables — any variables you need should either be defined as part of the effect's internal data structure that you define, or allocated locally to the procedure that is using them. This is important since there may be more than one process using the DLL at any given time.

If you follow the sample code, and use it as a starting point for your own effects module, then you can follow some of these guidelines. Use the DialogToStruct() function to copy data from your instance data structure to the dialog box. Use the StructToDialog() function to copy data from the dialog box back to your data structure. You can leave the CopyToDlgItem() and CopyFromDlgItem() functions essentially unchanged (except for inserting the name of your custom structure). These two functions copy the handles of your custom data structure, and to the COOLINFO structure to a dialog control so that the instance information can be kept for your instance. This way, if multiple copies of CoolEdit are using the same effects DLL, then the data will not be scrambled between them. The CopyToDlgItem() function is called during the dialog box's WM_INITDIALOG, and the CopyFromDlgItem() function is called before any of the data in the custom structure or COOLINFO structure is needed. Include the file xfmsdll.c to gain access to all the COOLINFO structure functions that are provided to you by CoolEdit.

All XFM functions should be declared as __declspec(dllexport) FAR PASCAL, if using Visual C/C++, or the equivalent export definition for any other C compiler.

- To support CoolEdit Pro, return 1160 from QueryXfm().
- To support CoolEdit 98 and CoolEdit Pro, return 1157 from QueryXfm(). Preview can be implemented in the module, but PreviewStart() will fail if being run from CoolEdit 98.

---

int QueryXfm(lpxq)	Required

XFMQUERY *lpxq	Structure to be filled with all information pertaining to the effect module

This function should fill the XFMQUERY structure with information about the effect module.

Returns: XFM_VALIDLABRARY if successful, zero otherwise.

The XFMQUERY data structure definition.

char	szName[80]	Text name of the effect that will appear in the menu bar
char	szCopyright[80]	Any copyright information you care to put in for your DLL
char	szToolHelp[80]	Text to be shown in toolbar quick help
WORD	wSupports	Combination of XFM_ flags for mono and stereo 8 or 16 bit
DWORD	dwFlags	Combination of the XF_ flags for describing module
DWORD	dwUserDataLength	Length of transform's internal data structure
char	szStructDef[96]	Array of chars representing types of data in internal data structure
char	szPresetDef[96]	Array of 'y' or 'n' representing whether or not to include data item in presets saving
WORD	wExtra	Number of extra bytes to follow

Constants for the wSupports field.

XFM_MONO8	Effect supports Mono 8-bit data
XFM_STEREO8	Effect supports Stereo 8-bit data
XFM_MONO16	Effect supports Mono 16-bit data
XFM_STEREO16	Effect supports Stereo 16-bit data
XFM_MONO32	Effect supports Mono 32-bit float data scaled to ±65536.0
XFM_STEREO32	Effect supports stereo 32-bit float data scaled to ±65536.0

Constants for the dwFlags field.

XF_TRANSFORM	This function works with highlighted audio, and belongs in the Transform menu
XF_GENERATE	This function generates new wave data, and belongs in the Generate menu
XF_ANALYZE	This function works with highlighted audio, and belongs in the Options menu. It does not modify any data, just examines existing data
XF_SYSTEM	This module should not appear in any menus, as it is a CoolEdit system DLL

XF_USESPRESETSAPI	Has a presets box (and uses the presets/scripts functions)
XF_USESGRAPHAPI	Has one or more graph controls (and uses the built-in graph control functions)
XF_USESFFTAPI	Uses CoolEdit's FFT functions
XF_PREVIEW	Uses Preview API for real-time interactive previewing of the effect
XF_USES32BITAPI	Uses new ReadDataEx() and WriteDataEx()
XF_MUSTHIGHLIGHT	If set, function only enabled if a selection is highlighted (most common)
XF_MUSTHAVECLIP	If set, function is grayed if nothing is on clipboard
XF_MODIFIESTOENDOFVIEW	Set this if function modifies data outside user's given selection to end of view
XF_MODIFIESTOENDOFFILE	Set if function modifies data outside user's given selection to end of file
XF_MODIFIESOUTSIDESEL	Set if function modifies data outside selection, and fill dwModifyLeft and dwModifyRight upon exiting XfmSetup() with the range of data that will be modified
XF_STRETCHES	Set if result is of a different size and time markers in that area should be proportionally stretched to match
XF_NOSINGLEEDIT	If set, function is grayed if editing one channel of a stereo waveform (do this if effect changes the size of the highlighted data, e.g., stretching)
XF_AUTOOK	During WM_INITDIALOG, this value should be checked, and if it is set, the dialog should send an IDOK message to itself to close the dialog
XF_AUTOCROSSFADE	If set, and user has global crossfade setting enabled under Settings/System/Data, then beginning and ending of selection is crossfaded with original audio (ideally used to prevent clicks occurring at boundaries of edits) (uses 32-bit API)
XF_COLORRED	If set, WM_CTLCOLOREDIT should color text red (used with Limits() to make text red if it is out of range)

Characters used in the structure definition (szStructDef) to tell CoolEdit the data types in the internal effect data structure.

'c'	char
'i'	16-bit int or WORD

`'l'`	32-bit `long` or `DWORD`
`'f'`	4-byte `float`
`'d'`	8-byte `double` floating point
`'g'`	`HANDLE` to a graph (please use the `WORD` type in your `struct` definition for graphs)
`'h'`	`HANDLE` to globally allocated memory
`'s'`	Array of 256 `chars`

`BOOL XfmInit(ci)` Required

`COOLINFO*ci`   CoolEdit info structure pointer

This function should do all initialization of the internal data structure that contains all data pertaining to the effect. The `hUserData` member of `ci` is a handle to this data. Basically, this function should lock the `hUserData` member, casting it to the internal data type, and fill all the members with default information, and finally unlock the handle. Data contained in handles to globally allocated memory are not saved in presets. If a handle is created, and the data pointed to by the handle needs to be saved with a preset, then this must be done by the caller. One way to do this is to save the filename of a file containing the data in a string in the structure, and then when the preset is called up, load the memory handle with the data from the file (and alternatively, save the data to this file when saving a preset).

Returns: `TRUE` if all was initialized OK; `FALSE` if there was an error.

`BOOL XfmDestroy(ci)` Required

`COOLINFO*ci`   CoolEdit info structure pointer

All internally allocated data should be freed (e.g., globally allocated handles) or destroyed (e.g., graphs). Use the `hUserData` member of `ci` to access any internal data.

Returns: `TRUE` if all was successfully destroyed; `FALSE` otherwise.

`BOOL XfmSetup(hWnd, hInst, ci)` Required

`COOLINFO*ci`	CoolEdit info structure pointer
`HWND    hWnd`	Handle to a parent window for your setup dialog box
`HINSTANCEhInst`	Instance handle for this DLL so you can access resources

This function should call up a Settings dialog. The dialog box template should be compiled with the DLL and can be accessed using the hInst parameter. Below is a sample XfmSetup() function that should suffice for any module. IDD_TRANSFORM is the dialog box ID, and the dialog box message proc is exported with ordinal 100. The lParam given to the dialog box proc during WM_INITDIALOG will be a far pointer to the COOLINFO structure.

```
BOOL FAR PASCAL __export XfmSetup(HWND hWnd, HINSTANCE hInst, COOLINFO *ci)
{   int nRc;
    FARPROC lpfnDIALOGMsgProc;
    lpfnDIALOGMsgProc = GetProcAddress(hInst,(LPCSTR)MAKELONG(100,0));
    nRc = DialogBoxParam( (HINSTANCE)hInst,
                          (LPCSTR)MAKEINTRESOURCE(IDD_TRANSFORM),
                          (HWND)hWnd, (DLGPROC)lpfnDIALOGMsgProc,(DWORD)ci);
    return nRc;
}
```

This set-up function is called when the user chooses this effect from the menu bar or wishes to call up the Settings dialog of the effect for any other reason. Sorry, there is no support for modeless dialog boxes yet. The dialog box routine should call End-Dialog() with TRUE if the user presses OK or FALSE if the they hit Cancel. Special care must be taken if more than one Settings dialog is open at one time (if two instances of CoolEdit each have your effect dialog open). See the example's Copy-ToDlgItem() and CopyFromDlgItem() functions, which copy the relevant instance data to a hidden dialog box text control.

Returns: TRUE to continue and call XfmDo() to run the function or FALSE if the user hit Cancel in the dialog box.

BOOL XfmDo(ci)   Required

COOLINFO*ci   CoolEdit info structure pointer

This function is what does all the actual work. The COOLINFO structure contains all information necessary for the effect, including the portion of the wave that was highlighted and a handle to this effect's specific internal data structure that should have been filled in during the Setup function.

Returns: TRUE if all went OK; FALSE if there was an error.

```
void DialogToStruct(ci, hWndDlg, lpVars)
void StructToDialog(ci, lpVars, hWndDlg)                    Recommended
```

COOLINFO*ci		CoolEdit info structure pointer
HWND	hWndDlg	Settings dialog window handle
EFFECTVARS*lpVars		Pointer to internally defined structure

These functions should transfer data from the dialog box to the user-defined data structure, and back. For example, an effect that has a Volume setting would get the text from the ID_VOLUME edit control in the dialog and copy it to the lpVars->wVolume parameter when calling DialogToStruct(). It would set the ID_VOLUME control text to a string containing a value derived from lpVars->wVolume when calling StructToDialog(). Since these operations are done frequently, it is best to have them as a separate function. StructToDialog() will be called during the WM_INITDIALOG call, and DialogToStruct() will be called during processing of IDOK .

```
BOOL DIALOGMsgProc(hWndDlg, Message, wParam, lParam)            Required
```

HWND	hWndDlg	Settings dialog window handle
UINT	Message	Message ID to window procedure
WPARAM	wParam	Windows wParam parameter
LPARAM	lParam	Windows lParam parameter

This is the main window procedure for the Settings dialog box. Please look at the sample code for a template of how to create the dialog box procedure. The main points are described below.

A STATE structure should be used to keep track of information local to this dialog box, such as the handle to the user data, a locked pointer to the data, and a pointer to the COOLINFO structure. This structure is allocated, locked, and then a pointer to it is set in the window user data area using SetWindowLong(). Upon entering the window proc, GetWindowLong() is used to retrieve the pointer to this structure. The STATE structure lloks like:

```
HANDLE  hThis
HANDLE  hTransform
TRANSFORM*pData
COOLINFO*ci
```

A sample main code fragment looks like:

```
typedef struct statinfo_t
{   HANDLE hThis;
    HANDLE hXfm;
    DISTORT *pXfm;
    COOLINFO *ci;
} STATE;


__declspec(dllexport) BOOL FAR PASCAL DIALOGMsgProc(HWND hWndDlg,
                    UINT Message, WPARAM wParam, LPARAM lParam)
{   STATE *pState= (STATE *)GetWindowLong(hWndDlg,GWL_USERDATA);
    if ((pState) && (pState->ci))
    {   if (DialogHook(pState->ci,hWndDlg,Message,wParam,lParam))
            return TRUE;
        if (GraphHook(pState->ci,pState->pXfm->hGraph,hWndDlg,
                                        Message,wParam,lParam))
            return TRUE;
    }
    switch(Message)
    {   case WM_INITDIALOG:
        {   MYXFM *pXfm = NULL;
            HANDLE hMyXfm = NULL;
            COOLINFO *ci=NULL;
            REGISTERGRAPH rgraph;
            REGISTERPRESETS rpinfo;
            HANDLE hState=GlobalAlloc(GMEM_MOVEABLE|GMEM_ZEROINIT,
                                sizeof(STATE));
            STATE *pState = (STATE *)GlobalLock(hState);

            pState->hThis=hState;
            pState->ci=(COOLINFO *)lParam;
            pState->hXfm= ci->hUserData;
            pState->pXfm=(MYXFM *)GlobalLock(pState->hXfm);
            SetWindowLong(hWndDlg,GWL_USERDATA,(LONG)pState);
            pState->ci->hWndDlg=hWndDlg;

            ...
```

```
        case WM_DESTROY:
        {    HANDLE h;
             STATE *pState= (STATE *)GetWindowLong(hWndDlg,GWL_USERDATA);
             h=pState->hThis;
             GlobalUnlock(h);
             GlobalFree(h);
             SetWindowLong(hWndDlg,GWL_USERDATA,0);
        }
```

The dialog box procedure should return the following in response to the various buttons that will close the dialog.

IDCANCEL	0	(User decided to cancel all changes made. All changes are forgotten after the dialog is closed)
IDOK	1	(User decided to accept changes and execute the transform)
IDCLOSE	2	(User decided to close the dialog, but keep the changes made to all settings)

COOLINFO Structure

The COOLINFO structure contains everything you may need for performing a transform effect on existing data or to create new data. Following are explanations of the various fields of COOLINFO, though they are not in the same order as declared in the xfms.h file. The procedure pointers are not mentioned (see the following section for details on all procedures).

WORD	wChannels	Number of channels in audio data (1 or 2)
WORD	wBitsPerSample	Bits per sample (8, 16, or 32)
WORD	wBlockAlign	Bytes per sample (1, 2, 4, or 8. Same as wChannels* wBitsPerSample/8)
long	lSamprate	Sample rate of given audio data
HANDLE	hUserData	Handle to user's internal data structure
HWND	hWndDlg	Handle to dialog box. This value must be set in WM_INITDIALOG
DWORD	dwLoSample	Starting sample offset to transform
DWORD	dwHiSample	Ending sample offset to transform
DWORD	dwRightSample	Rightmost visible sample on the current display
DWORD	dwLeftSample	Leftmost visible sample on the current display
DWORD	dwTimeOffset	If wave has a separate SMPTE offset, this is the number of samples to offset

FARPROC lpTestFunction		Nothing
XFMQUERY*cq		Structure returned by QueryXfm()
char	*szIniFile	INI file for storing data any data you might want to store
Char	*szHelpFile	HELP file for CoolEdit itself
HFONT	hFont	Handle to a small font for use in dialog boxes
BOOL	bReplacesHighlightedSelection	
		If set, when generating audio, newly generated audio will replace the highlighted selection
DWORD	dwInsertBlankSamples	
		If nonzero, this many blank samples will be inserted after OK is hit
BOOL	bHasCoprocessor	Obsolete (doesn't everybody have one now?). Set if user has a coprocessor
HINSTANCEhInst		Instance of this DLL
int	iScriptFile	Obsolete. -1 if no script running; else script is running
int	iScriptDialogStop	Obsolete. Nonzero if "Pause at Dialogs" set for Scripts
BOOL	*lpProgressCanceled	Pointer to progress canceled flag. If flag is set, user hit Cancel in progress meter
WORD	wInvalidEntry	Set to control ID of the last item that was invalid or out of range
WORD	wExtraReadFlags	Used automatically in xfmspdll.c to modify flags for reading
WORD	wExtraWriteFlags	Same as above, but for writing
DWORD	dwFileID	Used automatically in xfmspdll.c for reading and writing, or pass to ReadDataFileEx() and WriteDataFileEx()
int	iNumWaveforms	Number of waveforms if this is a multi-waveform transform
COOLWAVEINFO	*pWaveforms	Array of waveform information structures for multi-waveform transforms

The following are all functions defined in xfmspdll.c that wrap the the COOLINFO functions, organized by logical groupings.

Reading and Writing Audio Data

To make the most effective use of CoolEdit Pro, set the XF_USES32BITAPI flag and do not use the obsolete functions ReadData() and WriteData(). Audio can be read in as either 8-, 16-, or 32-bit float data, and even as mono or stereo regardless of the native format of the audio in CoolEdit. The wChannels and wBitsPerSample COOLINFO fields describe one of the requested formats [requested in the XFMQUERY structure's wSupports flags in QueryXfm()]. If you wish to optimize for 16- or 32-bit data specifically, then set both the XFM_MONO16|XFM_STEREO16 and XFM_MONO32|XFM_STEREO32 flags in QueryXfm(), and then poll wBitsPerSample for the ideal data format, and be sure to ask for data in this format.

ReadDataEx() and WriteDataEx() take a formatting flag as the last parameter, which is a combination of the RWEX_ flags. The RWEX_CLIP and RWEX_DITHER flags are set automatically by CoolEdit depending on the user's main data preferences and other criteria, but if used, data will forcibly be clipped or dithered. The 32-bit format is a 4-byte float, scaled to the range of ±32768.0 [basically the (float) representation of the corresponding short int 16-bit sample]. Other floating-point formats may be added in the future and the RWEX_ flags associated with them.

A transform of the data is performed in the XfmDo() function by reading data into a buffer, starting at the ci->dwLoSample offset, modifying the data in any way necessary, then writing the data back. The Reading and Writing method was chosen over Getting and Releasing a data buffer because of the flexibility of data types — a transform only needs to be written to handle 32-bit floating-point mono and stereo data and it will work flawlessly with 8-, 16-, or 32-bit data in CoolEdit. Plus, if the CoolEdit file is 16-bit, the 32-bit result can be dithered cleanly back to 16 bits if the user wishes. Of course, if the XFM only handles 16-bit data (as in the Distort example), it will still work with 32-bit audio data in CoolEdit.

The following flags can be OR'd together and used in the wFlags parameter to describe the data type you want to work with.

RWEX_MONO	RWEX_8BIT
RWEX_STEREO	RWEX_16BIT
	RWEX_32BIT

The buffer used to hold the data should be large enough to hold the number of samples necessary. To calculate, use the following macro or something similar:

```
dwBytes=dwLength*((wFlags&RWEX_STEREO)?2:1)
              *((wFlags&RWEX_16BIT)?2:((wFlags&RWEX_32BIT)?4:1))
```

DWORD ReadDataEx (ci, pData, lOffset, dwSize, wFlags)

COOLINFO*ci	CoolEdit info structure pointer
char *pData	Buffer to read into
long lOffset	Offset into caller's wave data
DWORD lSize	Number of samples to read
WORD wFlags	Data type to read as

Read dwSize samples into the pData buffer from the user's waveform data starting at lOffset samples into the CoolEdit waveform. CoolEdit tries to read the entire amount always, and will pad with silence where necessary. Remember that this function works in units of samples, not bytes, and that the formula for converting samples to bytes is Bytes = Samples*wBitsPerSample*wChannels/8.

Returns: Number of bytes read.

When working with multiple waveforms (when iNumWaveforms is non-zero and XF_MULTIWAVEFORM flag set), then the following function is called to determine whether or not this XFM module can handle the configuration of multiple files. Any number of input and output files may be specified, and any file may be flagged for input, output, or both. This example shows a transform that can handle two inputs and one output, and the output must not be the same as the inputs.

```
__declspec(dllexport) BOOL FAR PASCAL XfmMultiWaveformQuery(COOLINFO *ci)
{   DWORD dwDestFile=0;
    int t;
    int iNumSources=0;

    for (t=0;t<ci->iNumWaveforms;t++)
    {   if (ci->pWaveforms[t].dwFlags&WAVEINFO_DEST)
        {   if (dwDestFile)
                return FALSE; // more than one destination file
            dwDestFile=ci->pWaveforms[t].dwFileID;
            if (ci->pWaveforms[t].dwFlags&WAVEINFO_SOURCE)
                return FALSE;
                // dest file must not be one of the source files
        }
        else if (ci->pWaveforms[t].dwFlags&WAVEINFO_SOURCE)
            iNumSources++;
    }
```

```
    if (!dwDestFil)
        // This should never happen, but if no dest file, no dice
        return FALSE;


    if (iNumSources!=2)
        return FALSE;


    return TRUE;
}
```

DWORD WriteDataEx(ci, pData, lOffset, dwSize, wFlags)

COOLINFO*ci	CoolEdit info structure pointer
char *pData	Buffer to write out
long lOffset	Offset into caller's wave data
long lSize	Number of samples to write
WORD wFlags	Data type to read as

Write dwSize samples from the buffer pData to the caller's waveform data starting at lOffset samples into the data. Data at offsets below zero are ignored, and data at offsets greater than the length of the file extend the length of the CoolEdit file. Remember that this function works in units of samples, not bytes!

Returns: Number of samples written.

DWORD ReadDataFileEx (ci, dwFileID, pData, lOffset, dwSize, wFlags)
DWORD WriteDataFileEx (ci, dwFileID, pData, lOffset, dwSize, wFlags)

COOLINFO*ci	CoolEdit info structure pointer
DWORD dwFileID	ID of file to access
char *pData	Buffer to write out
long lOffset	Offset into caller's wave data
long lSize	Number of samples to write
WORD wFlags	Data type to read as

When working with multiple input and/or output files, the specific file id must be given to read from or write to. The file id is given in ci->pWaveforms[n].dwFileID, where n ranges from 0 to ci->iNumWaveforms-1. The functions work exactly like ReadDataEx() and WriteDataEx() in other respects.

Returns: Number of samples read or written.

BOOL DelayWriteInitEx(ci, lWriteBehindSize, lOffset, lFullLength, wFlags)

COOLINFO*ci		CoolEdit info structure pointer
long	lWriteBehindSize	Minimum amount of delay before writing
long	lOffset	Offset into caller's wave data to start writing
long	lFullLength	Total number of samples that will be written
WORD	wFlags	Data type to read as

Sometimes when writing data, the data written back will overlap data that is yet to be read, or you may wish to reread data and want to read the previous original data that was there, not the data that was just written back. This function will delay the actual writing of the data to the file by the number of samples given in lWriteBehindSize. The restriction to writing data this way is that writing must be sequential. Data will start being written at lOffset and continue sequentially until all lFullLength samples are written. Also, if working with multiple output streams, only one stream can be delay-written at a time.

Returns: TRUE if successful; FALSE if lWriteBehindSize is too large.

DWORD DelayWriteEx(ci, pData, lNumSamples)

COOLINFO*ci		CoolEdit info structure pointer
void	*pData	Data to be written
long	lNumSamples	Number of samples to write

Writes lNumSamples samples of data to the output stream. The offset has already been determined by the call to DelayWriteInitEx() and the previous calls to Delay-WriteEx().

Returns: The number of samples written.

Void DelayWriteDestroyEx(ci)

COOLINFO*ci	CoolEdit info structure pointer

Writes all pending data in the delay line to the output file, and frees all internal structures and memory allocated for the delay write process.

DWORD CutSamplesEx(ci, dwOffset, dwSamples)

COOLINFO*ci		CoolEdit info structure pointer
DWORD	dwOffset	Offset into caller's wave data
DWORD	dwSamples	Number of samples to remove

Removes samples from waveform file. This will shorten the caller's waveform file by the number of samples given.

Returns: Number of samples removed.

`DWORD InsertSamplesEx(ci, dwOffset, dwSamples, bClear)`

`COOLINFO*ci`	CoolEdit info structure pointer
`DWORD   dwOffset`	Offset into caller's wave data
`DWORD   dwSamples`	Number of samples to remove
`BOOL    bClear`	If set, clears the newly inserted data to silence

Insert samples from waveform file. This will lengthen the caller's waveform file by the number of samples given.

Returns: Number of samples inserted.

Progress Meter

If a function is going to take any length of time, a progress meter should be used so the user can see how long the function is going to take, and to provide a method for the user to cancel the operation. These functions should be used in the `XfmDo()` function, and anywhere else length operations are going to take place (such as in analyzing data, etc.).

`void ProgressCreate(ci, cszText, hwndParent)`

`COOLINFO*ci`	CoolEdit info structure pointer
`LPCSTR  cszText`	Progress message to display above meter
`HWND    hwndParent`	Parent window of meter display

Create a progress meter to show the user how much of the function has been processed. This meter automatically estimates the total time the function is anticipated to take. `cszText` contains text that will be displayed at the top of the meter window; for example, "Flanging Selection...". Set `hwndParent` to NULL to make the CoolEdit main window the parent (most common).

`BOOL ProgressMeter(ci, dwCurrent, dwTotal)`

`COOLINFO*ci`	CoolEdit info structure pointer
`DWORD   dwCurrent`	Measure of current amount done
`DWORD   dwTotal`	Measure of total amount to do

The progress bar is updated with the percentage 100*dwCurrent/dwTotal. Usually dwTotal is the total number of bytes that your function is going to process, and dwCurrent is the number of bytes processed so far when the function is called. The progress meter must have been created with ProgressCreate() before calling this function.

At the same time ProgressMeter() is called (or just after), you should also check the *ci->lpProgressCanceled flag, and if set, break out of the main loop and exit XfmDo().

```
ProgressMeter(ci,dwAmountProcessed,dwTotalAmount);
if (*(ci->lpProgressCanceled))
    break;
```

Returns: FALSE if user hit Cancel; TRUE if all is going OK.

BOOL SetProgressText(ci, szText)

COOLINFO*ci   CoolEdit info structure pointer
LPCSTR szText   Text to place inside progress meter window

Specific information about the transform being performed can be placed in the progress meter dialog itself. For example, the Click/Pop/Crackle Eliminator function reports the number of clicks found and repaired in this space.

Returns: TRUE if all went well.

---

void ProgressDestroy(ci)

COOLINFO*ci   CoolEdit info structure pointer

Close the progress meter box. Always remember to close the progress meter if you created one when processing is finished.

User Presets

User presets allow for saving and recalling of often-used settings. A set of pre-defined presets can also be generated by using the WritePrivateProfileString() in response to the PRESETN_NOPRESETS notification message. The bare minimum required to add preset capabilities to your XFM module is to call RegisterPresets() with a bounding rectangle for the presets controls during WM_INITDIALOG and respond to the PRESETN_ADDING and PRESETN_CHOSEN notification messages. The rest is automatic.

```
BOOL RegisterPresets(ci, prp)
```

COOLINFO	*ci	CoolEdit info structure pointer
REGISTERPRESETS *prp		Pointer to REGISTERPRESETS struct with initialization info

Fill the REGISTERPRESETS struct with pertinent information — all members must be filled out. This function should be called from within WM_INITDIALOG. The structure contains:

DWORD	dwSize	Just set to sizeof(REGISTERPRESETS)
char	szGroupName[64]	The group entry name in COOL.INI to place the presets under
HWND	hWndDlg	Set this to the main Settings dialog
UINT	uiListBox	ID of a List Box to hold names of presets, or zero for automatic
UINT	uiAddButton	ID of the "Add" button, or zero for automatic
UINT	uiDelButton	ID of the "Del" button, or zero for automatic
UINT	uiNotify	WM_COMMAND ID to receive notification messages
UINT	uiRect	Zero, or ID or bounding rectangle for automatic
char	szPresetHeading[32]	Empty string, or replacement name for "Presets" heading

Generally, you'll want to go the automatic route, by filling in zero for the uiList-Box, uiAddButton, and uiDelButton fields, and giving the ID of the bounding rectangle to uiRect in which the controls will go. Doing this will give your dialog a more standard look that will match with the rest of Cool Edit. If you wish to customize the locations of the list box and Add and Delete buttons, then you must fill in the three IDs accordingly.

Returns: TRUE if presets were started OK.

Notes: You will need to respond to a few of the notification messages for Presets to be fully functional. The WM_COMMAND message will have LOWORD(wParam)==uiNotify, and HIWORD (wParam)══Notification Message. To do this, set up the notification code as in the example WM_COMMAND entry below:

```
case ID_NOTIFY:
    switch (HIWORD(wParam))
    {   case PRESETN_NOPRESETS:
            WritePrivateProfileString(TRANSFORMNAME,"Item1",
                                "My Effect,3,2,0,0",ci->szIniFile);
            WritePrivateProfileString(TRANSFORMNAME,"Item2",
                                "Cool Thing,3,7,1,1",ci->szIniFile);
            break;
        case PRESETN_CHOSEN:
            StructToDialog(ci, lpAmp, hWndDlg);
            if (bPreviewing)
                PreviewUpdate(ci, PREVIEW_NORESTART);
            break;
        case PRESETN_ADDING:
            DialogToStruct(ci, hWndDlg, lpAmp);
            break;
    }
    break;
```

When a preset is chosen, the settings from the lpAmp structure (which is a locked-down pointer to the user-defined XFM data structure) are copied to the dialog box to update all the controls in PRESETN_CHOSEN. When a preset is about to be added, all the data from the controls is copied to the lpAmp structure first in PRESETN_ADDING.

PRESETN_ADDING	Sent before a preset is to be added to the presets list
PRESETN_ADDED	Sent after a preset has been added to the presets list (lParam══index of item added)
PRESETN_DELETING	Sent before a preset is to be deleted (lParam══index of item to delete)
PRESETN_DELETED	Sent after a preset has been deleted
PRESETN_CHOOSING	Sent before a preset is to be chosen (lParam══index of item being chosen)
PRESETN_CHOSEN	Sent after a preset has been chosen (lParam══index if item chosen)
PRESETN_NOPRESETS	Sent during registration if no presets are defined (define your defaults now)

PRESETN_INITIALIZING	Sent before presets are to be initialized
PRESETN_INITIALIZED	Sent after initializing is complete
PRESETREPLY_CANCEL	Return from _ADDING or _DELETING to cancel operation

If you want to manipulate the controls that are filled in automatically, the resource IDs for the new controls are 11500 (list box), 11501 (Del button), 11502 (Add button), and 11503 ("Presets" text).

BOOL DialogHook(ci, hWndDlg, wParam, lParam)

COOLINFO*ci	CoolEdit info structure pointer
HWND hWndDlg	Handle of Settings dialog
WPARAM wParam	wParam value of dialog proc
LPARAM lparam	lParam value of dialog proc

To make the presets work, CoolEdit needs to intervene and monitor the various messages coming in to your dialog box procedure. This is done via DialogHook(), which you simply place near the top of your dialog box code as shown:

```
STATE *pState= (STATE *)GetWindowLong(hWndDlg,GWL_USERDATA);


if ((pState) && (pState->ci))
{   if (DialogHook(pState->ci,hWndDlg,Message,wParam,lParam))
      return TRUE;
}
```

When CoolEdit completely handles the message, it will return TRUE from Dialog-Hook(), otherwise the message may need further processing from your dialog box procedure. Since the COOLINFO structure is needed to make the call to DialogHook(), it must be filled in and valid before making the call, which is the reason for the validity check in the earlier example. The GWL_USERDATA is zero by default when a dialog box is created, so DialogHook() will not be called then. GWL_USERDATA is set to the STATE structure in WM_INITDIALOG, so the DialogHook() function never gets called before WM_INITDIALOG. The DialogHook() function replaces all earlier HandleID_XXX functions from earlier versions of CoolEdit.

Graph and Display Objects

Common graphical entry and display objects are available for you to use, and using them brings more coherence between your XFM and the rest of CoolEdit. The graph object is the device where you click on points and drag them around to create a line that follows $y = f(x)$, where there is one and only one y for any x (i.e., the graph cannot loop back on itself). The Display object just displays a graph only, and has no user controls over it, but it can display multiple colored lines. Sometimes a single numerical value is not enough for some functions — in most cases, a graph control will do the trick.

`BOOL RegisterGraph(ci, pRegisterGraph)`

`COOLINFO *ci`	CoolEdit info structure pointer
`REGISTERGRAPH pRegisterGraph`	Pointer to `REGISTERGRAPH` structure for initialization

As many graph controls as needed may be used in a single Settings dialog. Call `RegisterGraph()` for each one. Each field must be filled out completely, with unused fields left as zero. The `REGISTERGRAPH` structure has the following members.

`DWORD`	`dwSize`	Just set to `sizeof(REGISTERGRAPH)`
`HWND`	`hWndDlg`	Set to the Settings dialog handle
`UINT`	`uiNotify`	`WM_COMMAND` ID that will be used for notification messages
`UINT`	`uiRect`	ID of bounding rectangle
`UINT`	`uiDisplayText`	ID of a static text control for displaying cursor location information
`WORD`	`hGraph`	Handle to the graph itself [created by calling `GraphCreate()` inside `XfmInit()` usually]

Notification messages will be sent to the `WM_COMMAND` ID given in `uiNotify` (with `LOWORD(wParam)`==`uiNotify` and `HIWORD(wParam)`==`Notification Message`). In all cases, the `lParam` is the index of the point that was clicked on, which will be between zero and `GraphCount() - 1`.

`GRAPHN_LBUTTONDOWN`	User clicked down on a point
`GRAPHN_RBUTTONDOWN`	User right-clicked down on a point
`GRAPHN_BUTTONUP`	User lifted up on button over a point
`GRAPHN_MOUSEMOVE`	User is moving a point with the mouse
`GRAPHN_DBLCLK`	User double-clicked on a point

Generally, you'll want to update the real-time preview during the `GRAPHN_LBUTTONUP`, and perhaps even the `GRAPHN_MOUSEMOVE` notifications to keep the audio in sync with what is being displayed.

Returns: `TRUE` if registered OK.

```
BOOL GraphHook(ci, hGraph, hWndDlg, wParam, lParam)
```

COOLINFO*ci		CoolEdit info structure pointer
WORD	hGraph	Handle to graph control
HWND	hWndDlg	Handle of Settings dialog
WPARAM	wParam	wParam value of dialog proc
LPARAM	lparam	lParam value of dialog proc

To make the graphs work properly, CoolEdit needs to intervene and monitor the various messages coming into your dialog box procedure. This is done via GraphHook(), which you simply place near the top of your dialog box for each graph control in your XFM as exemplified below.

```
STATE *pState= (STATE *)GetWindowLong(hWndDlg,GWL_USERDATA);


if ((pState) && (pState->ci))
{   if (GraphHook(pState->ci,pState->lpAmp->hGraph1,hWndDlg,
                                       Message,wParam,lParam))

      return TRUE;
   if (GraphHook(pState->ci,pState->lpAmp->hGraph2,hWndDlg,
                                       Message,wParam,lParam))

      return TRUE;

}
```

When CoolEdit completely handles the message, it will return TRUE from Graph-Hook(); otherwise, the message may need further processing from your dialog box procedure. Since the COOLINFO structure is needed to make the call to GraphHook() [just as in DialogHook() above], it must be filled in and valid before making the call, which is the reason for the validity check in the example above. GWL_USERDATA is zero by default when a dialog box is created, so GraphHook() will not be called then. GWL_USERDATA is set to the STATE structure in WM_INITDIALOG, so the GraphHook() function never gets called before WM_INITDIALOG. The GraphHook() function replaces all earlier GraphHandle_XXX functions from earlier versions of CoolEdit.

```
HANDLE GraphCreate(ci, iLeft, iRight, iMin, iMax, iLeftVal, iRightVal)
```

COOLINFO*ci		CoolEdit info structure pointer
int	iLeft	Leftmost graph coordinate (along x-axis)
int	iRight	Rightmost graph coordinate
int	iMin	Minimum graph value (along y-axis)
int	iMax	Maximum graph value
int	iLeftVal	Value of leftmost point (must be between iMin and iMax)
int	iRightVal	Value of rightmost point (must be between iMin and iMax)

This function returns a handle to a graph that is passed to subsequent graph functions. Graphs are usually created in the XfmInit() function, and destroyed in the XfmDestroy() function. A graph control allows the user to specify any number of points connected by a single line that has only one y for every x.

Returns: Handle to graph if successful, or NULL if no graph was created.

```
int GraphCount(ci, hGraph)
```

COOLINFO*ci	CoolEdit info structure pointer
HANDLE hGraph	Handle to graph given by GraphCreate()

The user can create new points in a graph at any time. This function returns the total number of points in the graph.

Returns: The number of points in the graph.

```
POINT GraphGetPoint(ci, hGraph, iIndex)
```

COOLINFO*ci	CoolEdit info structure pointer
HANDLE hGraph	Handle to graph given by GraphCreate()
int iIndex	Point to get [0 to GraphCount() - 1]

Returns a point structure (whose members are .x and .y) that contains the x and y coordinates of the point at the given index. Index values range from 0 for the left-hand point up to the number of points in the graph minus one for the rightmost point.

Returns: x and y values of point in graph.

`void GraphSetPoint(ci, hGraph, int iIndex, ptPoint, bEndPoint)`

`COOLINFO*ci`		CoolEdit info structure pointer
`HANDLE`	`hGraph`	Handle to graph given by `GraphCreate()`
`int`	`iIndex`	Index of point to set
`POINT`	`ptPoint`	Coordinates of point to add or set
`BOOL`	`bEndPoint`	`TRUE` if this point is the rightmost point (as in adding new points to the end)

Call this function to add new points or change the position of points in the graph. Not used very often. You can build a set of points for a graph by using this function, but usually the user is the one who creates the points by clicking and dragging. If the point you are adding is a new rightmost point, indicate by setting `bEndPoint` to `TRUE`.

`double GraphGetValueAt(ci, hGraph, fXValue)`

`COOLINFO*ci`		CoolEdit info structure pointer
`HANDLE`	`hGraph`	Handle to graph given by `GraphCreate()`
`double`	`fXValue`	x-coordinate on graph

Returns the y value at any given point along the x-axis of the graph. Values between points on the graph are linearly interpolated, or spline-wise interpolated, so no matter what x value is given (provided it is between zero and one) a valid value is returned. The function works with double-precision floating-point values since the y value is interpolated, and may not be an integral value. Only values between zero and one should be passed in. Zero corresponds to the leftmost side of the graph, while 1.0 corresponds to the right-hand side of the graph. Any x-axis coordinate must be scaled down to this range before calling `GraphGetValueAt()`. The return value will be in graph coordinates; that is, between the limits (`iMin` and `iMax`) given in the `GraphCreate()` call.

To convert the x-coordinate from graph values (between `iLeft` and `iRight`) and the normalized x value, use the following conversion:

`x = (double)(iValue-iLeft)/(double)(iRight-iLeft)`

Returns: The y value at the given x location in graph coordinates.

`void GraphDraw(ci, hGraph, hDC)`

`COOLINFO*ci`		CoolEdit info structure pointer
`HANDLE`	`hGraph`	Handle to graph given by `GraphCreate()`
`HDC`	`hDC`	Device context of dialog box for drawing graph into

This function should be called in the WM_PAINT routine of the Settings window. The hDC should be the same hDC that was returned from BeginPaint(). The code snippet below illustrates getting the pointer to the user-defined structure (lpMyVars), and then getting the dialog's device context, drawing the graph, and ending the paint routine. The graph is a HANDLE type in the user's structure, and has the code 'g' associated with it in the definition variable szStructDef.

```
case WM_PAINT:
{   PAINTSTRUCT ps;
    HDC hDC=BeginPaint(hWndDlg,&ps);
    GraphDraw(pState->ci,pState->lpAmp->hGraph,hDC);
    EndPaint(hWndDlg,&ps);
    return FALSE;
}
```

GraphDraw() may also need to be called at other times, for example during the StructToDialog() call, or any time you notice the graph is not being updated after making a particular change.

---

void GraphClear(ci, hGraph)

COOLINFO*ci	CoolEdit info structure pointer
HANDLE hGraph	Handle to graph given by GraphCreate()

Clear all the points in the graph except for the leftmost and rightmost endpoints.

---

void GraphInvert(ci, hGraph)

COOLINFO*ci	CoolEdit info structure pointer
HANDLE hGraph	Handle to graph given by GraphCreate()

Invert the graph if the graph is invertible. The graph is invertible if there is exactly one x point for every y point. When the graph is inverted, every x value is swapped with every y value. The property that there is only one y value for each x must still hold. Sorry, but no indication is returned if the function inverted the graph or not.

---

void GraphCopy(ci, hGraphDest, hGraphSource)

COOLINFO*ci	CoolEdit info structure pointer
HANDLE hGraphDest	Handle of graph to copy into
HANDLE hGraphSource	Handle of graph to copy from

Copy the contents of the source graph to the destination graph. Both graphs must have already been created using `GraphCreate()`.

```
void GraphSetDblClkScales(ci, hGraph, fScx, fOfx, fMagx, fScy, fOfy,
                          fMagy, wStyle)
```

`COOLINFO*ci`	CoolEdit info structure pointer
`HANDLE  hGraph`	Handle to graph given by `GraphCreate()`
`double  fScx`	Scaling factor for *x*-coordinate
`double  fOfx`	Offset for *x*-coordinate
`double  fMagx`	Precision of *x*-coordinate
`double  fScy`	Scaling factor for *y*-coordinate
`double  fOfy`	Offset for *y*-coordinate
`double  fMagy`	Precision of *y*-coordinate
`WORD    wStyle`	Style (see below)

Set the scaling factors that the actual points will be modified by before they are displayed in the text box. The text box whose ID was given in the `GraphSetDialog()` call will display the value of the point when a point is selected. The value displayed is not the same as the integer value of the point, but a value based on the formula: `DisplayValue = PointValue*fScx + fOfx`. The value of the point is multiplied by the scaling factor and added to the offset to get the final value displayed to the user. The magnitude/precision value tells how many decimal places to round the value off to. For example, `fMagx` of 10 will round values to one decimal place (e.g., 3.1), while `fMagx` of 100 will round values to two decimal places (e.g., 3.14). Values that are not powers of 10 will round according; for example, `fMagx` of 4 will give values such as "4", "4.25", "4.5", and "4.75".

The actual data displayed to the user will be this value along with the units given in the `GraphSetDblClkNames()` function. The `wStyle` parameter should always be zero. If it is set to 1, then the final value is converted to decibels (e.g., 0.5 becomes -3, and 2.0 becomes +3). The style can be any combination (or none) of the following:

`GRAPH_LOGXAXIS`	The graph is divided logarithmically from left to right (`iLeft` during `GraphCreate()` must not be zero)
`GRAPH_SPLINES`	If set, smooth spline curves are drawn between the points instead of straight linear lines.
`GRAPH_HASRULERS`	If set, horizontal and vertical rulers with grid are displayed as well.

```
void GraphSetDblClkNames (ci, hGraph, cszXText, cszXUnits,
                          cszYText, cszYUnits)
```

COOLINFO*ci	CoolEdit info structure pointer
HANDLE hGraph	Handle to graph given by GraphCreate()
LPCSTR cszXText	Name for values along x-axis (going left to right)
LPCSTR cszXUnits	Units for values along x-axis
LPCSTR cszYText	Name for values along y-axis (going up and down)
LPCSTR cszYUnits	Units for values along y-axis

Set the names for the values along the x- and y-axes, for example, "Input Signal Level" or "Frequency", and the names for the units of these, for example, "dB" or "Hz". The units given will also be used for labels in the rulers if rulers are displayed.

```
void GraphDestroy(ci, hGraph)
```

COOLINFO*ci	CoolEdit info structure pointer
HANDLE hGraph	Handle to graph given by GraphCreate()

Call this function to destroy the graph. This is usually done in XfmDestroy(). After the graph is destroyed, the handle is no longer valid, and all memory used by the graph is freed.

```
HWND DisplayInit(ci, hWndParent, wTemplateID, wID,
                 lpfnCallback, dwUserData)
```

COOLINFO*ci	CoolEdit info structure pointer
WORD wTemplateID	ID of bounding rectangle in dialog box
WORD wID	ID for this display object
DWFARPROClpfnCallback	Callback for getting display graph information
DWORD dwUserData	User data to be passed to the callback

This function creates a display object which is encapsulated in a window. The callback procedure is used to get the y values for a particular x value in the graph. Attributes of the graph can be changed via the WM_SETDISPLAYDATA message. The callback procedure should be defined as:

```
__declspec(dllexport) DWORD FAR PASCAL DisplayCallback(double fXpos,
                      DWORD dwUserData, WORD wChartNumber, double *pfYpos)
```

dwUserData should be set to the COOLINFO structure, or the pointer to your transform's data. The chart number is the same as that given in the DISPLAY_ADDCHART message, and fXpos ranges between the values given in the DISPLAY_LEFTVALUE and DISPLAY_RIGHTVALUE messages.

Once a display window has been initialized and created, messages can be sent to it through the WM_SETDISPLAYDATA message, with the graph number in the low word of wParam, and the message ID in the high word of wParam. The actual value to set the graph attribute to is in the lParam.

HIWORD(wParam)	lParam
DISPLAY_BAKCOLOR	RGB value of background
DISPLAY_GRIDCOLOR	RGB value of grid
DISPLAY_TEXTCONTROL	Dialog box ID of text control to display under the mouse location information
DISPLAY_XLABEL	Label for x-axis
DISPLAY_YLABEL	Label for y-axis
DISPLAY_SAMPLERATE	Sample rate of audio data (used in some conversions)
DISPLAY_ADDCHART	Add a new chart (colored line) to the display. See below for lParam values.
DISPLAY_LINECOLOR	RGB value of this line
DISPLAY_TOPVALUE	Topmost y value to display in chart
DISPLAY_BOTTOMVALUE	Bottom-most y value in chart
DISPLAY_LEFTVALUE	Leftmost x value in chart
DISPLAY_RIGHTVALUE	Rightmost x value in chart

For example, to create a graph of time in seconds vs. decibels, in the range of 0 to 12 seconds, and -40 to 0 dB, the following statements could be used:

```
double fLeft=0;
double fRight=12;
double fBottom=-40;
double fTop=0;
SendMessage(lpAmp->hWndDisplay,WM_SETDISPLAYDATA,MAKELONG(0,DISPLAY_XLABEL),
          (DWORD)(LPCSTR)"sec");
SendMessage(lpAmp->hWndDisplay,WM_SETDISPLAYDATA,MAKELONG(0,DISPLAY_YLABEL),
          (DWORD)(LPCSTR)"dB");
SendMessage(lpAmp->hWndDisplay,WM_SETDISPLAYDATA,
          MAKELONG(1,DISPLAY_ADDCHART),DISPLAY_MODE_NORMAL);
SendMessage(lpAmp->hWndDisplay,WM_SETDISPLAYDATA,
          MAKELONG(1,DISPLAY_LINECOLOR),(DWORD)RGB(0,255,0));
```

```
SendMessage(lpAmp->hWndDisplay,WM_SETDISPLAYDATA,
        MAKELONG(1,DISPLAY_TOPVALUE),(DWORD)&fTop);
SendMessage(lpAmp->hWndDisplay,WM_SETDISPLAYDATA,
        MAKELONG(1,DISPLAY_BOTTOMVALUE),(DWORD)&fBottom);
SendMessage(lpAmp->hWndDisplay,WM_SETDISPLAYDATA,
        MAKELONG(1,DISPLAY_LEFTVALUE),(DWORD)&fLeft);
SendMessage(lpAmp->hWndDisplay,WM_SETDISPLAYDATA,
        MAKELONG(1,DISPLAY_RIGHTVALUE),(DWORD)&fRight);
```

The DISPLAY_ADDCHART's value can be any of combination of the following:

DISPLAY_MODE_DB	Logarithmic scale on the y-axis.
DISPLAY_MODE_LOG	Logarithmic scale on the x-axis.
DISPLAY_MODE_NORMAL	Normal linear scale on the x-axis (do not use with DISPLAY_MODE_LOG).
DISPLAY_MODE_WRAP	If a value is too high or too low, the graph wraps around.

Real-Time Preview

If the processing in XfmDo() can be done faster than real time, then the result may be previewed in real time before pressing OK. Preview will basically call XfmDo(), but instead of the results being sent into a wave file, they are sent to the audio output. The preview can also adjust in real time to parameter changes in the Settings dialog by either automatically cancelling out of the XfmDo() procedure (via the Progress-Canceled flag) and restarting with new parameters in the user-defined structure or by calling XfmUpdate() and allowing XfmUpdate() to modify the parameters necessary so the main loop in XfmDo() does not have to exit.

DWORD PreviewStart(COOLINFO FAR *ci, DWORD dwStartFlags, WORD wCallbackID)

DWORD PreviewStart(ci, dwStartFlags, wCallbackID)

COOLINFO*ci	CoolEdit info structure pointer
DWORD dwStartFlags	Any of the flags listed below
WORD wCallbackID	WM_COMMAND ID to call back with notification messages

Call PreviewStart() to begin previewing, or to query whether or not the version of CoolEdit being used can do previews (CoolEdit shareware, for example, can not do real-time previews). PreviewStart() is generally called in response to pressing the "Preview" button. Possible flags to pass in are:

PREVIEW_RESTART_ALWAYS	If set, audio is played again from the beginning if any parameters were updated
PREVIEW_NORESTARTONLOOP	If set, XfmDo() sees an infinitely long data stream, which is a looped version of the original
PREVIEW_BUFSIZE125	Minimum buffer size to try is 125ms (may cause choppy playback on some systems)
PREVIEW_BUFSIZE250	Minimum buffer size is 250ms
PREVIEW_BUFSIZE500	Minimum buffer is 500ms (this is normal). Buffer size determines latency
PREVIEW_BUFSIZE1000	
PREVIEW_BUFSIZE1500	
PREVIEW_BUFSIZE2000	
PREVIEW_QUERY	Just query whether or not CoolEdit can do previews; if so, PreviewStart() returns TRUE

The following notification messages may be issued in the high word of wParam when sending a WM_COMMAND message with the given wCallbackID in the low word of wParam.

PREVN_STARTED	Preview mode was started successfully
PREVN_STOPOK	Preview mode ended normally
PREVN_STOPERR	Preview mode ended because of error; XfmDo() returned non-zero, and lParam contains the error number
PREVN_STOPNOW	You should call PreviewStop() in response to this notification

Following is an example of the WM_COMMAND handling for the preview notification callback ID:

```
case ID_PREVIEWNOTIFY:
    switch (HIWORD(wParam))
    {   case PREVN_STOPOK:
        case PREVN_STOPERR:
            bPreviewing=FALSE;
            SetDlgItemText(hWndDlg,IDC_PREVIEW,"&Preview");
            break;
        case PREVN_STARTED:
            SetDlgItemText(hWndDlg,IDC_PREVIEW,"&Stop");
            bPreviewing=TRUE;
            break;
```

```
        case PREVN_STOPNOW:
            PreviewStop(ci);
            break;
    }
```

This code snippet changes the text of the dual-purpose "Preview" button, and also sets a global flag that states whether or not we are currently previewing something. Elsewhere in the code, the bPreviewing flag can be queried, and XfmUpdate() called if we are previewing.

DWORD PreviewStop(ci)

COOLINFO*ci   CoolEdit info structure pointer

Stop previewing. The XfmDo() procedure is exited via setting *ci->lpProgressCanceled to TRUE.

DWORD PreviewUpdate(ci, dwFlags)

COOLINFO*ci   CoolEdit info structure pointer
DWORD   dwFlags   Any of the flags listed below

If audio is currently being played for the preview, and a Settings dialog parameter changes, PreviewUpdate() can be called to update the various parameters without interrupting playback.

If dwFlags is set to PREVIEW_NORESTART then XfmUpdate() is called to update parameters, and the *ci->lpProgressCancelled flag is not set. XfmUpdate() should handle all parameter updating necessary so that XfmDo() does not need to be exited and called again. If dwFlags is zero, then XfmDo() is cancelled, and called again with the updated parameters.

_declspec(dllexport) BOOL FAR PASCAL XfmUpdate(ciSource, ciDest, dwCursor)

COOLINFO*ciSource   CoolEdit info structure to copy data from
COOLINFO*ciDest   CoolEdit info structure to copy data to [currently in use by XfmDo()]
DWORD   dwCursor   Current sample offset being played (for updating properly if data varies over time)

XfmUpdate() is used to update live parameters while they are still being used in XfmDo(). If done properly, preview audio can very closely follow the user's motions in moving the controls in the Settings dialog. Sometimes an internal runtime structure is created during XfmDo() which contains all the local variables that control the transform's progress. This runtime structure can be updated to change XfmDo()'s internal processing variables if necessary. See the Distort sample file for somewhat of an example.

FFTs and Convolution

Some routines are used over and over again in digital audio. One of the most popular is the FFT, as well as the most popular use for an FFT— digital convolution. Both these functions are provided in double-precision floating-point accuracy for the XFM module developer. The FFT will transform time-domain audio data to a frequency domain set of amplitudes and phases. Convolution will basically multiply every sample in an impulse by every sample in the user's data stream, and is a popular method for implementing large FIR filters. More information on FFTs and convolution can be found in DSP-related literature.

```
void LFFT(ci, hpReData, hpImData, lSize, iDirection)
```

COOLINFO*ci	CoolEdit info structure pointer
double huge*hpReData	Pointer to array of lSize doubles representing the Real components
double huge*hpImData	Pointer to array of lSize doubles representing the Imaginary components
long iSize	Size of FFT (number of data points)
int iDirection	1 for FFT; -1 for inverse FFT

Perform an FFT (Fast Fourier transform) on the data. The largest-sized FFT supported depends on the amount of RAM the user has, which will be evident when the huge arrays are allocated by your DLL. On average, FFT sizes up to 262,144 points can be performed successfully on most systems. The data is zero-based, so hpReData[0] is the first real value part of the first data point for example. The number of data points should only have the factors 2, 3, 4, and 5. So an FFT of size 10,000 is fine, since 10,000 factors to $5 \times 5 \times 5 \times 5 \times 4 \times 4$. So, any powers of 2, 3, 4, or 5 are also valid by this rule.

After the FFT is performed, the data can be manipulated, or viewed, or whatever. If it is manipulated in some way (e.g., changing the phase or amplitude of various frequency components), then the inverse FFT can be performed by setting iDirection to -1. You must provide any windowing necessary on this data, as the only "Large" FFT function provided is this one.

It is usually necessary to window data first before performing an FFT on it. A triangular window, for example, is just multiplying the data going into the FFT, sample by sample, by a triangular-shaped set of data — that is, the beginning and ending elements are multiplied by near zero, and slowly each element is multiplied by a slightly larger value until at the center point, the element is being multiplied by 1.0, then the multiplier goes linearly back down to zero again by the last sample.

```
void ConvoluteEx(ci, wBitsPerSample, wChannels, wBlockAlign, dwLoSample,
                 dwHiSample, bCombine, wNumImpulses, hImpulse,
             lImpulseSize, lDelaySamples, szProgressText, dwUser,
          lpfnReadData, lpfnWriteData, fLeftAmp, fRightAmp)
```

COOLINFO*ci		CoolEdit info structure pointer
WORD	wBitsPerSample	Must be either 16 or 32 bits per sample.
WORD	wChannels	Number of channels of source data will be convolved
WORD	wBlockAlign	Block alignment of source data (bytes per sample)
DWORD	dwLoSample	First sample to convolve
DWORD	dwHiSample	Last sample to convolve [samples offsets from dwLoSample to dwHiSample will be passed into the lpfnReadData() and lpfnWriteData() functions]
BOOL	bCombine	If source audio is stereo, combine into mono data before performing convolution
WORD	wNumImpulses	Number of impulses interleaved in hImpulse
HANDLE	hImpulseIn	Handle to globally allocated memory block of lImpulseSize*wNumImpulses doubles
long	lImpulseSize	Number of samples represented in the impulse block
long	lDelaySamples	Number of samples to skip before returning valid data
LPCSTR	szProgressText	If given, a progress meter with this text will show progress of convolution
DWORD	dwUser	This value is passed on into the read and write callbacks
FARPROC	lpfnReadData	int ReadData(char huge *data, DWORD dwOffset, int iAmount)
FARPROC	lpfnWriteData	int WriteData(char huge *data, DWORD dwOFfset, int iAmount)
double	fLeftAmp	Final amplification of left channel (1.0 to remain unchanged)
double	fRightAmp	Amplification multiplier for right channel

One of the most popular uses for FFTs is in convolution. Since this function is so useful, it has been included here so any Transform function may use it. Impulses of up to about 100,000 samples may be given. The impulse data consists of an interleaved (e.g., Left0, Right0, Left1, Right1, Left2, Right2, …) array of doubles generally with values between -1.0 and 1.0. The number of impulses being interleaved (valid values are 1, 2, or 4) must be given in wNumImpulses. The following types of convolution may be performed [with the symbol (X) standing for convolution]:

Source Channels	Number of Impulses	Combine	Action Performed
1	1	TRUE/ FALSE	Out = In (X) Impulse
2	1	FALSE	OutLeft = InLeft (X) Impulse OutRight = InRight (X) Impulse
2	1	TRUE	OutLeft = ((InLeft + InRight)/2) (X) Impulse OutRight = ((InLeft + InRight)/2) (X) Impulse
2	2	FALSE	OutLeft = InLeft (X) Impulse1 OutRight = InRight (X) Impulse2
2	2	TRUE	OutLeft = ((InLeft + InRight)/2) (X) Impulse1 OutRight = ((InLeft + InRight)/2) (X) Impulse2

2	4	FALSE	OutLeft = InLeft (X) Impulse1
			    + InRight (X) Impulse3
			OutRight = InRight (X) Impulse2
			    + InLeft (X) Impulse4

A custom ReadData() and WriteData() should be passed in (see below for a good example). This function can be a bit tricky to use, but it is extremely powerful. Convolution entails multiplying every sample in the source by every sample in the impulse. If the impulse is symmetric, then the lDelaySamples should be set to 1/2 the length of the impulse. For example, a certain filter may have a 200-point impulse symmetric about 100. In this case, set lDelaySamples to 100, otherwise the audio will be delayed by 100 samples. If the impulse is the impulse of a reverb or echo, then no delay is necessary, except perhaps the delay to the primary (first) echo. Please refer to text on digital signal processing for more information about impulses and convolution.

The Read and Write routines take offsets and amounts in bytes, so they must be converted to samples before calling the Read and Write routines of CoolEdit. The dwUser parameter below is the same as that passed into ConvoluteEx(), and should be the file ID (`ci->dwFileID`). Below are examples of Read and Write routines that can be passed into `ConvoluteEx()`.

```
__declspec(dllexport) DWORD FAR PASCAL ConvReadData(DWORD dwUser, char *pData,
    long lOffset, long lAmount)
{   COOLINFO *ci;
    WORD wFlags=RWEX_CLIP;
    ci=(COOLINFO *)dwUser;
    wFlags|=(ci->wChannels==1)?RWEX_MONO:RWEX_STEREO;
    wFlags|=(ci->wBitsPerSample==32)?RWEX_32BIT:RWEX_16BIT;
    return ReadDataEx(ci,pData,lOffset/ci->wBlockAlign,lAmount/ci
                    ->wBlockAlign,wFlags)*ci->wBlockAlign;
}


__declspec(dllexport) DWORD FAR PASCAL ConvWriteData(DWORD dwUser, char *pData,
    long lOffset, long lAmount)
{   COOLINFO *ci;
    WORD wFlags=RWEX_CLIP;
    ci=(COOLINFO *)dwUser;
    wFlags|=(ci->wChannels==1)?RWEX_MONO:RWEX_STEREO;
    wFlags|=(ci->wBitsPerSample==32)?RWEX_32BIT:RWEX_16BIT;
    return WriteDataEx(ci,pData,lOffset/ci->wBlockAlign,lAmount/ci
                    ->wBlockAlign,wFlags)*ci->wBlockAlign;
}
```

Miscellaneous Helper Functions

The following functions are some little extras. The first three are for calling other XFMs from within your XFM module. The `CenterDialog()` function can always be called from within `WM_INITDIALOG`.

HANDLE CreateXfmVars(ci, cszXfmName)

`COOLINFO*ci`	CoolEdit info structure pointer
`LPCSTR cszXfmName`	Name of the other effects module

Other effects modules can be used from within your effects module if they exist. For example, you can possibly use CoolEdit's Filter or Amplify functions from within your own. But since the internal structures for these functions may change over time, you are better off just using other XFM modules that you have created. When called, the external structure is initialized to its defaults [from the other module's `XfmInit()` function].

Returns: Handle to be used in subsequent calls for identifying the external effect module's data structure.

---

void SetXfmVar(ci, hVars, iOffset, iLength, lpvMem)

COOLINFO	*ci	CoolEdit info structure pointer
HANDLE	hVars	Handle for use with external transforms [from `CreateXfmVars()`]
int	iOffset	Offset into transform's structure
int	iLength	Length of data in transform's structure
void	*lpvMem	Memory to copy to transform's structure

Set some data in another effect module's user-defined data structure. After the data is set, the effect may be called using `CallXfm()`.

---

long CallXfm(ci, cszXfmName, hVars, dwLoSamp, dwHiSamp, bShowMeter)

COOLINFO	*ci	CoolEdit info structure pointer
LPCSTR	cszXfmName	Name of the other effects module
HANDLE	hVars	Handle for use with external transforms [from `CreateXfmVars()`]
DWORD	dwLoSamp	Low sample value to work with
DWORD	dwHiSamp	High sample value to work with
BOOL	bShowMeter	Should the called effect show its progress meter?

Actually calls the other effect. The effect will work with samples from `dwLoSamp` on up to `dwHiSamp`. If your function is displaying a progress meter, you will want to set `bShowMeter` to `FALSE` so that the effect you are calling doesn't try to display its own progress meter as well (which would end in total confusion and chaos).

Returns: Value depends on the function being called. It is the return value from `XfmDo()` of the effect being called.

---

void GetTempName(ci, szThree, szFilename)

COOLINFO	*ci	CoolEdit info structure pointer
LPSTR	szThree	Three-letter identifier for type of temporary file
LPSTR	szFilename	Complete filename of temporary file

This function returns the name of a valid, unique temporary filename in szFilename. The user's choice of temporary directory in CoolEdit is used as the directory for this file.

```
void CenterDialog(ci, hWndDlg)
```

COOLINFO*ci	CoolEdit info structure pointer
HWND hWndDlg	Handle of Settings dialog

Call this function in the WM_INITDIALOG section of your dialog's Settings window to center the dialog with respect to the CoolEdit window that called it.

Definitions File

xfmspro.h

```
#define XFM_MONO8        1
#define XFM_STEREO8      2
#define XFM_MONO16       4
#define XFM_STEREO16     8
#define XFM_MONO32       16
#define XFM_STEREO32     32


//#define XFM_HELPFILE "waycool.hlp"


typedef DWORD (CALLBACK*     DWFARPROC)();
typedef void  (CALLBACK*     VFARPROC)();


typedef struct xfmquery_tag
{   char szName[80];            // Appears in Menu (without elipsis...)
    char szCopyright[80];
    char szToolHelp[80];
    WORD wSupports;
            // OR's of XFM_MONO16, XFM_STEREO16, XFM_MONO8, XFM_STEREO8
    DWORD dwFlags;          // Any of the XF_ flags OR'd together
                            // Must have either XF_TRANSFORM or XF_GENERATE
                            // May have XF_HASPRESETS, XF_HASSCRIPTS
    DWORD dwUserDataLength;        // Length of transform's internal data
    char szStructDef[96];
            // Array of chars representing types in user data area
    char szPresetDef[96];
            // Array of chars 'y' to include in presets, 'n' to ignore
    WORD wExtra;                   // Number of extra bytes to follow
} XFMQUERY;


#define XFM_VALIDLIBRARY 1160
#define XFM_VALIDLIBRARY98 1157
```

```
#define XF_TRANSFORM              0x0001
    // This function transforms highlighted audio
#define XF_GENERATE               0x0002
    // This function generates new pcm wave data
#define XF_USESPRESETSAPI         0x0004
    // Has a presets box (and uses the presets/scripts API)
#define XF_USESGRAPHAPI           0x0008
    // Uses the built in graph control functions
#define XF_MUSTHIGHLIGHT          0x0010
    // If set, function only enabled if a selection is highlighted (normal)
#define XF_MUSTHAVECLIP           0x0020
    // If set, function is grayed if nothing on clipboard
#define XF_MODIFIESTOENDOFVIEW    0x0040
    // Set this if function modifies data outside user's given selection
#define XF_MODIFIESTOENDOFFILE    0x0080
    // Set this if function modifies data outside user's given selection
#define XF_USESFFTAPI             0x0100      // Uses Cool's FFT functions
#define XF_NOSINGLEEDIT           0x0200
    // Do not include if editing only one channel
#define XF_ANALYZE                0x0400      // This function analyzes data
#define XF_MODIFIESOUTSIDESEL     0x2000
    // Set this if function modifies data outside selection, and fill
    // dwModifyLeft and dwModifyRight upon exiting XfmSetup
#define XF_SYSTEM                 0x4000
    // System function - does not show up in drop-down list
#define XF_PREVIEW                0x8000      // Preview API is being used
#define XF_USES32BITAPI           0x10000
    // Uses lpReadEx and lpWriteEx and NOT lpReadData/lpWriteData
#define XF_AUTOCROSSFADE          0x20000
    // Start and End (as defined by dwLoSample/dwHiSample) will be
    // crossfaded automatically roughly 200 samples (Uses 32-bit API)
#define XF_MULTIWAVEFORM          0x40000
    // Set if this is an XFM that accepts more than one wave stream
#define XF_AUTOOK                 0x80000000L
    // If set, wm_initdialog should send "OK" message to close
#define XF_COLORRED               0x40000000L
    // If set, WM_CTLCOLOREDIT should color text red
#define XF_STRETCHES              0x20000000L
    // Set if result is of a different size and time markers
    // in that area should be proportionally stretched to match
```

```
//#define COOLCALLBACK     _far _pascal
//typedef int (COOLCALLBACK*COOLINTPROC)();

// Flags for new Read and Write functions lpReadEx and lpWriteEx
#define RWEX_MONO                0x01
#define RWEX_STEREO              0x02
#define RWEX_8BIT                0x08
#define RWEX_16BIT               0x10
#define RWEX_32BIT               0x20
#define RWEX_CLIP                0x1000
#define RWEX_DITHER              0x2000


#define WAVEINFO_SOURCE     1     // for dwFlags of COOLWAVEINFO
#define WAVEINFO_DEST       2


typedef struct waveinfo_tag
    // Information for a particular waveform if processing multiple waves
{   DWORD     dwFlags;
    WORD      wChannels;
    WORD      wBitsPerSample;
    long      lSampleRate;
    DWORD     dwLoSample;
    DWORD     dwHiSample;
    DWORD     dwFileID;
} COOLWAVEINFO;


// Old legacy COOLINFO members are italicized


typedef struct coolinfo_tag
{    WORD wChannels;          // Number of channels
    WORD wBitsPerSample;      // Bit size, 8 or 16 for now
    WORD wBlockAlign;         // Bytes per sample (eg stereo 16-bit = 4)
    long lSamprate;          // Sample rate (8000,11025,22050, etc.)
    HANDLE hUserData;
        // Handle to specialized transform data, depends on DLL
    DWORD dwLoSample;         // Lowest sample to transform
```

```
    DWORD dwHiSample;          // Highest sample
    FARPROC lpTestFunction;
    XFMQUERY FAR *cq;          // Pointer to query struct
    // General-Purpose Tools
    FARPROC lpCenterDialog;
        // (HWND hWndDlg, int iUnused)  Center the dialog box
    // Reading and Writing audio data
    // OLD CODE - use lpReadEx, and lpWriteEx now.
    FARPROC lpReadData;        // (char huge *data,long offset,long amount)
    FARPROC lpWriteData;       // (char huge *data, DWORD offset,DWORD nbytes)
    // Progress Meter
    BOOL FAR *lpProgressCanceled;
        // if TRUE, user hit Cancel button, you must stop processing NOW
    FARPROC lpProgressMeter;
        // (DWORD dwCurrent, DWORD dwTotal)
        // percent done is 100*dwCurrent/dwTotal
    FARPROC lpProgressCreate;
        // (LPCSTR szText) szText is message to indicate type of processing
    FARPROC lpProgressDestroy;      // (void)  Call to remove progress meter
    // Presets
    char far *szIniFile;            // Ini file
    FARPROC lpPresetsInit;          // (HWND hWndDlg,LPSTR szGroupName)
    FARPROC lpHandleID_PRESETNAME;
    FARPROC lpHandleID_PRESETS;
    FARPROC lpHandleID_ADD;
    FARPROC lpHandleID_DEL;
    // Graph control
    FARPROC lpGraphCreate;
    FARPROC lpGraphCount;
    FARPROC lpGraphGetPoint;
    FARPROC lpGraphGetValueAt;
    FARPROC lpGraphDraw;
    FARPROC lpGraphClear;
    FARPROC lpGraphInverse;
    FARPROC lpGraphDestroy;
    // OLD CODE -- call lpGraphControlHook in main msg proc for each graph
    FARPROC lpGraphHandleWM_LBUTTONDOWN;
    FARPROC lpGraphHandleWM_LBUTTONUP;
```

```
FARPROC lpGraphHandleWM_LBUTTONDBLCLK;
FARPROC lpGraphHandleWM_MOUSEMOVE;
FARPROC lpGraphSetDblClkScales;
FARPROC lpGraphSetDblClkNames;
FARPROC lpGraphSetDialog;
short iScriptFile;
short iScriptDialogStop;
HFONT hFont;                            // smaller font for dialogs
// for XF_GENERATE type, must be filled out during XfmSetup
BOOL    bReplacesHighlightedSelection;
    // if TRUE, highlighted selection is deleted before inserting blanks
DWORD   dwInsertBlankSamples;      // the number of samples to generate
BOOL    bHasCoprocessor;
    // if TRUE, use fp, otherwise use table lookups and other speedups
// OLD CODE - should do windowing in XFM module itself,
// and call lpLFFT or lpConvolute
FARPROC lpFFT;            // perform n-point FFT (radix-2 for now)
FARPROC lpWindowFFT;
FARPROC lpIWindowFFT;
FARPROC lpSetWindowType;
    // 0..5: Triangular, von Hann, Hamming, Blackman,
    // Cosine^2, Blackman-Harris
FARPROC lpGetTempName;             // get temporary file name
FARPROC lpSetFFT16bit;
FARPROC lpSetFFT8bit;
FARPROC lpSetStereoFFT16bit;
FARPROC lpSetStereoFFT8bit;
FARPROC lpSetStereoFFT16bitInterleaved;
FARPROC lpSetStereoFFT8bitInterleaved;
FARPROC lpGetStereoFFT16bitInterleaved;
FARPROC lpGetStereoFFT8bitInterleaved;
FARPROC lpCreateXfmVars;
FARPROC lpSetXfmVar;
DWFARPROC lpCallXfm;
// OLD CODE - use lpDelayWriteEx, etc.
// Delayed writing
VFARPROC lpDelayWrite;
VFARPROC lpDelayWriteInit;
```

```
VFARPROC lpDelayWriteDestroy;
HINSTANCE hInst;
FARPROC lpGraphCopy;
FARPROC lpCutData;
FARPROC lpInsertBlankData;
FARPROC lpDeleteXfmVars;
DWORD dwRightSample;          // Current viewing screen rightmost sample
DWORD dwLeftSample;           // and leftmost sample
FARPROC lpGraphSetPoint;
DWORD dwExtraFlags;
FARPROC lpLFFT;
DWFARPROC lpConvolute;
DWORD dwModifyLeft;
DWORD dwModifyRight;
// Used for Preview mode
HINSTANCE hInstThisLibrary;
DWFARPROC lpPreviewStart;
DWFARPROC lpPreviewStop;
DWFARPROC lpPreviewUpdate;
DWFARPROC lpDisplayInit;
DWORD dwTimeOffset;
DWFARPROC lpLimits;
WORD wInvalidEntry;
HWND hWndDlg;
DWFARPROC lpReadEx;
DWFARPROC lpWriteEx;
WORD wExtraReadFlags;
WORD wExtraWriteFlags;
DWORD dwFileID;
DWFARPROC lpConvoluteEx;
DWFARPROC lpCutSamplesEx;
DWFARPROC lpInsertSamplesEx;
DWFARPROC lpGetMinMax;
DWFARPROC lpDelayWriteInitEx;
DWFARPROC lpDelayWriteEx;
VFARPROC  lpDelayWriteDestroyEx;
char far *szHelpFile;         // Help file for main CoolEdit
```

```
    FARPROC lpProgressSetText;
        // (LPCSTR szText) to change progress text at any time
    FARPROC lpGraphControlHook;
    FARPROC lpDialogHook;
    int iNumWaveforms;
    COOLWAVEINFO *pWaveforms;
    FARPROC lpRegisterPresets;
    FARPROC lpRegisterGraph;
    FARPROC lpGraphSetProperty;
} COOLINFO;

typedef struct registergraph_tag
{   DWORD dwSize;
    HWND hWndDlg;
    UINT uiNotify;
    UINT uiRect;
    UINT uiDisplayText;
    WORD hGraph;
} REGISTERGRAPH;

#define GRAPHN_LBUTTONDOWN    201
#define GRAPHN_RBUTTONDOWN    202
#define GRAPHN_BUTTONUP       203
#define GRAPHN_MOUSEMOVE      204
#define GRAPHN_DBLCLK         205


// Graph Properties
//#define GRAPHP_SPLINES     1     // non-zero=splines, 0=linear
//#define GRAPHP_RULERS      2     // non-zero=draw rulers
//#define GRAPHP_GRID        3     // non-zero=draw grid

// Struct passed to RegisterPresets
typedef struct registerpresets_tag
{   DWORD dwSize;                  // size of REGISTERPRESETS struct
    char szGroupName[64];
    HWND hWndDlg;
    UINT uiListBox;
```

```
    UINT uiAddButton;
    UINT uiDelButton;
    UINT uiNotify;
    UINT uiRect;
    char szPresetHeading[32];
} REGISTERPRESETS;


// Sent if using new presets format, RegisterPresets and PresetsInitEx
#define PRESETN_ADDING          101
#define PRESETN_ADDED           102
#define PRESETN_DELETING        103
#define PRESETN_DELETED         104
#define PRESETN_CHOOSING        105
#define PRESETN_CHOSEN          106
#define PRESETN_NOPRESETS       107
#define PRESETN_INITIALIZING    108
#define PRESETN_INITIALIZED     109
#define PRESETREPLY_CANCEL       86


// Used in PreviewStart, just to see if the version of
// Cool supports Preview or not
#define PREVIEW_QUERY            16384



// Used in PreviewUpdate - if set, XfmDo is not called during update,
// but XfmUpdate IS called, otherwise XfmDo is called every time
// PreviewUpdate is called
#define PREVIEW_NORESTART         1



// Used in PreviewStart - if set, PREVIEW_NORESTART is ignored,
// plus XfmDo is called with data starting at the beginning of the
// highlight instead of data at the current location.  This flag
// makes sense only if PREVIEW_NORESTARTONLOOP is not specified
// in PreviewStart.
#define PREVIEW_RESTART_ALWAYS    4
```

```
// Used in PreviewStart - if set, XfmDo sees an infinite signal
// which is the looped version of the highlighted selection,
// otherwise XfmDo is called repeatedly for each loop
#define PREVIEW_NORESTARTONLOOP     2


// If 0 is passed to all Preview functions for the flags, XfmDo is
// called whenever Update is called, and every time the selection loops.
// During update, read pointer is not reset to the start of
// selection, but stays wherever it left off before the call to Update.
#define PREVIEW_BUFSIZE125          0x10
#define PREVIEW_BUFSIZE250          0x00
#define PREVIEW_BUFSIZE500          0x20
#define PREVIEW_BUFSIZE1000         0x30
#define PREVIEW_BUFSIZE1500         0x40
#define PREVIEW_BUFSIZE2000         0x50


// Do not use, it's the mas for the above sizes
#define PREVIEW_BUFSIZEMASK         0x70


// Flags that are used in the HIWORD(wParam) of the callback command message
// To use preview, you must fill the ci->hWndDlg during WM_INITDIALOG
#define PREVN_STARTED 1         // Preview mode was started successfully
#define PREVN_STOPOK  2         // Preview mode ended normally
#define PREVN_STOPERR 3
    // Preview mode ended because of error - XfmDo() returned non-zero
    // and lParam contains the error number.
#define PREVN_STOPNOW 4
    // You must call PreviewStop in response to this message!!!


// Display Analysis Control
#define WM_SETDISPLAYDATA WM_USER+100
#define DISPLAY_ADDCHART        1
#define DISPLAY_BAKCOLOR        2
#define DISPLAY_LINECOLOR       3
#define DISPLAY_GRIDCOLOR       4
#define DISPLAY_TOPVALUE        5
#define DISPLAY_BOTTOMVALUE     6
#define DISPLAY_LEFTVALUE       7
```

```
#define DISPLAY_RIGHTVALUE     8
#define DISPLAY_TEXTCONTROL    9
#define DISPLAY_SAMPLERATE    10
#define DISPLAY_XLABEL        11
#define DISPLAY_YLABEL        12


#define DISPLAY_MODE_DB        1
#define DISPLAY_MODE_LOG       2
#define DISPLAY_MODE_WRAP      4
```

Converting from 16-Bit Transforms

Following are the major differences between 16-bit and 32-bit effects modules:

- The __export declarator should be changed to __declspec(dllexport) in all function definitions.
- LibMain() and _WEP have been replaced by DllMain().
- All occurrences of int should be changed to short.
- Message procedures are now of the form MsgProc(HWND, UINT, WPARAM, LPARAM), and all functions using wParam and lParam should be verified.
- EXETYPE WINDOWS can be removed from the DEF file.
- The user structure should be bracketed with #pragma pack(1) and #pragma pack().
- The valid library return value is 1157 for use with both CoolEdit 98 and CoolEdit Pro, or 1160 for just CoolEdit Pro compatibility only.

Obsolete Functions

All of the following functions have been superseded by updated versions. Most will still function for backwards compatibility, but the new functions should be used instead as soon as possible. The only functions that have no new version are the windowing functions for working with FFTs. If you are working with FFTs, then you no doubt know about windowing functions, and can window the data from within the XFM module directly.

void CopyToDlgItem(hWndDlg, iControl, ci, hVars, lpVars)	Obsolete
void CopyFromDlgItem(hWndDlg, iControl, lpci, phVars, plpVars)	

HWND	hWndDlg	Settings dialog window handle
int	iControl	ID of control to store information into (control should be text type and hidden)

`COOLINFO*ci`	CoolEdit info structure pointer
`COOLINFO**lpci`	Pointer to CoolEdit info structure pointer
`HANDLE  hVars`	Handle to internally defined structure
`HANDLE  *phVars`	Pointer to handle to internally defined structure
`EFFECTVARS*lpVars`	Pointer to internally defined structure
`EFFECTVARS**plpVars`	Pointer to pointer to internally defined structure

These functions are obsolete. See the `STATE` structure under `DIALOGMsgProc()`.

These obsolete functions basically were used to copy the handles and pointers used most frequently to a dialog control, and to copy data back from the control to the appropriate handles and pointers. During `WM_INITDIALOG`, the internal data is locked and the pointer to the data is saved in a control's text field. When any other messages are processed that need this information, it is copied back from the control to the locally declared pointers and handles. Following is the most common implementation for these two functions. The control ID passed into these functions should represent a hidden text field in the dialog box.

```
void CopyToDlgItem(HWND hWndDlg, int iControl, COOLINFO *ci,
                HANDLE hMyVars, COMPRESS *lpMyVars)
{   char m[80];
    wsprintf(m,"%ld,%d,%ld",lpAmp,hCompress,ci);
    SetDlgItemText(hWndDlg,iControl,m);
}


void CopyFromDlgItem(HWND hWndDlg, int iControl,  COOLINFO **ci,
                HANDLE *hMyVars, COMPRESS **lpMyVars)
{   char m[80];
    char *cursor;
    GetDlgItemText(hWndDlg,iControl,m,80);
    if (!m[0])
    {   *lpMyVars=NULL;
        *hMyVars=0;
        *ci=NULL;
        return;
    }
    cursor=m;
    *lpMyVars=(EFFECTVARS *)longfromstring(&cursor);
    *hMyVars=(HANDLE)longfromstring(&cursor);
    *ci=(COOLINFO *)longfromstring(&cursor);
}
```

int ReadData(ci, hpData, lOffset, lSize)		Obsolete

COOLINFO*ci	CoolEdit info structure pointer
char huge*hpData	Buffer to read into
long lOffset	Offset into user's wave data
long lSize	Number of bytes to read

This function is obsolete (but will still work). Please use ReadDataEx() instead. Read lSize bytes into the buffer hpData from the user's waveform data starting at lOffset bytes into user's data. The entire amount is always read. Data at offsets less than zero, or greater than the length of the file are filled with zeros for 16-bit data, or 128s for 8-bit data. Remember that this function works in units of bytes, not samples, and that the formula for converting samples to bytes is Bytes = Samples*wBitsPerSample*wChannels/8.

Returns: 0 if all went OK; otherwise, an error occurred.

int WriteData(ci, hpData, lOffset, lSize)		Obsolete

COOLINFO*ci	CoolEdit info structure pointer
char huge*hpData	Buffer to write out
long lOffset	Offset into user's wave data
long lSize	Number of bytes to write

This function is obsolete (but still works). Please use WriteDataEx() instead. Write lSize bytes from the buffer hpData to user's waveform data starting at lOffset bytes into the user's data. The entire amount is always written. Data at offsets below zero is ignored, and data at offsets greater than the length of the file extends the length of the user's file. Remember that this function works in units of bytes, not samples, and that the formula for converting samples to bytes is Bytes = Samples*wBitsPerSample*wChannels/8.

Returns: 0 if all went OK; otherwise, an error occurred.

BOOL PresetsInit(ci, hWndDlg, cszGroupName)		Obsolete

COOLINFO*ci	CoolEdit info structure pointer
HWND hWndDlg	Handle of Settings dialog
LPCSTR cszGroupName	Name of function

This should be called in the WM_INITDIALOG section of your dialog's Settings window if presets are used. cszGroupName identifies a unique name for your function which is used as a key in the INI file for keeping the preset information. This is usually the same as the name of the function given in the XFMQUERY structure. The szStructDef array of XFMQUERY contains an array of characters representing each of the data items in the user-defined structure. szPresetDef contains an array of 'y' and 'n' characters that specify whether or not to include the associated user data item in the preset. For example, a user-defined structure with 4 integers (szStructDef="iiii"), of which only the first 3 should be saved when added to the presets, would have szPresetDef="yyyn".

The dialog box must have the following IDs for various preset controls:

```
#define ID_PRESETS      1005  // Single selection sorted list box with
                              // Notify flag set
#define ID_PRESETNAME   1006  // Edit control for user to enter new preset
                              // name to add or delete
#define ID_ADD           325  // Button with text "Add"
#define ID_DEL          1109  // Button with text "Del"
```

Returns: 0 if no presets exist, or the number of presets if some do exist. (You can cast the BOOL return type to int.)

BOOL HandleID_PRESETS(ci, hWndDlg, cszGroupName, lParam)	Obsolete

COOLINFO*ci	CoolEdit info structure pointer
HWND hWndDlg	Handle of Settings dialog
LPCSTR cszGroupName	Name of function
LPARAM lParam	lParam passed to Settings dialog

Call this function in response to the WM_COMMAND message when wParam is ID_PRESETS. The function returns TRUE if the user chose a new preset item (double-clicked on a name in the Presets box). The StructToDialog() procedure should be called if this is the case to update all the dialog controls with the new data.

```
case ID_PRESETS:
    if (HandleID_PRESETS(ci, hWndDlg, TRANSFORMNAME, lParam))
    {   StructToDialog(ci, lpMyVars, hWndDlg);
    }
    break;
```

Returns: TRUE if the user chose a new preset item; FALSE otherwise.

void HandleID_ADD(ci, hWndDlg, cszGroupName)	Obsolete

COOLINFO*ci	CoolEdit info structure pointer
HWND hWndDlg	Handle of Settings dialog
LPCSTR cszGroupName	Name of function

Call this function in response to the WM_COMMAND message when wParam is ID_ADD. The DialogToStruct() function should be called before calling HandleID_ADD() to fill the COOLINFO structure with the information in the dialog box controls.

```
case ID_ADD:
    DialogToStruct(ci, hWndDlg, lpMyVars);
    HandleID_ADD(ci, hWndDlg, TRANSFORMNAME);
    break;
```

void HandleID_DEL(ci, hWndDlg, cszGroupName)	Obsolete

COOLINFO*ci	CoolEdit info structure pointer
HWND hWndDlg	Handle of Settings dialog
LPCSTR cszGroupName	Name of function

Call this function in response to the WM_COMMAND message when wParam is ID_DEL.

```
case ID_DEL:
    HandleID_DEL(ci, hWndDlg, TRANSFORMNAME);
    break;
```

void HandleID_PRESETNAME(ci, hWndDlg)	Obsolete

COOLINFO*ci	CoolEdit info structure pointer
HWND hWndDlg	Handle of Settings dialog

Call this function in response to the WM_COMMAND message when wParam is ID_PRESET-NAME.

```
case ID_PRESETNAME:
    HandleID_PRESETNAME(ci, hWndDlg);
    break;
```

void GraphSetDialog (ci, hGraph, hWndDlg, uiControl, uiDisplay)	Obsolete

COOLINFO*ci	CoolEdit info structure pointer
HANDLE hGraph	Handle to graph given by GraphCreate()
HWND hWndDlg	Handle of Settings dialog
UINT uiControl	ID of rectangle control defining boundaries of graph
UINT uiDisplay	ID of static text control for displaying graph point data

This function is obsolete. The SetDialog() portion gets called automatically during RegisterGraph() now. This should be called in the WM_INITDIALOG section of your dialog's Settings window to attach dialog controls to the graph. The graph must know of a rectangle it can be drawn inside of, and of a text control for displaying graph relevant information, and of the handle to the dialog box itself. uiControl is the ID of a Frame-type picture control in the dialog box, while uiDisplay is the ID of a left-aligned text control in the dialog box. The graph will automatically fit itself inside the bounds of the rectangle.

int GraphHandleWM_LBUTTONDOWN(ci, hGraph, ptCursor)	Obsolete

COOLINFO*ci	CoolEdit info structure pointer
HANDLE hGraph	Handle to graph given by GraphCreate()
POINT ptCursor	Coordinate of mouse down point (gained from lParam)

Call this function in response to the WM_LBUTTONDOWN message in the Settings dialog. The lParam parameter can be converted to a point since the LOWORD of lParam is the *x*-coordinate and the HIWORD of lParam is the *y*-coordinate.

```
case WM_LBUTTONDOWN:
{    POINT here;
     int whichpoint;
     CopyFromDlgItem(hWndDlg, IDC_STORAGE, &ci, &hMyVars, &lpMyVars);
     here.x=LOWORD(lParam);
     here.y=HIWORD(lParam);
     whichpoint=GraphHandleWM_LBUTTONDOWN(ci, lpMyVars->hGraph, here);
}
```

Returns: The index of the point that is under the mouse. A new point may have been created if the user clicked in an empty area of the graph. This index will always range from 0 to one minus the number of points in the graph.

int GraphHandleWM_LBUTTONUP(ci, hGraph, ptCursor)	Obsolete

COOLINFO*ci	CoolEdit info structure pointer
HANDLE hGraph	Handle to graph given by GraphCreate()
POINT ptCursor	Coordinate of mouse up point (gained from lParam)

Call this function in response to the WM_LBUTTONUP message in the Settings dialog. See GraphHandleWM_LBUTTONDOWN for more specifics.

Returns: The index of the point that is under the mouse.

int GraphHandleWM_LBUTTONDBLCLK(ci, hGraph, ptCursor)	Obsolete

COOLINFO*ci	CoolEdit info structure pointer
HANDLE hGraph	Handle to graph given by GraphCreate()
POINT ptCursor	Coordinate of mouse double-click point (gained from lParam)

Call this function in response to the WM_LBUTTONDBLCLK message in the Settings dialog. See GraphHandleWM_LBUTTONDOWN for more specifics.

Returns: The index of the point that is under the mouse.

int GraphHandleWM_MOUSEMOVE(ci, hGraph, ptCursor)	Obsolete

COOLINFO*ci	CoolEdit info structure pointer
HANDLE hGraph	Handle to graph given by GraphCreate()
POINT ptCursor	Coordinate of mouse point (gained from lParam)

Call this function in response to the WM_MOUSEMOVE message in the Settings dialog. See GraphHandleWM_LBUTTONDOWN for more specifics.

Returns: The index of the point that is under the mouse.

void SetWindowType(ci, wType)	Obsolete

| COOLINFO*ci | CoolEdit info structure pointer |
| WORD   wType | Type of window |

Sets the window type when windowing data for an FFT [when using WindowFFT() and IWindowFFT()]. Window types can be any one of the following:

0	Triangular
1	Von Hann
2	Hamming
3	Blackman
4	Cosine squared
5	Blackman–Harris

The values go from narrowest band, but widest side lobes and gentlest slopes, to wider bands with smaller side lobes and steeper slopes. In general, Hamming is used most often, but Blackman will give a little steeper slopes.

---

void WindowFFT(ci, lpData, iSize, bStereo)		Obsolete
COOLINFO*ci		CoolEdit info structure pointer
float	*lpData	Pointer to FFT-prepared data (prepare with the Set-FFT*x*bit functions)
int	iSize	Size of FFT (number of data points)
BOOL	bStereo	TRUE for stereo; FALSE for mono

Once data is converted from 16- or 8-bit audio to floating-point data, the floating-point data can be windowed before performing the FFT. Windowing the data consists of multiplying each sample by the window values. Samples near the center are multiplied by 1.0, while samples near the end are multiplied by smaller values down to but not including 0. Windowing the data will give a more precise measurement of the frequencies present, since the data is assumed to be circular. Please read about FFTs in other material related to Digital Signal Processing for more information.

---

void IWindowFFT(ci, lpData, iSize, bStereo)		Obsolete
COOLINFO*ci		CoolEdit info structure pointer
float	*lpData	Pointer to FFT-prepared data (prepare with the SetFFT... functions)
int	iSize	Size of FFT (number of data points)
BOOL	bStereo	TRUE for stereo; FALSE for mono

Perform the inverse window on FFT data. Calling IWindowFFT() right after Window-FFT() will result in no change in the original data. Instead of multiplying the data by the window coefficients, the data is divided by the window coefficients.

void FFT (ci, lpData, iSize, iDirection)		Obsolete
COOLINFO*ci	CoolEdit info structure pointer	
float *lpData	Pointer to FFT-prepared data (prepare with the Set-FFT... functions)	
int iSize	Size of FFT (number of data points)	
int iDirection	1 for FFT; -1 for inverse FFT	

Perform an FFT (Fast Fourier Transform) on the prepared data. The floating-point values will then contain pairs of {real, imaginary} data starting at index 1, so lpData[1] contains the real part of the first bin, lpData[2] contains the imaginary part of the first bin, and so on. The data going into the FFT should have been prepared from the original audio data using one of the SetFFT... functions. The largest size of FFT that should be used with this function is a 4096 point FFT, so iSize should never be more than 4096, and must be a power of 2.

After the FFT is performed, the data can be manipulated, or viewed, or whatever. If it is manipulated in some way (e.g., changing the phase or amplitude of various frequency components), then the inverse FFT can be performed by setting iDirection to -1, then an inverse Window can be performed by calling IWindowFFT() and finally calling the GetFFT... functions to convert the floating-point data back to audio data. Please read books on Digital Signal Processing for more information on the FFT.

void SetFFT16bit(ci, lpFloats, lpiAudio, iSize)		
void SetFFT8bit(ci, lpFloats, lpcAudio, iSize)		Obsolete
COOLINFO *ci	CoolEdit info structure pointer	
float *lpFloats	Destination array of floating-point values for use with FFT functions	
int *lpiAudio	Source array of 16-bit samples	
usigned char *lpcAudio	Source array of 8-bit samples	
int iSize	Number of samples to prepare or copy	

Fill an array of complex floating-point data with integer audio data. There will be twice as many floating-point numbers generated as iSize since complex numbers require two floats (one for the real part and one for the imaginary part). The floats array must be large enough to hold iSize*2 + 1 floating-point values (the +1 is because the array is 1 based, not zero based). iSize integer samples from lpiAudio will be converted to iSize complex floating-point values in lpFloats.

If preparing 8-bit audio for an FFT, the SetFFT8bit() function is used, and character data is read instead of integer data.

```
void SetStereoFFT16bit(ci, lpFloats, lpiLeft, lpiRight, iSize)
void SetStereoFFT8bit(ci, lpFloats, lpcLeft, lpcRight, iSize)        Obsolete
```

COOLINFO	*ci	CoolEdit info structure pointer
float	*lpFloats	Destination array of floating-point values for use with FFT functions
int	*lpiLeft	Source array of 16-bit samples for left channel
int	*lpiRight	Source array of 16-bit samples for right channel
usigned char	*lpcLeft	Source array of 8-bit samples for left channel
unsigned char	*lpcRight	Source array of 8-bit samples for right channel
int	iSize	Number of samples to prepare or copy

Two channels can be converted to one set of floating point data at a time for more efficiency. This way, two FFTs can be performed in parallel.

```
void SetStereoFFT16bitInterleaved(ci, lpFloats, lpiAudio, iSize)
void SetStereoFFT8bitInterleaved(ci, lpFloats, lpcAudio, iSize)       Obsolete
```

COOLINFO	*ci	CoolEdit info structure pointer
float	*lpFloats	Destination array of floating-point values for use with FFT functions
int	*lpiAudio	Source array of 16-bit samples, left/right interleaved
usigned char	*lpcAudio	Source array of 8-bit samples, left/right interleaved
int	iSize	Number of samples to prepare or copy

Two channels can be converted to one set of floating point data at a time for more efficiency. This way, two FFTs can be performed in parallel. The source data is interleaved left sample 0, right sample 0, left sample 1, right sample 1, . . .

```
void GetStereoFFT16bitInterleaved(ci, lpFloats, lpiAudio, iSize, iOp)
void GetStereoFFT8bitInterleaved(ci, lpFloats, lpcAudio, iSize, iOp)
                                                                    Obsolete
```

COOLINFO	*ci	CoolEdit info structure pointer
float	*lpFloats	Source array of floating-point values for use with FFT functions
int	*lpiAudio	Destination array of 16-bit samples, left/right interleaved
usigned char	*lpcAudio	Destination array of 8-bit samples, left/right interleaved
int	iSize	Number of samples to prepare/copy
int	iOp	Operation — set this to zero

Convert floating-point FFT data (data gained after doing an inverse FFT) back to interleaved audio.

To convert FFT data back that was set using `SetFFT16bit()`, simply convert every other floating-point value starting with offset 1. So `IntegerSample[0] = (int)lpFloats[1]`, `IntegerSample[1] = (int)lpFloats[3]`, `IntegerSample[2] = (int)lpFloats[5]`, and so on until you have `IntegerSample[iSize-1] = (int)lpFloats[(iSize-1)*2+1]`.

Index

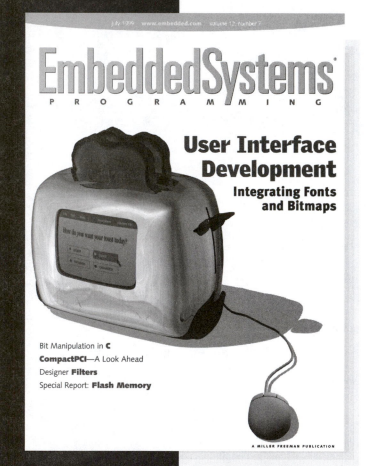

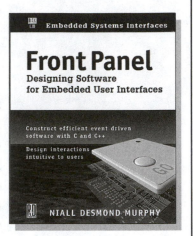
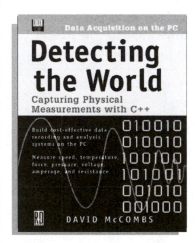

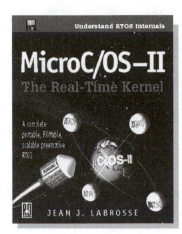

Writing Windows WDM Device Drivers

by Chris Cant

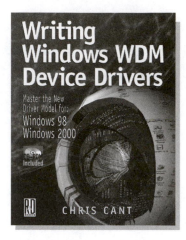

Master the new Windows Driver Model (WDM) that is the common hardware interface for Windows 98 and Windows 2000. You get overview of driver development, including hardware and software interface issues, appropriate application of driver types, and descriptions of where drivers fit in the new 'layer' model of WDM. Instructions on writing device drivers, including installation and debugging, are demonstrated with code provided on the companion CD-ROM. 568pp, ISBN 0-87930-565-7

RD3121   **$49.95**

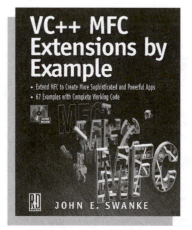

RD3250   **$49.95**

VC++ MFC Extensions by Example

by John E. Swanke

Extend MFC to create more sophisticated and powerful applications. You get 67 examples — each fully annotated and ready to insert into applications. This book features a menu of advanced techniques across the entire range of Windows functions that complement the author's earlier title, *Visual C++ MFC Programming by Example*. The CD contains working projects in Visual C++ V5.0 & V6.0 and the author's own *SampleWizard* utility that facilitates adding these examples into users' applications. CD and MFC Quick Reference Guide included, 528 pp, ISBN 0-87930-588-6

What's on the CD-ROM?

The CD-ROM that accompanies **Digital Audio Processing** contains complete C++, C, and assembly source code for all of the applications discussed in the book, as well as some bonus code the reader should find useful. Project files for DevStudio rev 6 are in the appropriate directories along with the source to make it easy to get started. The CD is a simple copy of various project directories from the author's machine at the time the book was written. There is no dangerous or unpredictable installation program; you simply copy whatever is wanted from the CD to your hard disk and then use Windows Explorer to remove the read-only attributes the CD-burning process adds to all files. Besides source code, the CD contains audio plugin development tools and documentation for the DirectX and CoolEdit standards.

The CD contains full source and pre-built executable code for:

- The WavEd waveform/music editor discussed in the book, with hundreds of reusable C++ classes for DSP transforms, signal generation, analysis, and DSP-specialized graphical user interfaces.
- WavEdPro, an enhanced version with example modifications by the author.
- Fast Intel assembly modules for DSP and often needed utilities in both source and .obj form.
- The FilterDraw ActiveX control which allows graphical user input and plotting.
- The Presets ActiveX control which makes it easy to add presets to applications.
- An example plotting program used to create some of the book plots.
- A DevStudio AppWizard that generates CoolEdit transform plugins.
- Example programs that help the reader explore the S-Domain intuitively.
- An arbitrary IIR filter designer for the DirectX plugin format.
- Various reusable C++ classes for benchmarking and Windows UI.
- An optimized Fast Fourier Transform (FFT) that can create bitmaps for fast screen drawing.
- Example code for m-Law and ADPCM coding and decoding.

For more information, see the WavEd Programmer's Guide and the WavEd User's Guide on the CD-ROM and Appendix A and B of the book.
